COOPERATIVITY AND REGULATION IN BIOCHEMICAL PROCESSES

COOPERATIVITY AND REGULATION IN BIOCHEMICAL PROCESSES

Arieh Ben-Naim
The Hebrew University of Jerusalem
Jerusalem, Israel

Kluwer Academic / Plenum Publishers
New York, Boston, Dordrecht, London, Moscow

Library of Congress Cataloging-in-Publication Data

Ben-Naim, Arieh, 1934–
Cooperativity and regulation in biochemical processes/Arieh Ben-Naim.
p. cm.
Includes bibliographical references and index.

1. Cooperative binding (Biochemistry) 2. Statistical mechanics. 3. Physical biochemistry. I. Title.

QP517.C66 .B46 2000
572′.43—dc21
00-021561

ISBN 978-1-4419-3336-2

233 Spring Street, New York, New York 10013

http://www.wkap.nl/

10 9 8 7 6 5 4 3 2 1

A C.I.P. record for this book is available from the Library of Congress.

Printed in the United States of America

To Alicia
and
To the Memory of My Mother

Preface

This book evolved from a graduate course on applications of statistical thermodynamics to biochemical systems. Most of the published papers and books on this subject used in the course were written by experimentalists who adopted the phenomenological approach to describe and interpret their results. Two outstanding papers that impressed me deeply were the classical papers by Monod, Changeux, and Jacob (1963) and Monod, Wyman, and Changeux (1965), where the allosteric model for regulatory enzymes was introduced. Reading through them I felt as if they were revealing one of the cleverest and most intricate tricks of nature to regulate biochemical processes.

In 1985 I was glad to see T. L. Hill's volume entitled *Cooperativity Theory in Biochemistry, Steady State and Equilibrium Systems*. This was the first book to systematically develop the molecular or statistical mechanical approach to binding systems. Hill demonstrated how and why the molecular approach is so advantageous relative to the prevalent phenomenological approach of that time. On page 58 he wrote the following (my italics):

> The naturalness of Gibbs' grand partition function for binding problems in biology is evidenced by the *rediscovery* of what is *essentially* the grand partition function for this particular type of problem by various physical biochemists, including E. Q. Adams, G. S. Adair, H. S. Simms, K. Linderstrom-Lang, and, especially, J. Wyman. These treatments, however, were empirical or thermodynamic in content, that is, expressed from the outset in terms of thermodynamic equilibrium constants. The advantage of the explicit use of the *actual* grand partition function is that it is more general: it includes everything in the empirical or thermodynamic approach, plus providing, when needed, the background molecular theory (as statistical mechanics always does).

Indeed, there are two approaches to the theory of binding phenomena. The first, the older, and the more common approach is the thermodynamic or the phenomenological approach. The central quantity of this approach is the *binding polynomial* (BP). This polynomial can easily be obtained for any binding system by viewing each step of the binding process as a chemical reaction. The mass action law of thermodynamics associates an equilibrium constant with such a reaction. The BP is constructed in terms of these constants (see an example in Section 2.3). It has the general form

$$BP = 1 + \beta_1 C + \beta_2 C^2 + \beta_3 C^3 + \cdots \quad (1)$$

where β_i are products of the equilibrium constants K_i, and C is the ligand concentration in a reservoir at equilibrium with the binding system.

The statistical mechanical approach starts from more fundamental ingredients, namely, the molecular properties of all the molecules involved in the binding process. The central quantity of this approach is the *partition function* (PF) for the entire macroscopic system. In particular, for binding systems in which the adsorbent molecules are independent, the partition function may be expressed as a product of partition functions, each pertaining to a single adsorbent molecule. The latter function has the general form

$$PF = Q(0) + Q(1)\lambda + Q(2)\lambda^2 + \cdots \quad (2)$$

where $Q(i)$ is the so-called canonical partition function of a single adsorbent molecule having i bound ligands, and λ is the absolute activity of the ligand in the reservoir being at equilibrium with the binding systems.

Both the BP and PF are polynomials of degree m for a system having m binding sites. However, the PF is the more *fundamental*, the more *general*, and the more *powerful* quantity of the two functions. It is more fundamental in the sense that it is based on the basic molecular properties of the molecules involved in the system. Therefore, from the PF one can obtain the BP. The reverse cannot, in general, be done. It is more general in the sense that it is applicable for *any* ligand concentration* in the reservoir. The BP, based on the mass action law, is valid only for ligand reservoirs in which the ligand concentration is very low, such that $\lambda = \lambda_0 C$, i.e., an ideal-dilute with respect to the ligand.

Thermodynamics cannot provide the extension to the BP for nonideal systems (with respect to either the ligands or the adsorbent molecules). The statistical mechanical approach can, in principle, provide corrections for the nonideality of the system. An example is worked out in Appendix D.

Finally, it is more powerful in its interpretative capability. In particular, the central concept of the present book—the cooperativity—may be interpreted on a molecular level. All possible sources of cooperativity may be studied and their relative importance estimated. None of this can be done with the phenomenological

*In general, the statistical mechanical approach may also be applied to systems where the adsorbent molecules are not necessarily independent. However, in this book we shall always assume independence of the adsorbent molecules.

approach. The BP can give the general form of the binding curve.* In spite of this limited interpretative power of the BP, it is astonishing to see so many formal manipulations applied to it or to its derivative, the binding isotherm (BI). They range from rearrangements of the BI and plotting it in different forms, differentiating the BP followed by integration, taking the roots and rewriting the BI as a product of linear factors, or "cutting" and "pasting" the cuts. None of these manipulations can enhance or improve the interpretative power of the BP.

Returning to the quotation from Hill, I fully agree with its content except for the word "rediscovery," which he uses to describe the BP, referring to it as "*essentially* the grand partition function," while the PF as cited in Eq. (2) is referred to as "the *actual* grand partition function."

A genuine rediscovery of the PF should provide the functional dependence of the coefficients of the BP in terms of the molecular properties of the system. This has never been done independently since Gibbs' discovery. Therefore, one should make a clear-cut distinction between the phenomenological BP on the one hand and the molecular PF on the other. Unfortunately, the distinction between the two quantities is often blurred in the literature, the two terms sometimes being used as synonyms.

The main objective of this book is to understand the molecular origin of cooperativity and its relation to the actual function of biochemical binding systems.

The term *cooperativity* is used in many branches of science. Two atoms cooperate to form molecules, molecules cooperate to build up a living cell, cells cooperate to construct a living organism, men and women cooperate in a society, and societies and nations cooperate or do not cooperate in peace and war. In all of these situations, cooperation is achieved by exchanging signals between the cooperating units. The signals may be transmitted electromagnetically, chemically, or verbally. In this book we confine ourselves to one kind of cooperativity—that between two (or more) ligands bound to a single adsorbent molecule. The type of information communicated between the ligands is simple: which sites are occupied and which are empty. The means of communication are varied and intricate and are explored herein, especially in Chapters 4, 5, and 9.

Even when the term *cooperativity* is confined to binding systems, it has been defined in a variety of ways. This has led to some inconsistencies and even to conflicting results.

In this book, we define cooperativity in probabilistic terms. This is not the most common or popular definition, yet it conveys the spirit and essence of what researchers mean when they use this term. Since the partition function embodies the probabilities of the occupancy events, the definition of cooperativity can

*This is true only for ideal systems with respect to both the ligand and the adsorbent molecules (see Appendix D).

immediately be translated in terms of molecular properties of the system. Thus, the sequence of concepts leading to cooperativity is the following: molecular parameters → molecular events (which sites are occupied) → correlation between molecular events → cooperativity between bound ligands.

The term *interaction* is sometimes used almost synonymously with cooperativity. In this book we reserve the term *interaction* to mean *direct* interaction energy between two (or more) particles. Indeed, sometimes interaction, in the above sense, is the sole source of cooperativity, in which case the two terms may be used interchangeably. However, in most cases of interest in biochemistry, *interaction* in the above sense is almost negligible, such as in two oxygen molecules in hemoglobin. Cooperativity in such systems is achieved by *indirect* routes of communication between the ligands.

The practice of using the term *interaction* (or related terms such as interaction parameters, interaction free energy, etc.), though legitimate, can lead to misinterpretation of experimental results. An example is discussed in Chapter 5.

The contribution of the direct interaction to cooperativity is easy to visualize and understand. On the other hand, the indirect part of cooperativity is less conspicuous and more difficult to grasp. There are two "lines of indirect communication" between the ligands: one through the adsorbent molecule and the other through the solvent. Both depend on the ability of the ligands to induce "structural changes" in either the adsorbent molecule or the solvent. The relation between the induced structural changes and the resulting cooperativity is not trivial. Nevertheless, by using very simplified models of adsorbent molecules we can obtain explicit relations between cooperativity and molecular parameters of these simplified models. The treatment of the more difficult communication through the solvent is left to Chapter 9, where we outline the complexity of the problem rather than derive explicit analytical results.

While there are several books that deal with the subject matter of this volume, the only one that develops the statistical mechanical approach is T. L. Hill's monograph (1985), which includes equilibrium as well as nonequilibrium aspects of cooperativity. Its style is quite condensed, formal, and not always easy to read. The emphasis is on the effect of cooperativity on the form of the PF and on the derived binding isotherm (BI). Less attention is paid to the sources of cooperativity and to the mechanism of communication between ligands, which is the main subject of the present volume.

There are three books that review the experimental aspects of cooperativity using the phenomenological-theoretical approach. Levitzki (1978) develops the binding isotherms for various allosteric models, based on the relevant mass action laws. Imai (1982) describes the function of hemoglobin as an oxygen carrier in living systems, emphasizing experimental methods of measuring binding oxygen to hemoglobin and ways of analyzing the obtained experimental data. Perutz (1990) emphasizes structural aspects of hemoglobin and other allosteric enzymes. Perutz also raises some fundamental questions regarding the exact molecular mechanism of the allosteric model.

Two more recent books by Wyman and Gill (1990) and by DiCera (1996) present the phenomenological approach in much greater detail. Wyman and Gill describe a large number of binding systems, illustrating various experimental aspects of the binding data, but the theoretical treatment is cumbersome, sometimes confusing. They treat the BP as equivalent to the PF. The concept of cooperativity is introduced in several different ways, without showing their formal equivalence. This inevitably leads to some ambiguous statements regarding the cooperativity of specific systems.

DiCera's book starts with the construction of the PF of the system, then switches to the BP based on the mass action law, but still refers to it as the PF of the system. Much of the remainder of the book contains lengthy lists of mass-action-law equations for binding reactions and the corresponding equilibrium constants. This is followed by lengthier lists of contracted BPs (referred to as contracted PFs). The contracted BPs (or PFs) do not provide any new information that is not contained in, or can be extracted from, the PF of the binding system, nor do they possess any new interpretive power.

In summary, although each of the aforementioned books does touch upon some aspects of cooperativity in binding systems, none of them explores the details of the mechanisms of cooperativity on a molecular level. In this respect I feel that the present book fills a gap in the literature. I hope it will help the reader to gain insight into the mechanism of cooperativity, one of the cleverest and most intricate tricks that nature has evolved to regulate biochemical processes.

This volume is addressed mainly to anyone interested in the life sciences. There are, however, a few minimal prerequisites, such as elementary calculus and thermodynamics. A basic knowledge of statistical thermodynamics would be useful, but for understanding most of this book (except Chapter 9 and some appendices), there is no need for any knowledge of statistical mechanics.

The book is organized in nine chapters and eleven appendices. Chapters 1 and 2 introduce the fundamental concepts and definitions. Chapters 3 to 7 treat binding systems of increasing complexity. The central chapter is Chapter 4, where all possible sources of cooperativity in binding systems are discussed. Chapter 8 deals with regulatory enzymes. Although the phenomenon of cooperativity here is manifested in the kinetics of enzymatic reactions, one can translate the description of the phenomenon into equilibrium terms. Chapter 9 deals with some aspects of solvation effects on cooperativity. Here, we only outline the methods one should use to study solvation effects for any specific system.

Many students and friends have contributed to my understanding of the binding systems discussed in this book. In particular, I am most grateful to Dr. Harry Saroff, who introduced me to this field and spent so much time with me describing some of the experimental binding systems. I am also grateful to Drs. Robert Mazo,

Mihaly Mezei, Wilse Robinson, Jose Sanchez-Ruiz, and Eugene Stanley for reading parts of the manuscript and sending me their comments and suggestions. The entire manuscript was typed by Ms. Eva Guez to whom I am deeply grateful for her efforts in deciphering my handwriting and preparing the first, second, and third drafts.

Finally, I wish to express my thanks and admiration to Wolfram Research for creating the *Mathematica* software. I have used *Mathematica* for simplifying many mathematical expressions and for most of the graphical illustrations.

Arieh Ben-Naim

Jerusalem
October 2000

Contents

Chapter 1

Introducing the Fundamental Concepts

1.1. Correlation and Cooperativity 1
1.2. The Systems of Interest 9
1.3. States of the System and Their Energies 12
1.4. Construction of the Partition Function 17
1.5. Probabilities 20

Chapter 2

The Binding Isotherm

2.1. The General Form of the Binding Isotherm 25
2.2. The Intrinsic Binding Constants 29
2.3. The Thermodynamic Binding Constants 34
2.4. The Simplest Molecular Model for the Langmuir Isotherm 38
2.5. A Few Generalizations 40
2.5.1. Mixture of Two (or More) Types of Adsorbing Molecules . . 40
2.5.2. Mixture of Two (or More) Ligands Binding to the Same Site 41
2.6. Examples 43
2.6.1. Normal Carboxylic Acids 43
2.6.2. Normal Amines 47

Chapter 3

Adsorption on a Single-Site Polymer with Conformational Changes Induced by the Binding Process

3.1. Introduction 51

3.2. The Model and Its Partition Function 52
3.3. The Binding Isotherm 56
3.4. Induced Conformational Changes 57
3.5. Spurious Cooperativity 60

Chapter 4

Two-Site Systems: Direct and Indirect Cooperativity

4.1. Introduction 67
4.2. The General Definition of Correlation and Cooperativity in a Two-Site System 68
4.3. Two Identical Sites: Direct Correlation 73
4.4. Two Different Sites: Spurious Cooperativity 77
4.5. Two Sites with Conformational Changes Induced by the Ligands: Indirect Correlations 82
4.6. Spurious Cooperativity in Two Identical-Site Systems 91
4.7. Two Sites on Two Subunits: Transmission of Information across the Boundary between the Subunits 100
4.7.1. The Empty System 100
4.7.2. The Binding Isotherm 104
4.7.3. Correlation Function and Cooperativity 105
4.7.4. Induced Conformational Changes in the Two Subunits . . . 107
4.7.5. Two Limiting Cases 112
4.8. Binding of Protons to a Two-Site System 114
4.8.1. Introduction, Notation, and Some Historical Perspectives . . . 114
4.8.2. Two Identical Sites: Dicarboxylic Acids and Diamines . . . 119
4.8.3. Two Different Sites: Amino Acids 121
4.8.4. Maleic, Fumaric, and Succinic Acids 122
4.8.5. A Fully Rotating Electrostatic Model 127
4.8.6. Spurious Cooperativity in Some Alkylated Succinic Acids . . 131
4.8.7. Conclusion 141

Chapter 5

Three-Site Systems: Nonadditivity and Long-Range Correlations

5.1. Introduction 143
5.2. General Formulation of the Partition Function 143
5.3. Direct Interaction Only 145
5.4. Three Strictly Identical Sites: Nonadditivity of the Triplet Correlation 147

5.5. Three Different, Linearly Arranged Sites: Long-Range Correlations 151
5.6. Three Linearly Arranged Subunits: Correlation Transmitted across the Boundaries between the Subunits 155
5.7. A Simple Solvable Model 159
5.8. A Measure of the Average Correlation in a Binding System . . . 164
5.8.1. Introduction and Historical Background 164
5.8.2. Definition of the Average Correlation in Any Binding System 166
5.8.3. Some Numerical Illustrations 171
5.9. Correlations Between Two and Three Protons 173
5.10. Binding of Proteins to DNA 177
5.10.1. Introduction 177
5.10.2. Sources of Long-Range and Nonadditivity of the Correlation Functions 179
5.10.3. Processing the Experimental Data on Binding of the λ Repressor to the Operator 184
5.10.4. Conclusions 187

Chapter 6

Four-Site Systems: Hemoglobin

6.1. Introduction 193
6.2. The General Theoretical Framework 193
6.3. The Linear Model 197
6.4. The Square Model 199
6.5. The Tetrahedral Model 200
6.6. The Average Cooperativity of the Linear, Square, and Tetrahedral Models: The "Density of Interaction" Argument 202
6.7. Benzene-Tetracarboxylic Acids 204
6.8. Hemoglobin—The Efficient Carrier of Oxygen 207
6.8.1. Introduction and a Brief Historical Overview 207
6.8.2. A Sample of Experimental Data 212
6.8.3. Utility Function under Physiological Conditions 218

Chapter 7

Large Linear Systems of Binding Sites

7.1. The Matrix Method 223
7.2. Correlation Functions 230

7.3. 1-D System with Direct Correlations Only 239
7.4. A System of m Linearly Arranged Subunits 242

Chapter 8

Regulatory Enzymes

8.1. Introduction and Historical Perspective 255
8.2. The Connection between the Kinetic Equation and the Binding Isotherm 258
8.3. The Regulatory Curve and the Corresponding Utility Function . . 261
8.4. The Competitive Regulation 263
8.5. A Simple Allosteric Regulation 264
8.6. One Active and Two Regulatory Sites 267
8.7. One Active and m Regulatory Sites 269
8.8. A Cyclic Model for Allosteric Regulatory Enzymes 272
8.9. Aspartate Transcarbamoylase (ATCase) 277

Chapter 9

Solvent Effects on Cooperativity

9.1. Introduction 281
9.2. Solvation Effect on the Equilibrium Constants 282
9.3. Solvent Effect on the Ligand–Ligand Pair Correlation 287
9.4. Decomposition of the Solvation Gibbs Energy of Macromolecules 293
9.5. Effect of Size on the Cooperativity 298
9.6. Some Specific Solvent Effects 302

Appendices

A. Pair and Triplet Correlations between Events 309
B. Localization of the Adsorbent Molecules and Its Effect on the Binding Isotherm 311
C. Transition from Microstates to Macrostates 313
D. First-Order Correction to Nonideality of the Ligand's Reservoir 317
E. Relative Slopes of Equilibrated and "Frozen-in" BIs in a Multimacrostate System 320
F. Spurious Cooperativity in Single-Site Systems 322

G. The Relation between the Binding Isotherm and the Titration Curve for Two-Site Systems 328
H. Fitting Synthetic Data 330
I. A Comment on the Nomenclature 332
J. Average Binding Constants and Correlation Functions 335
K. Utility Function in a Binding System 337

Abbreviations Used in the Text 341

References 343

Index 347

1

Introducing the Fundamental Concepts

1.1. CORRELATION AND COOPERATIVITY

Let us consider a system of m binding sites. Each site can be in one of two states: empty or occupied by a ligand **L**. First, we treat the case where the system contains a *fixed* number of ligands, n. In thermodynamic terms, we refer to such a system as a *closed* system with respect to the ligands. We are interested in asking probabilistic questions about this system. To this end we imagine a very large collection, an ensemble of such systems, all of which are identical in the sense that each has a *fixed* number of n ligands occupying n of the m sites ($n \leq m$). If the sites are distinguishable but the ligands are indistinguishable, then altogether we have

$$\binom{m}{n} = \frac{m!}{n!(m-n)!} \tag{1.1.1}$$

distinguishable *configurations* of such a system.

Figure 1.1 shows all of these configurations for $m = 4$ and $n = 2$. In this example there are altogether $\binom{4}{2} = 6$ distinguishable configurations. Note that if the sites were indistinguishable, there would be only *one* configuration for such a system. On the other hand, if the ligands were labeled (say, blue and red in the case of Fig. 1.1) then the number of distinguishable configurations would be $m!/(m-n)!$, or 12 in the case of Fig. 1.1. In general, the $\binom{m}{n}$ configurations will have different probabilities, i.e., different frequencies of occurrence in an ensemble of such systems. For instance, if two ligands attract each other, then a configuration for which the two ligands are closer will have higher probability. For the moment, we assume that each of these configurations has equal probability. Since there are $\binom{m}{n}$ distinguishable configurations, the probability of finding a system in one *specific* configuration is $\binom{m}{n}^{-1}$.

What is the probability of finding a *specific site*, say the ith site, occupied? The answer can be given by using the so-called *classical* definition of probability

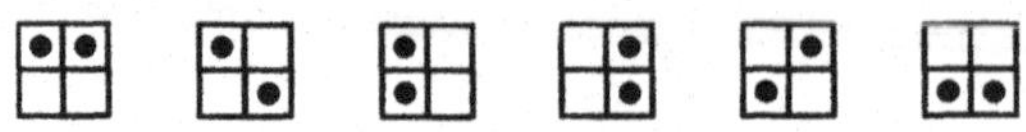

Figure 1.1. The six distinguishable configurations of a four-site system ($m = 4$) with two bound ligands ($n = 2$).

(Feller, 1957; Papoulis, 1965), namely,

$$P_i(1) = \frac{\binom{m-1}{n-1}}{\binom{m}{n}} = \frac{n}{m} = \theta \tag{1.1.2}$$

In the denominator we have the total number of configurations; in the numerator we have the number of configurations that fulfill the requirement "site i is occupied." To count the latter, we simply place one ligand at site i and count the number of ways of arranging the remaining $(n - 1)$ ligands on the remaining $(m - 1)$ sites, $\binom{m-1}{n-1}$. Clearly, since all the sites are identical the same result holds for any specific i.

Next, we seek the probability of finding two specific sites i and j simultaneously occupied. This can be calculated again by the classical definition of probability,

$$P_{ij}(1, 1) = \frac{\binom{m-2}{n-2}}{\binom{m}{n}} = \frac{n(n-1)}{m(m-1)} = \theta^2 \frac{(1 - 1/n)}{(1 - 1/m)} \tag{1.1.3}$$

As in Eq. (1.1.2) we have the total number of configurations in the denominator, and the number of configurations that fulfill the condition "site i *and* site j are occupied" in the numerator.

Two events $\mathcal{A}$ and $\mathcal{B}$ are said to be independent, if and only if the probability of the joint event $\mathcal{A} \cdot \mathcal{B}$ (read: $\mathcal{A}$ *and* $\mathcal{B}$)* is the product of the probabilities of the two events, i.e.,

$$P(\mathcal{A} \cdot \mathcal{B}) = P(\mathcal{A})P(\mathcal{B}) \tag{1.1.4}$$

If $\mathcal{A}$ is the event "site i is occupied" and $\mathcal{B}$ is "site j is occupied," then clearly

$$P(\mathcal{A} \cdot \mathcal{B}) = P_{ij}(1, 1) = \theta^2 \left(\frac{1 - 1/n}{1 - 1/m}\right) \neq \theta^2 = P_i(1)P_j(1) = P(\mathcal{A})P(\mathcal{B}) \tag{1.1.5}$$

i.e., the two events $\mathcal{A}$ and $\mathcal{B}$ are *not* independent.†

*Another notation for $\mathcal{A} \cdot \mathcal{B}$ is $\mathcal{A} \cap \mathcal{B}$, referred to as the intersection of the two events $\mathcal{A}$ and $\mathcal{B}$.

†We use the notation $P(\mathcal{A} \cdot \mathcal{B})$ for any two events $\mathcal{A}$ and $\mathcal{B}$; $P_{ij}(1, 1)$ is used for the two specific events "site i is occupied" and "site j is occupied."

For any two events $\mathcal{A}$ and $\mathcal{B}$, we define the correlation function by the ratio*

$$g(\mathcal{A}, \mathcal{B}) = \frac{P(\mathcal{A} \cdot \mathcal{B})}{P(\mathcal{A})P(\mathcal{B})} \tag{1.1.6}$$

We say that the correlation is *positive* whenever $g - 1 > 0$, or $g > 1$, and *negative* whenever $g - 1 < 0$, or $g < 1$. For independent events $g = 1$, i.e., the correlation is unity.†

We see from Eq. (1.1.5) that for any $m > 1$ and $1 < n \leq m$, the correlation between the two events "site i is occupied" and "site j is occupied" is negative, i.e., $g < 1$. Recall that we have assumed that all of the $\binom{m}{n}$ configurations have equal probabilities. This is usually the case when there are no interactions between the ligands occupying different sites.

The conditional probability of an event $\mathcal{A}$ given the occurrence of event $\mathcal{B}$ is defined by

$$P(\mathcal{A}/\mathcal{B}) = \frac{P(\mathcal{A} \cdot \mathcal{B})}{P(\mathcal{B})} = \frac{P(\mathcal{A})P(\mathcal{B})g(\mathcal{A}, \mathcal{B})}{P(\mathcal{B})} = P(\mathcal{A})g(\mathcal{A}, \mathcal{B}) \tag{1.1.7}$$

and similarly

$$P(\mathcal{B}/\mathcal{A}) = \frac{P(\mathcal{A} \cdot \mathcal{B})}{P(\mathcal{A})} = P(\mathcal{B})g(\mathcal{A}, \mathcal{B}) \tag{1.1.8}$$

Thus, the correlation $g(\mathcal{A}, \mathcal{B})$ measures the extent of the difference between the *conditional* probability and the *unconditional* probability.‡

Returning to $P_{ij}(1, 1)$, we find that the correlation function is always negative, i.e.,

$$g_{ij}(1, 1) = \frac{1 - 1/n}{1 - 1/m} \leq 1 \tag{1.1.9}$$

Usually the *conditional* probability of "site i is occupied" given that "site j is occupied" is different from the *unconditional* probability of "site i is occupied," whenever there exists some kind of "communication" between the sites, i.e., when a ligand at site i "knows" or "senses" the state of occupation of site j. This book is devoted to the study of various mechanisms for transmitting such information

*This definition of correlation differs from the common definition of correlation in the mathematical theory of probability (Feller, 1957; Papoulis, 1965). The latter is defined, up to a normalization constant, as the difference $P(\mathcal{A} \cdot \mathcal{B}) - P(\mathcal{A})P(\mathcal{B})$.

†When $g(\mathcal{A}, \mathcal{B}) > 1$ it is often said that $\mathcal{A}$ supports $\mathcal{B}$ (and vice versa), and when $g(\mathcal{A}, \mathcal{B}) < 1$, $\mathcal{A}$ does not support $\mathcal{B}$ (and vice versa).

‡Strictly speaking, every probability is a conditional probability. For instance, $P_i(1)$ is the probability of finding site i occupied *given* the condition that the *experiment* of selecting site i and determining its state of occupation has been performed. In general, the condition "the experiment . . . has been performed" is suppressed whenever this is the only condition. However, whenever the experiment may be performed in various ways, one must specify the exact manner in which it is performed, since this could affect the probabilities of the various outcomes.

between different sites. Usually, the type and extent of communication between two sites depends on the specific sites i and j. For instance, if two ligands attract each other, then the correlation between two ligands at sites i and j would depend on the distance between the sites; the larger the distance, the weaker, in general, the correlation (a detailed example is discussed in Section 4.3). There are also examples where the correlation between the two ligands does not depend on the distance between the sites, but only on the *type* of the sites i and j (an example is the model treated in Section 4.5). The correlation written in Eq. (1.1.9) does not depend on the specific sites i and j. It is the same for any pair of sites.

Clearly, the correlation function computed in Eq. (1.1.9) is a result of our choice of the *fixed* and *finite* values of n and m. If we let $n \to \infty$ and $m \to \infty$ in such a way that the ratio n/m remains constant, we obtain $g_{ij}(1, 1) \to 1$ and

$$P_{ij}(1, 1) = P_i(1)P_j(1) = \theta^2 \tag{1.1.10}$$

In this limit the two events become independent. This is what we expect from a system where no "communication" between the sites exists. In the example of Fig. 1.1, the conditional probability of finding "site i is occupied given that site j is occupied" differed from the unconditional probability, only because of the *finite* values of n and m. If one site is occupied, then there remain only $(n - 1)$ ligands to be arranged at the $(m - 1)$ sites. This is the only reason for the correlation between the sites. One site "knows" that another site is occupied only due to the fact that the number of arrangements has changed from $\binom{m}{n}$ to $\binom{m-1}{n-1}$. Clearly, in this example it does not matter which site is i and which site is j $(i \neq j)$. When $n \to \infty$ and $m \to \infty$, this "communication" between the sites is lost, i.e., $g \to 1$.

In this book we examine various types of correlations that arise from (direct or indirect) "communication" between the ligands at different sites. We require that the correlation functions be unity whenever the two sites are physically independent. This excludes the type of correlation we found in Eq. (1.1.9). Yet, we wish to study systems with *small* values of m. This can be achieved by *opening* the system with respect to the ligands. We still keep m fixed, but now the ligands bound to the system are in equilibrium with a reservoir of ligands at a fixed chemical potential, or at a fixed density (see also Section 1.2).

Once we open the system to allow exchange of ligands between the sites and the reservoir, the number of occupancy states of our system is not $\binom{m}{n}$ (or 6 in the case of Fig. 1.1), but 2^m (or $2^4 = 16$ as in Fig. 1.2). This is so because any site can be either empty or occupied, i.e., 2 states for each site, hence 2^m states for the m sites. Clearly, in an open system these 2^m configurations are not equally probable. For calculating the probabilities of the various events statistical mechanics provides a general recipe which differs from the classical method used above. The latter is applicable only when there are Ω equally probable events (say, six outcomes of casting a die with probability 1/6 for each outcome).

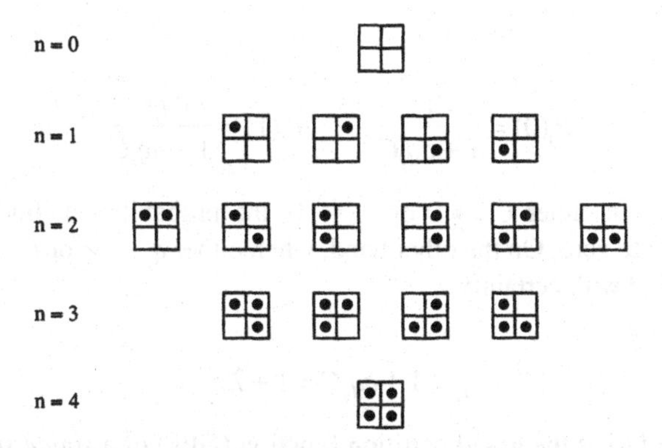

Figure 1.2. All of the sixteen configurations for a system with four sites, arranged in five groups with $n = 0, 1, 2, 3, 4$.

We shall present here an intuitively plausible argument on how to construct the probabilities of the various events of our new system, having *m independent* sites, opened with respect to the ligand. Each site, say *j*, can be in one of two states: empty, with probability $P_j(0)$, and occupied, with probability $P_j(1)$. Since the system is at equilibrium with the ligand at some fixed chemical potential, it is reasonable to assume that the probability ratio $P_j(1)/P_j(0)$ is proportional to two factors: one that measures the affinity of the site to the ligand, which we denote by q_j, and the second that depends on the concentration C of the ligand in the reservoir, i.e.,

$$\frac{P_j(1)}{P_j(0)} = aq_jC \tag{1.1.11}$$

where a is a constant. The rationale for this choice of probability ratio is that the stronger the attraction between the site and the ligand, the larger the probability ratio. In addition, the site is exposed to incessant collisions by ligand molecules. The larger the number of such collisions, the larger the probability ratio. Statistical mechanics provides more general and more accurate recipes to compute such ratios. We shall discuss this in Section 1.5. Instead of being proportional to the attractive energy, the theory tells us that the probability ratio is proportional to $\exp(-\beta U_j)$, where $\beta = (k_BT)^{-1}$ and U_j is the interaction energy. Instead of being proportional to the ligand *concentration* C, the theory tells us that it should be proportional to the absolute activity (which is a monotonic increasing function of C). Hence, one should identify aC with λ, the absolute activity of the ligand (see also Section 1.2). For the purpose of this section we retain the form (1.1.11), and add only the requirement

$$P_j(0) + P_j(1) = 1 \tag{1.1.12}$$

to obtain

$$P_j(0) = \frac{1}{1 + aq_jC}, \qquad P_j(1) = \frac{aq_jC}{1 + aq_jC} \tag{1.1.13}$$

Clearly, when either $C \to 0$ or $q_j \to 0$, the probability of finding the site occupied tends to zero. On the other hand, when either $q_j \to \infty$ or $C \to \infty$, the site will be occupied with certainty.

The quantity

$$\xi_j = 1 + aq_jC = 1 + \lambda q_j \tag{1.1.14}$$

will be referred to as the grand partition function (GPF) of a *single site*. We shall see later how to construct the GPF for more general systems. Here, we extend our qualitative argument to construct the GPF of an adsorbent molecule **P**, having *m identical* and *independent* sites, opened with respect to the ligand at some fixed chemical potential or absolute activity (see also Section 1.2 for more details).

For a system of *m identical* and *independent* sites, the probability of finding k specific sites (say, $j = 1, 2, 3, \ldots, k$) occupied and the remaining sites ($j = k + 1, k + 2, \ldots, m$) empty is

$$P_s(k) = \left(\frac{aqC}{1 + aqC}\right)^k \left(\frac{1}{1 + aqC}\right)^{m-k} = \frac{(\lambda q)^k}{(1 + \lambda q)^m} \tag{1.1.15}$$

Note that since all the sites are identical, we have $q_j = q$ for all $j = 1, 2, \ldots, m$. From the assumption of independence we constructed $P_s(k)$ by taking a product of k factors $P_j(1)$ and $m - k$ factors $P_j(0)$. We stress that $P_s(k)$ refers to a *specific* set of k sites. The probability of finding *any* k sites occupied and the remaining sites empty is

$$P(k) = \binom{m}{k} P_s(k) \tag{1.1.16}$$

In calculating $P(k)$ we simply sum the probabilities of the $\binom{m}{k}$ disjoint events, each of which has the same probability $P_s(k)$.

We denote the denominator in Eq. (1.1.15) by ξ, and refer to this quantity as the GPF of a single adsorbent molecule,

$$\xi = (1 + \lambda q)^m = \sum_{k=0}^{m} \binom{m}{k} (\lambda q)^k \tag{1.1.17}$$

Clearly, each term in this sum represents one configuration of the system. There are altogether

$$\sum_{k=0}^{m} \binom{m}{k} = (1+1)^m = 2^m \tag{1.1.18}$$

configurations (see Fig. 1.2 for $m = 4$), collected in $m + 1$ groups ($k = 0, 1, 2, \ldots, m$). Each term in the GPF is also proportional to the probability of the "event" it represents. As we shall see in Section 1.4, this is a very general property of the GPF.

In Eq. (1.1.17) we derived the GPF of a system having m *independent* sites. Statistical mechanics provide the recipe for constructing the GPF for more general systems. This is discussed in Section 1.4. Here, we present the general form of the GPF of a single adsorbent molecule with m (identical or different) binding sites, namely,

$$\xi = \sum_{k=0}^{m} Q(k)\lambda^k = \sum_{k=0}^{m} \binom{m}{k} Q_s(k)\lambda^k \tag{1.1.19}$$

In the first sum, $Q(k)$ is referred to as the canonical partition function (CPF) of the system having a fixed number of k bound ligands. This quantity is itself a sum over terms, each of which represents one arrangement of the k ligands at the m sites. The terms could be different or equal, depending on whether the sites are different or identical. If all the sites are identical* then we can take one representative, denoted by $Q_s(k)$, and multiply it by the number of such terms $\binom{m}{k}$. In this case, the second equality on the right-hand side (rhs) of Eq. (1.1.19) holds.

The rule for reading the probabilities of the various events (here the events are the occurrence of a specific configuration of the k ligands; we discuss other derived events in Section 1.5) is

$$P_s(k) = \frac{Q_s(k)\lambda^k}{\xi} \tag{1.1.20}$$

Compare this with Eq. (1.1.15). In general, the probabilities $P_s(k)$ cannot be factorized into a product of probabilities each pertaining to a single site [as in Eq. (1.1.15)]. We define the pair correlation function by

$$g_{ij}(1, 1) = \frac{P_{ij}(1, 1)}{P_i(1)P_j(1)} \tag{1.1.21}$$

*In Section 2.2 we shall distinguish between sites that are identical in a strict or in a weak sense. Here, "identical" means that all $Q_s(k)$ are equal, independently of the specific set of k sites.

which measures the extent of dependence between two ligands occupying the two specific sites. Similarly, one defines the triplet correlation functions by

$$g_{ijl}(1, 1, 1) = \frac{P_{ijl}(1, 1, 1)}{P_i(1)P_j(1)P_l(1)} \tag{1.1.22}$$

and so on, for higher-order correlations. Note that the correlation function is defined for a *specific* set of sites. In general, different sites might be differently correlated. We say that a specific set of sites is uncorrelated whenever the corresponding correlation function is unity. It is positively or negatively correlated when the correlation function is greater or smaller than unity.*

The term *correlation* is used throughout this book as a measure of the extent of dependence between *any* two (or more) events pertaining to a binding system. The term *interaction* is frequently used in the literature also as a measure of dependence. We shall refrain from such usage since this might lead to some misinterpretations. An example is discussed in Section 5.10. Instead, we shall reserve the term *interaction* to mean *interaction energy*. Two (or more) particles are said to be interacting with each other whenever there exists a potential energy change in the process of bringing these particles from infinite separation to their final configuration *in vacuum*. Usually, the existence of *interaction* between two ligands occupying two sites also implies the occurrence of correlation between the corresponding events (unless there exists an accidental cancellation by an indirect correlation, see Chapter 4). The reverse is, in general, not true. Two ligands occupying two sites may be correlated but not interacting with each other. These correlations will be the subject of most of this book, beginning in Chapter 4.

The term *cooperativity* will be used almost synonymously with *correlation*, except for restricting its usage to a particular type of event, namely, "site i is occupied and site j is occupied." In Eq. (1.1.21), we defined the *pair* correlation between two such events. In Eq. (1.1.22), we defined the triplet correlation among three such events.

It should be noted that the existence (or nonexistence) of one type of correlation does not, in general, imply the existence (or nonexistence) of another type of correlation. For instance, a system can be pairwise correlated but not triply correlated. In Appendix A, we present two simple probabilistic examples where there exist pair correlations but not triple correlations, and vice versa.

*In the theory of probability the term *correlation* is normally applied to two random variables, in which case correlation means that the average of the product of two random variables X and Y is the product of their averages, i.e., $\langle X \cdot Y\rangle = \langle X\rangle\langle Y\rangle$. Two independent random variables are necessarily uncorrelated. The reverse is usually not true. However, when the term *correlation* applies to *events* rather than to random variables, it becomes equivalent to *dependence* between the events.

1.2. THE SYSTEMS OF INTEREST

As in any treatment of a thermodynamic system one must first describe the system, the properties of which are to be examined. The typical system to be studied in this book consists of M adsorbent molecules, **P**, each having m sites. Each site can accommodate a single ligand molecule **L**. There are only *two* occupancy states of the site: empty or occupied.* For the entire molecule **P** there are $m+1$ occupancy states. For instance, in hemoglobin, the occupancy states are 0, 1, 2, 3, 4, according to the number of bound oxygen molecules.

The real system, consisting of **P** and **L** molecules, is usually dissolved in a solvent denoted by w (w can be a pure one-component liquid, say water, or any mixture of solvents and other solutes) and maintained at some fixed temperature T and pressure p.

Figure 1.3 depicts a series of systems in which the *real* system is reduced to a more simplified system that is more manageable for theoretical study.

First, we remove the solvent and consider only the system of adsorbent and ligand molecules. We make this simplification not because solvent effects are unimportant or negligible. On the contrary, they are very important and sometimes can dominate the behavior of the systems. We do so because the development of the theory of cooperativity of a binding system in a solvent is extremely complex. One could quickly lose insight into the molecular mechanism of cooperativity simply because of notational complexity. On the other hand, as we shall demonstrate in subsequent chapters, one can study most of the aspects of the theory of cooperativity in unsolvated systems. What makes this study so useful, in spite of its irrelevance to real systems, is that the basic formalism is unchanged by introducing the solvent. The theoretical results obtained for the unsolvated system can be used almost unchanged, except for reinterpretation of the various parameters. We shall discuss solvated systems in Chapter 9.

Second, we *define* our system (whether in a solvent or not) as the system of M (M being very large) adsorbent molecules, including any ligands that are bound to them. The new system is at equilibrium with a very large reservoir of ligand molecules at a *fixed* chemical potential μ. Thermodynamically, our system is now closed with respect to **P** but open with respect to **L**. The free ligand molecules are not considered as part of the system but rather part of the environment. Like a thermostat that maintains a fixed temperature T, the ligand reservoir maintains a fixed chemical potential μ.

*We shall never discuss a continuous state of occupation. For instance, a ligand **L** approaching a site might interact with **P** according to some interaction potential $U(R)$, where R is the distance (and, in general, also the relative orientation) between the ligand and the site. One can, in principle, define the (continuous) state of occupation with respect to the distance R or the interaction energy $U(R)$. In this book we assume that the site is either empty or occupied, and no intermediate states are considered.

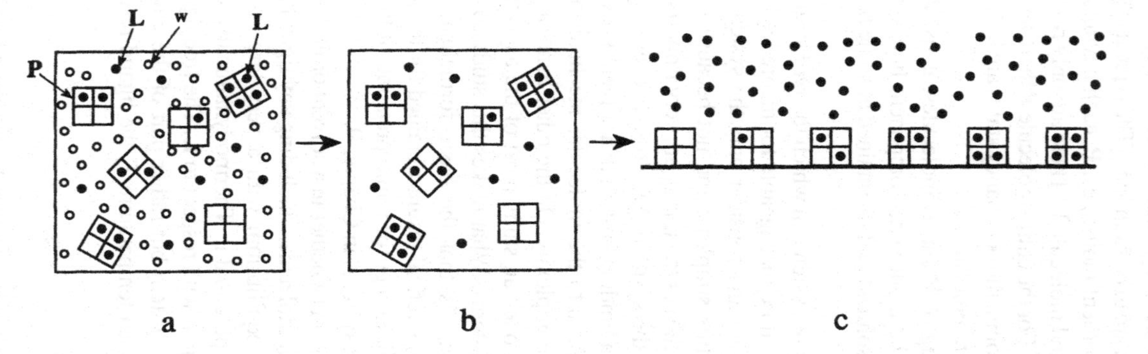

Figure 1.3. Three stages in the process of simplification of the thermodynamic system under consideration. (a) The original system consists of M adsorbent molecules (**P**), each of which has four binding sites, solvent (w) and ligand (**L**) molecules, all in a volume V and at temperature T. (b) The solvent is removed. (c) The final system consists of M *localized* adsorbent molecules opened with respect to the ligands.

In general, the chemical potential of any species **L** can be written as[†]

$$\mu = \mu^* + k_B T \ln C\Lambda^3 \tag{1.2.1}$$

where k_B is the Boltzmann constant ($1.3807\ 10^{-23}$JK^{-1}), T is the absolute temperature, and C is the number density $C = N/V$ (where N and V can be either fixed or average quantities, depending on the type of ensemble); Λ^3 is the momentum partition function, or the de Broglie thermal wavelength, and is given by

$$\Lambda^3 = \left(\frac{h}{\sqrt{2\pi m_L k_B T}}\right)^3 \tag{1.2.2}$$

where h is Planck's constant (6.626×10^{-34} Js) and m_L is the mass of a single molecule. We note that Λ^3 has dimensions of V^{-1}, hence $C\Lambda^3$ is a dimensionless quantity. We also note that for classical systems, for which Eq. (1.2.1) is valid, $C\Lambda^3 << 1$. The quantity μ^* is referred to as the pseudo-chemical potential. In general, μ^* depends on the density C, but for our purposes we shall always assume that μ^* is independent of C. This is true either when the ligand is in an ideal gas phase, or when it is in a very dilute solution in a solvent.

We define the absolute activity of the ligand by

$$\lambda = \exp(\beta\mu) = \lambda_0 C \tag{1.2.3}$$

where $\beta = (k_B T)^{-1}$. Note that whatever the dependence of μ^* on the density C, thermodynamic stability requires that μ or λ be a monotonically increasing function of C. In our special case discussed above, we assume that λ_0 (or μ^*) is independent of C. Hence λ is simply proportional to the density C.

The third step of our simplification is to "freeze-in" the translational and rotational degrees of freedom of the entire **P** molecules. Clearly, our system is now different and all the thermodynamic functions, such as energy, entropy, etc., are changed. However, being interested in the binding properties of the system, it can be shown that if the ligands are very small compared with the adsorbed molecules, then the binding isotherm, hence the cooperativities, will be almost unchanged by this simplification.[‡] We refer the reader to Appendix B for further discussion of this step.

Finally, we reduce the multitudinal number of energy levels of each molecule, **P** or **L**, to a very few, enough to obtain insight into the mechanism of communication between the sites. Once this insight is gained, it is easy to reintroduce all the original energy levels into the final results. The more general results are obtained by reinterpreting the various parameters involved in the simplified models. This is very much the same as we do by eliminating and reintroducing the solvent.

[†]See any textbook on statistical thermodynamics, such as, Hill (1960) or Ben-Naim (1992).
[‡]An exception to this assumption is discussed in Section 5.10.

1.3. STATES OF THE SYSTEM AND THEIR ENERGIES

Having defined the system to be studied, we proceed to characterize the *states* of the system and the corresponding *energies*. These are the fundamental building blocks for constructing the grand partition function (GPF) of the system (Section 1.4).

The *states** of our macroscopic system **P** are numerous and unknown, so we shall always consider only a few macrostates and their corresponding energies. The transition from the microstates of the system to the macrostates used throughout the book is described in Appendix C. Also, it is shown there that the "energy levels" of the macrostates when applied to the real system are actually free-energy levels. The free-energy character of these "energy levels" is suppressed, first, for convenience, otherwise there is no way to proceed with the theory using the entire set of true energy levels, and second to gain insight into the way *free energies* emerge as a result of the averaging process. We shall encounter two types of averaging process in the following chapters.

For convenience, we shall distinguish two types of states.

1. *Conformational states.* Since **P** is either a macromolecule such as hemoglobin, or even a relatively small molecule such as succinic acid, it has a very large number of conformations. We shall reduce these to a very few conformations and refer to them as the macrostates, or simply as the states of **P**. In most cases the ligand will be considered to be in a single macrostate. Only in one case (Section 5.10) we shall allow different conformations for ligand **L**.

2. *Occupancy states.* When the sites are different, we shall need to specify which of the sites are occupied and which are empty. On the other hand, when the sites are identical it is often sufficient to specify only the *number* of occupied sites.

In order to compute the binding isotherm (Section 2.1) of any system, one must know all the microstates of the system. This cannot be done for even the smallest binding system. However, in order to understand the origin of cooperativity and the mechanism by which ligands cooperate, it is sufficient to consider simple models having only a few macrostates. This understanding will be helpful for the selection of methods to extract information from experimental data, and for the meaningful interpretation of this information.

Once we have enumerated all the states of our system, we *assign* energies to these states. If we have only two macrostates, say L and H, we shall assign the corresponding energies E_L and E_H. These are not *derived* from either experiment or from computation. They will be used in the theory as parameters of the model. We shall then examine how the cooperativity, or the correlation function, depends on these parameters.

*These are the solutions ψ_i of the time-independent Schrödinger equation $\mathbf{H}\psi_i = E_i\psi_i$, where **H** is the Hamiltonian operator of the system and E_i is the energy corresponding to the state ψ_i.

When a ligand binds to a site j, we assign to the process a binding energy U_j. This is essentially the work required to bring the ligand from rest at some fixed point infinitely far from the site, to the empty site j. As with the parameters E_L and E_H, U_j is also a free energy (Appendix C), but it will be convenient to suppress the free-energy character of this quantity. Of course, for different ligands and different sites we must assign different binding energies.

Two ligands situated at different sites will usually interact via an intermolecular potential function $U(R)$. We shall refer to this interaction as *direct* interaction. By *direct* interaction we mean the work, or the change in energy, for the process of bringing the two ligands from infinite separation to the final distance (and possibly also orientation), in *vacuum*, i.e., the work arises only from the *interaction between the two ligands*. Since we assume discrete sites, and only a single state for the ligand, the function $U(R)$ reduces to a single number, say $U(R_d)$, where R_d is the ligand–ligand distance at the two sites. Sometimes we shall use the notation $U(1, 1)$ or $U_{ab}(1, 1)$ when the two occupied sites are of the same or different kind.

The triplet *direct* interaction between three ligands is defined as the work, or the energy change, associated with the process of bringing the three particles from infinite separation between each other to the final configuration, say $(\mathbf{R}_1, \mathbf{R}_2, \mathbf{R}_3)$, in *vacuum*. Again, this work arises from the *direct interactions* among the three particles. A similar definition applies for any number of ligands being at some specified configuration.

We shall always assume that the *direct* interaction among any group of ligands is *pairwise additive*. For example, for three ligands at $(\mathbf{R}_1, \mathbf{R}_2, \mathbf{R}_3)$ we write

$$U(\mathbf{R}_1, \mathbf{R}_2, \mathbf{R}_3) = U(\mathbf{R}_1, \mathbf{R}_2) + U(\mathbf{R}_1, \mathbf{R}_3) + U(\mathbf{R}_2, \mathbf{R}_3) \tag{1.3.1}$$

which means that the work associated with the process of bringing the three particles from infinite separation to the final configuration $(\mathbf{R}_1, \mathbf{R}_2, \mathbf{R}_3)$ is the sum of three works, each associated with the process of bringing a pair of particles to the final configuration $(\mathbf{R}_1, \mathbf{R}_2)$, $(\mathbf{R}_1, \mathbf{R}_3)$, and $(\mathbf{R}_2, \mathbf{R}_3)$.

In subsequent chapters we shall see that part of the ligand–ligand correlation in a binding system is due to the *direct* interaction between two (or more) ligands. This will be referred to as the *direct* correlation. However, there is a second part of the correlation referred to as *indirect*, which is not related to the *direct* interactions between the ligands. The indirect correlations are sometimes far more important than the direct correlations. We shall devote the remainder of this section to examine some properties of the direct interactions. The indirect part of the correlations will be the subject of most of the subsequent chapters.

Consider first two point dipoles. The interaction energy between them, as a function of the distance $R = |\mathbf{R}_2 - \mathbf{R}_1|$ and of the orientations (θ_1, ϕ_1) and (θ_2, ϕ_2), is

$$U(1,2)=\frac{\mu^2}{R^3}\left[\sin\theta_1\sin\theta_2\cos(\phi_2-\phi_1)-2\cos\theta_1\cos\theta_2\right] \tag{1.3.2}$$

where μ is the dipole moment* (assumed identical for the two dipoles). The various angles are shown in Fig. 1.4 ($\phi_{12}=\phi_2-\phi_1$).

Now, suppose we fix the orientations $\theta_1=\theta_2=\frac{1}{2}\pi$, $\phi_1=\phi_2=0$, which will be referred to as up-up, and for which

$$U_{uu}(1,2)=+\frac{\mu^2}{R^3} \tag{1.3.3}$$

This is a repulsive interaction that changes with distance as R^{-3}. On the other hand, fixing the orientation $\theta_1=\frac{1}{2}\pi$, $\theta_2=-\frac{1}{2}\pi$, $\phi_1=\phi_2=0$, referred to as up-down, we find

$$U_{ud}(1,2)=-\frac{\mu^2}{R^3} \tag{1.3.4}$$

This is an attractive interaction, again changing as R^{-3}. Now, suppose we average over all possible orientations of the two dipoles, giving *equal weight* to each orientation. We find

$$U=\frac{\int U(1,2)d\Omega_1 d\Omega_2}{\int d\Omega_1 d\Omega_2}=0 \tag{1.3.5}$$

where $\int d\Omega_i=\int_0^\pi \sin\theta_i d\theta_i \int_0^{2\pi} d\phi_i$. This average is always zero, for any distance R. The reason for this result is simple. For any orientation of the two dipoles there exists another orientation obtained by inverting the direction of one dipole. Since we have given *equal probability* to each orientation, the sum of the attractive interactions will exactly cancel out the sum of the repulsive interactions and the net result is zero.

The actual average interaction between two dipoles is not zero, however. At equilibrium, the different configurations are weighted by the Boltzmann distribution

$$P(\Omega_1,\Omega_2)=\frac{\exp[-\beta U(\Omega_1,\Omega_2)]}{\int \exp[-\beta U(\Omega_1,\Omega_2)]d\Omega_1 d\Omega_2} \tag{1.3.6}$$

and the average interaction under equilibrium conditions is

$$U_{eq}=\int P(\Omega_1,\Omega_2)U(\Omega_1,\Omega_2)d\Omega_1 d\Omega_2 \tag{1.3.7}$$

*Not to be confused with the chemical potential.

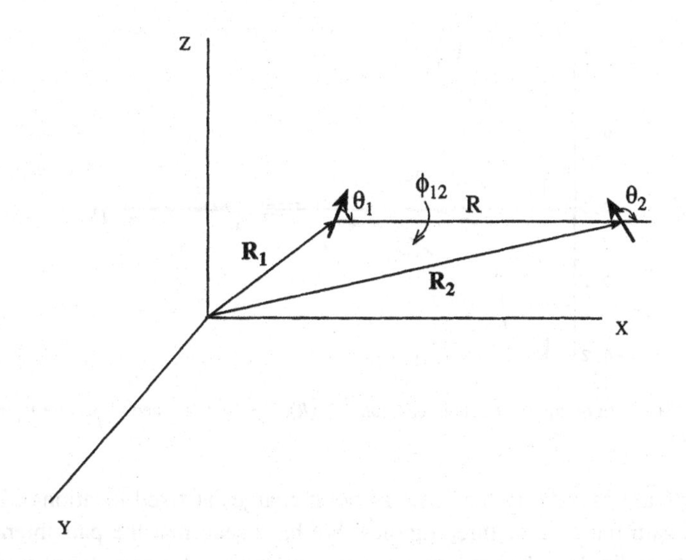

Figure 1.4. Two dipoles at distance $R = |\mathbf{R}_2 - \mathbf{R}_1|$, where $\mathbf{R}_i$ is the locational vector of the center of the dipole i. The orientation of the dipoles is defined by the two angles θ_i and ϕ_i. Only the relative dihedral angle $\phi_{12} = \phi_2 - \phi_1$ enters into the dipole–dipole interaction [Eq. (1.3.2)].

This result is *always attractive*. The reason is again very simple. At any orientation for which $U(\Omega_1, \Omega_2)$ is positive, $U > 0$, there exists another orientation (obtained by inverting the direction of one dipole) for which $U < 0$. The weights given by the Boltzmann factors are now *different*; we have $\exp(-\beta U) > 1$ for $U < 0$, and $\exp(-\beta U) < 1$ for $U > 0$. The average is therefore biased in favor of the attractive orientations, therefore the net result is $U_{eq} < 0$. We note also that when either the distance or the temperature increases so that $|\beta U| << 1$, we can expand the exponent in Eq. (1.3.7) to obtain

$$U_{eq} \approx \frac{\int [1 - \beta U(\Omega_1, \Omega_2)] U(\Omega_1, \Omega_2) d\Omega_1 d\Omega_2}{\int [1 - \beta U(\Omega_1, \Omega_2)] d\Omega_1 d\Omega_2} \approx -\frac{2}{3k_B T}\frac{\mu^4}{R^6} \tag{1.3.8}$$

i.e., the distance dependence is now as R^{-6}. Note also that for any given R, when $T \to \infty$ we have $U_{eq} \to U \to 0$. At this limit the Boltzmann factor becomes independent of Ω and we obtain the average result (1.3.5). Figure 1.5 shows $U_{uu}(R)$, $U_{ud}(R)$, $U(R)$, and the approximate limit of U_{eq} [see Eq. (1.3.8)]. Once we appreciate the effect of averaging on the interaction between two particles, it is easy to understand the effect of averaging on the interaction between three or more particles.

The additivity assumption, as written in Eq. (1.3.1), is exact for hard-sphere particles, point charges, or point dipoles at fixed orientation (point dipoles at fixed

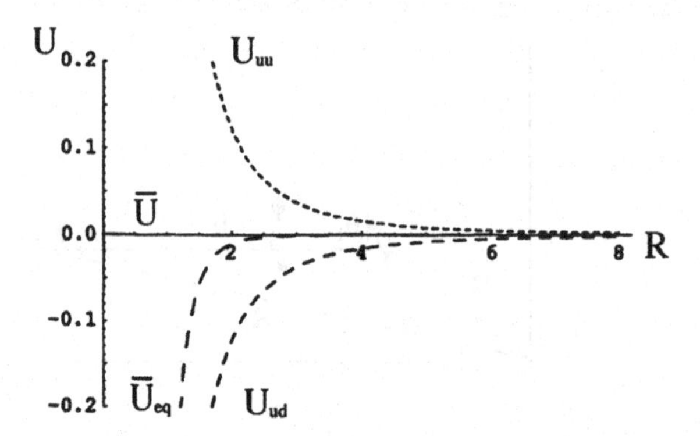

Figure 1.5. The form of dipole–dipole interactions $U_{uu}(R)$, $U_{ud}(R)$, $U(R)$, and $U_{eq}(R)$ for $\mu_1 = \mu_2 = 1$ and $\beta = (k_B T)^{-1} = 1$.

orientations are essentially the same as point charges at fixed locations). Let us examine again the case of three dipoles. We have seen that the pair interaction between two dipoles changes upon averaging over all orientations with the Boltzmann distribution (1.3.6). It is easily realized that when a third dipole is present, the average pair interaction $U_{eq}(1, 2)$ will be modified, since the Boltzmann distribution is now dependent on Ω_3, i.e.,

$$P(\Omega_1, \Omega_2/\Omega_3) = \frac{\exp[-\beta U(\Omega_1, \Omega_2, \Omega_3)]}{\int \exp[-\beta U(\Omega_1, \Omega_2, \Omega_3)] d\Omega_1 d\Omega_2} \tag{1.3.9}$$

hence

$$\begin{aligned} U_{eq}(1, 2/3) &= \int P(\Omega_1, \Omega_2/\Omega_3) U(\Omega_1, \Omega_2) d\Omega_1 d\Omega_2 \\ &\neq \int P(\Omega_1, \Omega_2) U(\Omega_1, \Omega_2) d\Omega_1 d\Omega_2 \\ &= U_{eq}(1, 2) \end{aligned} \tag{1.3.10}$$

As an extreme example suppose we have two dipoles, as before. The equilibrium average interaction between them is always *attractive* [see Eq. (1.3.7) or (1.3.8)]. We now introduce a point charge of any sign as shown in Fig. 1.6. If the dipole–charge interaction is strong, it could orient the dipoles in such a way that the average dipole–dipole interaction will become *repulsive*. Clearly, similar effects will be observed when a third dipole is present. In this case the additivity assumption (1.3.1) becomes invalid.

In spite of what we have seen above, we shall always assume the validity of Eq. (1.3.1) for three or more ligands. In most cases this is indeed a good approxi-

Figure 1.6. Two dipoles in the presence of a point charge *e*.

mation, e.g., for two inert molecules, or polar molecules at large distances. It is sometimes exact (e.g., for a fixed distribution of point charges). The nonadditivity in the triplet interactions discussed above results from averaging over the degrees of freedom (orientations in the case of dipoles) of the particles *involved* in the interactions. We shall encounter other types of correlations—referred to as *indirect*—that arise from averaging over degrees of freedom of molecules *other* than the particles involved in the interactions. This is the case when we average over the *states* of the adsorbent molecule (Chapters 5–8) or over configurations of the solvent molecules (Chapter 9).

We conclude this section by adding one more type of interaction that will appear in some models discussed in this book. These are interactions between subunits in a multisubunit model for adsorbent molecules. The subunit–subunit interaction will in general depend on the conformational states of the subunit. For instance, two subunits in states α, β will be assigned interaction energy $E_{\alpha\beta}$. As with ligand– ligand interaction, subunit–subunit interaction is modified when we *average* over all states of the subunits.

1.4. CONSTRUCTION OF THE PARTITION FUNCTION

Below we present the rules for constructing the partition function (PF) of a binding system.

1. Enumerate all the microstates of the molecule. Each microstate of the molecule, having *k* ligands bound to *k specific sites*, is characterized by an energy level $E_i(k)$. We usually combine many microstates into one macrostate denoted by α, and write the corresponding *canonical* PF as

$$Q_{s,\alpha}(k) = \sum_{i \in \alpha} \exp[-\beta E_i(k)] \tag{1.4.1}$$

where the sum is over all the microstates belonging to the macrostate α (see Appendix C). We shall usually consider only very few macrostates; say one or two (except for Section 4.8, where we take the continuous rotation about a C–C bond in succinic acid).

2. Sum over all macrostates α of the molecule to obtain the canonical PF of the molecule with k ligands bound to k *specific sites* (s),

$$Q_s(k) = \sum_{\alpha} Q_{s,\alpha}(k) \tag{1.4.2}$$

3. Sum over all *specific* sites s to obtain the canonical PF of the molecule with k bound ligands,

$$Q(k) = \sum_{s} Q_s(k) = \binom{m}{k} Q_s(k) \tag{1.4.3}$$

In some cases where the canonical PF of all the $\binom{m}{k}$ *specific* configurations are equal, we say that the system has m identical sites in the *strict sense*. Only in this case does the last equality on the rhs of Eq. (1.4.3) hold. Thus, from Eq. (1.4.1) to (1.4.3) we have proceeded from $Q_{s,\alpha}(k)$ to $Q_s(k)$, the latter being the canonical PF of an adsorbent molecule with k (unspecified) sites occupied by ligands.

4. The grand PF (GPF) of the molecule is now constructed by

$$\xi = \sum_{k=0}^{m} Q(k)\lambda^k \tag{1.4.4}$$

where $\lambda = \exp(\beta\mu)$ is the absolute activity of the ligand. The sum in Eq. (1.4.4) is over all occupation numbers k ($k = 0, 1, 2, \ldots, m$). Normally we shall assume only a few macrostates, say L and H, to which we shall ascribe corresponding "energy levels" E_L and E_H (see also Appendix C). In these cases the construction of the GPF is quite simple.

As an example, consider a system with two conformations L and H, each having two different sites, say a and b. The canonical PFs of the systems are

$$Q(0) = Q_L(0) + Q_H(0) = Q_L + Q_H \tag{1.4.5}$$

$$\begin{aligned} Q(1) &= Q_L(a, 0) + Q_L(0, b) + Q_H(a, 0) + Q_H(0, b) \\ &= Q_L q_{La} + Q_L q_{Lb} + Q_H q_{Ha} + Q_H q_{Hb} \end{aligned} \tag{1.4.6}$$

and

$$Q(2) = Q_L(a, b) + Q_H(a, b) = Q_L q_{La} q_{Lb} S_{LL} + Q_H q_{Ha} q_{Hb} S_{HH} \tag{1.4.7}$$

where $Q_\alpha = \exp(-\beta E_\alpha)$, $q_{\alpha i} = \exp(-\beta U_{\alpha i})$, and $S_{\alpha\alpha} = \exp[-\beta U_{\alpha\alpha}(1, 1)]$. Thus, each $Q(k)$ is a sum over all the states corresponding to a fixed occupation number k. The PF $Q(0)$ has only two states, L and H; $Q(1)$ has four, L occupied at a, L occupied at b, H occupied at a, and H occupied at b; and $Q(2)$ has two states, L fully occupied and H fully occupied. Each of the terms in $Q(k)$ consists of a product of a factor Q_α ($\alpha = L, H$) for the conformational states of **P**, one factor $q_{\alpha i}$ ($\alpha = L, H, i = a, b$) for each ligand occupying the site i, and a factor $S_{\alpha\alpha}$ for the *direct* interaction $[U_{\alpha\alpha}(1, 1)]$ between the two ligands occupying the sites a, b when the state is α ($\alpha = L, H$).

We shall treat more complicated cases, such as systems with a larger number of identical or different sites, and also cases of more than one type of ligand. But the general rules of constructing the canonical PF, and hence the GPF, are the same. The partition functions, either Q or ξ, have two important properties that make the tool of statistical thermodynamics so useful. One is that, for macroscopic systems, each of the partition functions is related to a thermodynamic potential. For the particular PFs mentioned above, these are

$$A = -k_B T \ln Q \tag{1.4.8}$$

and

$$pV = k_B T \ln \Xi \tag{1.4.9}$$

where Q and Ξ are the canonical PF and GPF for macroscopic systems, A is the Helmholz energy, V is the volume, and p is the pressure of the system.

The second property is that each term of the PF is proportional to the probability of occurrence of the particular state it represents when the system is at equilibrium. We shall use mainly the second property of the PF. The next section is devoted to this aspect of the theory. Once we have the probabilities of all possible "events" we can compute average quantities pertaining to the system at equilibrium. Of these, the average occupation number, or the binding isotherm, will be the central quantity to be examined and analyzed in this book.

The skeptical reader may reasonably ask from where we have obtained the above rules and where is the proof for the relation with thermodynamics and for the meaning ascribed to the individual terms of the PF. The ultimate answer is that there is no proof. Of course, the reader might check the contentions made in this section by reading a specialized text on statistical thermodynamics. He or she will find the "proof" of what we have said. However, such proof will ultimately be derived from the fundamental postulates of statistical thermodynamics. These are essentially equivalent to the two properties cited above. The fundamental postulates are statements regarding the connection between the PF and thermodynamics on the one hand (the famous Boltzmann equation for entropy), and the probabilities of the states of the system on the other. It just happens that this formulation of the postulates was first proposed for an isolated system—a relatively simple but uninteresting system (from the practical point of view). The reader interested in the subject of this book but not in the foundations of statistical thermodynamics can safely adopt the rules given in this section, trusting that a "proof" based on some more basic postulates has already been established. The ultimate proof of any physical theory lies in its capacity for predicting results that agree with experiments. Such "proofs" have been abundantly obtained ever since the establishment of statistical thermodynamics almost a century ago.

1.5. PROBABILITIES

The quantity most referred to in this book is the binding isotherm (BI). This is simply the *average* number of bound ligands (per site or per molecule) as a function of the ligand activity (or concentration, or partial pressure). To compute any average quantity at equilibrium one needs to know the probabilities of all the events that contribute to that average. Some of the probabilities can be read directly from the GPF, others may be derived by using elementary rules for calculating probabilities of sums (or unions) and products (or intersections) of events. We shall encounter many examples throughout the book. Here, we present a few examples of these rules.

Consider again the example of a system having two conformational states, L and H, and two different sites, a and b. The relevant canonical PFs are given in Eqs. (1.4.5)–(1.4.7). The event "the system is in state L, having a ligand bound on site a, and empty on site b" is denoted by $(L; a, 0)$. The probability of this event is proportional to the corresponding term in the GPF, i.e.,

$$P(L; a, 0) = \frac{Q_L q_{La} \lambda}{\xi} \tag{1.5.1}$$

where the proportionality constant is simply ξ^{-1}. This guarantees that when we sum over all possible states (both conformational and occupancy) we shall obtain unity, as expected from any distribution. When one speaks of the probability of an event, referring to a *single* molecule, one can interpret this as being equivalent to the mole fraction of finding that event in a very large ensemble ($M \to \infty$) of such molecules under the same conditions of T and λ (and of the composition of any solvent, if present).

Two events are said to be disjoint or mutually exclusive if the occurrence of one excludes the occurrence of the other. We write this as $\mathcal{A} \cdot \mathcal{B} = \varnothing$ where $\varnothing$ denotes the event having zero probability. The product $\mathcal{A} \cdot \mathcal{B}$ (or $\mathcal{A} \cap \mathcal{B}$) denotes the occurrence of $\mathcal{A}$ *and* $\mathcal{B}$. If $\mathcal{A}$ and $\mathcal{B}$ are two disjoint events, then the probability of the event "either $\mathcal{A}$ or $\mathcal{B}$," denoted by $\mathcal{A} + \mathcal{B}$ (or $\mathcal{A} \cup \mathcal{B}$), is given by

$$P(\mathcal{A} + \mathcal{B}) = P(\mathcal{A}) + P(\mathcal{B}) \tag{1.5.2}$$

In the example of Section 1.4, the system can be either in conformational state L or in H. Therefore, for any state of occupation, the two events "L" and "H" are disjoint. Hence the probability of finding the system empty, in either state L or H, is simply

$$P(0,0) = P(L;0,0) + P(H;0,0) = \frac{Q_L}{\xi} + \frac{Q_H}{\xi} \tag{1.5.3}$$

Note that whenever we sum over all possible states of one variable, we eliminate the notation of that variable from the resulting probability. In Eq. (1.5.3), we refer to $P(0,0)$ as the probability of the event "site a and site b are empty." This is obtained by summing over all possible states (here, L and H) of the empty system.

Similarly, the event "the system is either singly or doubly occupied" is obtained by the sum

$$P[(a,0)+(0,b)+(a,b)] = P(a,0) + P(0,b) + P(a,b)$$

$$= \frac{\Sigma_\alpha Q_\alpha q_{\alpha a}\lambda}{\xi} + \frac{\Sigma_\alpha Q_\alpha q_{\alpha b}\lambda}{\xi} + \frac{\Sigma_\alpha Q_\alpha q_{\alpha a} q_{\alpha b} S_{\alpha\alpha}\lambda^2}{\xi} \tag{1.5.4}$$

where the sum over α is over the two states L and H. Clearly, the three events "$(a, 0)$," "$(0, b)$," and "(a, b)" are disjoint, and the probability of the sum is the sum of their probabilities. Note that the sign "+" on the left-hand side (lhs) of Eq. (1.5.4) stands for the *union* of two events. The "+" on the rhs stands for the *addition* of two numbers.

When the two events $\mathcal{A}$ and $\mathcal{B}$ are not disjoint, i.e., when the occurrence of one event does not exclude the occurrence of the other, we have the relation

$$P(\mathcal{A}+\mathcal{B}) = P(\mathcal{A}) + P(\mathcal{B}) - P(\mathcal{A}\cdot\mathcal{B}) \tag{1.5.5}$$

Compare this with Eq. (1.5.2). Here, we must subtract the probability of the product (or the intersection, also denoted $\mathcal{A}\cap\mathcal{B}$) of the two events.

The two events "the system is in state L" *and* "the system is singly occupied" are clearly not disjoint. The occurrence of one does not exclude the occurrence of the other. In order to construct the probability of the sum of these two events, we need the following three ingredients:

$$P(L) = \frac{\sum_{k=0}^{2} Q_L(k)\lambda^k}{\xi} \tag{1.5.6}$$

$$P(1) = P[(a,0) + P(0,b)] = \frac{[Q_L(a,0) + Q_L(0,b) + Q_H(a,0) + Q_H(0,b)]\lambda}{\xi} \tag{1.5.7}$$

and

$$P(L\ \textbf{and}\ singly\ occupied) = \frac{[Q_L(a,0) + Q_L(0,b)]\lambda}{\xi} \tag{1.5.8}$$

Hence the probability of the required event "either L or singly occupied" is given by

$$P(either\ L\ \textbf{or}\ singly\ occupied) = P(L) + P(1) - P(L\ \textbf{and}\ singly\ occupied)$$

$$= \frac{\sum_{k=0}^{2} Q_L(k)\lambda^k + [Q_H(a,0) + Q_H(0,b)]\lambda}{\xi}$$

$$= \frac{Q_L + [Q_L(1) + Q_H(1)]\lambda + Q_L(2)\lambda^2}{\xi} \tag{1.5.9}$$

Note that if we do not subtract Eq. (1.5.8) from the sum of Eqs. (1.5.6) and (1.5.7), we would have counted twice the term $[Q_L(a, 0) + Q_L(0, b)]\lambda/\xi$. The last result on the rhs of Eq. (1.5.9) can also be obtained by directly collecting all the relevant terms from the GPF; first, take all the terms having subscript L [i.e., Q_L, $Q_L(1)$ and $Q_L(2)$], then add the term with subscript H pertaining to single occupation only [i.e., $Q_H(1)$].

Two events are said to be independent if and only if the probability of their product (or intersection) is equal to the product of their probabilities, i.e.,

$$P(\mathcal{A} \cdot \mathcal{B}) = P(\mathcal{A} \cap \mathcal{B}) = P(\mathcal{A}) \cdot P(\mathcal{B}) \tag{1.5.10}$$

The independence of two events means that the occurrence of one event does not affect the probability of occurrence of the second. We define the *conditional* probability of occurrence of $\mathcal{A}$, given that the event $\mathcal{B}$ occurred, by

$$P(\mathcal{A}/\mathcal{B}) \equiv \frac{P(\mathcal{A} \cdot \mathcal{B})}{P(\mathcal{B})} \tag{1.5.11}$$

and similarly

$$P(\mathcal{B}/\mathcal{A}) \equiv \frac{P(\mathcal{A} \cdot \mathcal{B})}{P(\mathcal{A})} \tag{1.5.12}$$

Clearly, when $\mathcal{A}$ and $\mathcal{B}$ are independent it follows from Eq. (1.5.10) that the conditional probability is equal to the unconditional probability,* i.e.,

*Strictly speaking, any probability is "conditional" in the sense that an experiment has been performed. Normally we suppress this condition in our notation.

$$P(\mathcal{A}/\mathcal{B}) = \frac{P(\mathcal{A}) \cdot P(\mathcal{B})}{P(\mathcal{B})} = P(\mathcal{A}) \tag{1.5.13}$$

The events "the system is singly occupied" and "the system is in state L" are clearly not independent,

$$P(singly\ occupied) = P(1) = P[(a, 0) + (0, b)] \tag{1.5.14}$$

$$P(L) = P(L; 0, 0) + P(L; a, 0) + P(L; 0, b) + P(L; a, b) \tag{1.5.15}$$

The conditional probability is [from Eqs. (1.5.7) and (1.5.8)]

$$P(L/singly\ occupied) = \frac{Q_L(a, 0) + Q_L(0, b)}{Q_L(a, 0) + Q_L(0, b) + Q_H(a, 0) + Q_H(0, b)} \tag{1.5.16}$$

On the other hand [from Eqs. (1.5.8) and (1.5.15)]

$$P(singly\ occupied/L) = \frac{[Q_L(a, 0) + Q_L(0, b)]\lambda}{Q_L + [Q_L(a, 0) + Q_L(0, b)]\lambda + Q_L(a, b)\lambda^2} \tag{1.5.17}$$

Replacing the condition "L" by "H" will result in a different conditional probability,

$$P(singly\ occupied/H) = \frac{[Q_H(a, 0) + Q_H(0, b)]\lambda}{Q_H + [Q_H(a, 0) + Q_H(0, b)]\lambda + Q_H(a, b)\lambda^2} \tag{1.5.18}$$

Whenever two events are not independent, i.e., when Eq. (1.5.10) is not fulfilled, we define a correlation function by

$$g(\mathcal{A}, \mathcal{B}) = \frac{P(\mathcal{A} \cdot \mathcal{B})}{P(\mathcal{A}) \cdot P(\mathcal{B})} \tag{1.5.19}$$

or, equivalently, by

$$P(\mathcal{A}/\mathcal{B}) = g(\mathcal{A}, \mathcal{B})P(\mathcal{A}) \tag{1.5.20}$$

and

$$P(\mathcal{B}/\mathcal{A}) = g(\mathcal{A}, \mathcal{B})P(\mathcal{B}) \tag{1.5.21}$$

*The definition of correlation functions in this book differs from the definition of the correlation *coefficient* in the theory of probability. The difference is essentially in the normalization, i.e., whereas $g(\mathcal{A}, \mathcal{B})$ can be any positive number $0 \leq g \leq \infty$, the correlation coefficient varies within $[-1, 1]$. We have chosen the definition of correlation as in Eq. (1.5.19) or (1.5.20) to conform with the definition used in the theory of liquids and solutions.

The correlation function measures the extent of deviation from independence.*
Clearly, when the two events are independent then $g(\mathcal{A}, \mathcal{B}) = 1$. Note that independence is defined symmetrically with respect to $\mathcal{A}$ and $\mathcal{B}$. Hence, also $g(\mathcal{A}, \mathcal{B})$ is symmetrical with respect to $\mathcal{A}$ and $\mathcal{B}$. Thus, whenever the occurrence of $\mathcal{A}$ affects the probability of occurrence of event $\mathcal{B}$, also the occurrence of $\mathcal{B}$ will affect the probability of $\mathcal{A}$. This is also clearly seen from Eqs. (1.5.20) and (1.5.21). We shall say that the events $\mathcal{A}$ and $\mathcal{B}$ are *positively* correlated whenever $P(\mathcal{A}/\mathcal{B}) > P(\mathcal{A})$ or, equivalently, $g(\mathcal{A}, \mathcal{B}) > 1$. They are *negatively* correlated whenever $P(\mathcal{A}/\mathcal{B}) < P(\mathcal{A})$ or, equivalently, $0 < g(\mathcal{A}, \mathcal{B}) < 1$.

We have defined the correlation function for any two events. Of particular interest in this book will be correlations between events such as "site a is occupied" and "site b is occupied." For the specific system described in the previous section we have

$$P(\textit{site } a \textit{ is occupied}) = P[(a, 0) + (a, b)]$$

$$= \frac{[Q_L(a, 0) + Q_H(a, 0)]\lambda + [Q_L(a, b) + Q_H(a, b)]\lambda^2}{\xi} \tag{1.5.22}$$

$$P(\textit{site } b \textit{ is occupied}) = P[(0, b) + (a, b)]$$

$$= \frac{[Q_L(0, b) + Q_H(0, b)]\lambda + [Q_L(a, b) + Q_H(a, b)]\lambda^2}{\xi} \tag{1.5.23}$$

$$P(\textit{site } a \textbf{ and } \textit{site } b \textit{ are occupied}) = P(a, b)$$

$$= \frac{[Q_L(a, b) + Q_H(a, b)]\lambda^2}{\xi} \tag{1.5.24}$$

The correlation function for these two events is clearly a ratio of two polynomials in λ. We shall need only the $\lambda \to 0$ limit of this correlation function, which in this case is

$$\overset{\circ}{g}(a, b) = \lim_{\lambda \to 0} g(a, b)$$

$$= \frac{[Q_L(a, b) + Q_H(a, b)][Q_L + Q_H]}{[Q_L(a, 0) + Q_H(a, 0)][Q_L(0, b) + Q_H(0, b)]} \tag{1.5.25}$$

This, and similar quantities defined for more than two sites, will be studied extensively in the following chapters.

2

The Binding Isotherm

2.1. THE GENERAL FORM OF THE BINDING ISOTHERM

The binding isotherm (BI) of any binding system was originally referred to as a curve of the amount of ligands adsorbed as a function of the concentration or partial pressure of the ligand at a fixed temperature. A typical curve of this kind is shown in Fig. 2.1. Numerous molecular models have been studied that simulate particular experimental BIs.

The formal derivation of the BI from the GPF is based on the thermodynamic relation

$$\overline{N} = \left(\frac{\partial(pV)}{\partial\mu}\right)_{T,V} \tag{2.1.1}$$

where $\overline{N}$ is the average number of molecules in a macroscopic open system, V is the volume of the system, and p is the partial pressure. The connection with the

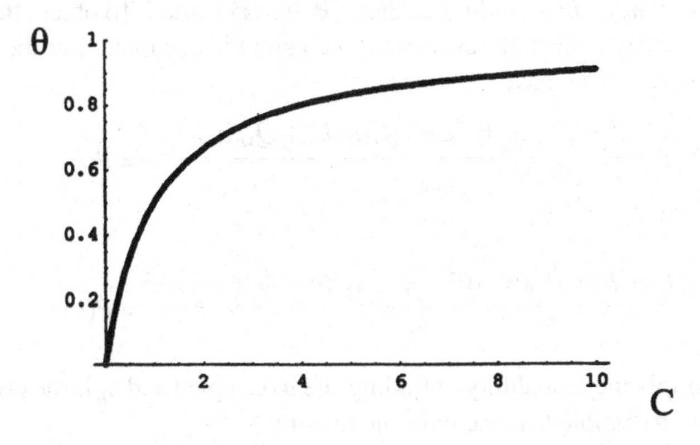

Figure 2.1. A typical binding isotherm, θ, as a function of the concentration, C, for $k = 1$.

GPF of the system [Eq. (1.4.9)] leads to

$$\overline{N} = k_B T \left(\frac{\partial \ln \Xi}{\partial \mu} \right)_{T,V} \tag{2.1.2}$$

Since we have assumed that our macroscopic system consists of M independent adsorbent molecules, we can use either $\Xi = \xi^M / M!$ or $\Xi = \xi^M$ (see Appendix B) to obtain

$$\bar{n} = \frac{\overline{N}}{M} = \lambda \left(\frac{\partial \ln \xi}{\partial \lambda} \right)_T = \frac{\Sigma k Q(k) \lambda^k}{\xi} = \sum_{k=1}^{m} k P(k) \tag{2.1.3}$$

where we have used the relation $\mu = k_B T \ln \lambda$ to convert the derivative with respect to the chemical potential into a derivative with respect to the absolute activity λ; $\bar{n}$ is the average number of bound ligands per adsorbent molecule. The last equality in Eq. (2.1.3) confirms this interpretation of $\bar{n}$; $P(k)$ is the probability of finding the adsorbent molecule with k bound ligands; hence $\Sigma k P(k)$ is the average number of bound ligands.

Thus, accepting the rules for assigned probabilities discussed in Section 1.5 is sufficient for calculating the average number of bound ligands, or any other average relevant to our system. The function $\bar{n} = \bar{n}(T, \lambda)$ is the equation of state of the bound ligands. Following $\bar{n}$ as a function of λ (or concentration) provides the required *binding isotherm*. Normally, when all the sites are identical, one follows the average number of ligands per site. For an m-site system, this is simply defined by

$$\theta = \frac{\bar{n}}{m} \tag{2.1.4}$$

The study of BIs per site is important when the sites are different, in which case we have a different *individual* binding isotherm θ_i for each site i. To obtain these from the GPF we simply collect all terms for which site i is occupied. For the example given in Section 1.4, we have

$$\theta_a = \frac{[Q_L(a, 0) + Q_H(a, 0)]\lambda + [Q_L(a, b) + Q_H(a, b)]\lambda^2}{\xi} = P(a) \tag{2.1.5}$$

and

$$\theta_b = \frac{[Q_L(0, b) + Q_H(0, b)]\lambda + [Q_L(a, b) + Q_H(a, b)]\lambda^2}{\xi} = P(b) \tag{2.1.6}$$

Thus, θ_a is simply the probability of finding site a occupied and θ_b is the probability of finding site b occupied, so we have the identity

$$\bar{n} = \theta_a + \theta_b \tag{2.1.7}$$

and, in general, for m different sites

$$\bar{n} = \Sigma\, \theta_i \tag{2.1.8}$$

It should be noted that although θ_i is the *probability* of finding "site i occupied," the sum $\bar{n}$ is *not* the probability of the sum (or union) of these events. In fact, $\bar{n}$ is not a probability at all. What we actually sum over are not probabilities but average quantities. To see this, we reinterpret $P(a)$ [which is a genuine probability $0 \le P(a) \le 1$] as an average quantity,

$$\theta_a = 0 \cdot [1 - P(a)] + 1 \cdot P(a) = P(a) \tag{2.1.9}$$

and

$$\theta_b = 0 \cdot [1 - P(b)] + 1 \cdot P(b) = P(b) \tag{2.1.10}$$

Since site a can be either empty [with probability $1 - P(a)$] or occupied [with probability $P(a)$], θ_a is the average occupation number for the site a. Clearly, $0 \le \theta_a \le 1$. When forming the sum $\bar{n}$ in Eq. (2.1.7) or (2.1.8), we sum over all *average* quantities θ_i and obtain the average occupation number for the entire molecule. Clearly $0 \le \bar{n} \le m$, and in general $\bar{n}$ is not a probability.

There are many other averages that can be defined by collecting the relevant probabilities from the GPF. We mention here two more types of individual BIs. Before doing so, we rewrite $P(a)$ and $P(b)$ as $P(a, \_)$ and $P(\_, b)$, respectively. The blank "_" denotes an unspecified state. Thus, $P(a)$ is the probability of finding "site a occupied," which is the same as the probability of finding "site a occupied and the state of site b unspecified." We can now construct two *conditional* probabilities, hence individual BIs for which the state of site b *is specified*.

The first is denoted by $P(a/b = 0)$* which, for the model of Section 1.4, is

$$P(a/b = 0) = \frac{P(a, 0)}{P(b = 0)} = \frac{(Q_L q_{La} + Q_H q_{Ha})\lambda}{Q_L + Q_H + (Q_L q_{La} + Q_H q_{Ha})\lambda}$$

$$= \frac{(X_L^0 q_{La} + X_H^0 q_{Ha})\lambda}{1 + (X_L^0 q_{La} + X_H^0 q_{Ha})\lambda} \tag{2.1.11}$$

Clearly, this is a simple Langmuir (see Section 2.4) isotherm with a binding constant for site a that is an average of the binding constants for site a in a system at states L and H, with weights X_L^0 and X_H^0, respectively. Similarly, we define the conditional probability of finding site a occupied given that site b is occupied which, for the

*$a/b = 0$ means: a is occupied *given* that b is empty. $a/b = 1$ means: a is occupied *given* that b is occupied.

model of Section 1.4, is

$$P(a/b=1)=\frac{P(a,b)}{P(b=1)}=\frac{(Q_L q_{La} q_{Lb} S_{LL}+Q_H q_{Ha} q_{Hb} S_{HH})\lambda}{(Q_L q_{Lb}+Q_H q_{Hb})+(Q_L q_{La} q_{Lb} S_{LL}+Q_H q_{Ha} q_{Hb} S_{HH})\lambda}$$

$$=\frac{(X_L^{(b)} q_{La} S_{LL}+X_H^{(b)} q_{Ha} S_{HH})\lambda}{1+(X_L^{(b)} q_{La} S_{LL}+X_H^{(b)} q_{Ha} S_{HH})\lambda} \tag{2.1.12}$$

This is again a simple Langmuir (see Section 2.4) isotherm with a binding constant to site *a* which is the average of the binding constants to site *a* when the system is in states *L* and *H*. The quantities $X_L^{(b)}$ and $X_H^{(b)}$ are the mole fractions of states *L* and *H* given that site *b* is occupied.

Clearly, both $P(a/b=0)$ and $P(a/b=1)$ can be interpreted as individual BIs for site *a* (and similar definitions apply to site *b*). It should be noted that all three individual BIs defined above can, in principle, be measured experimentally. The conditions of the experiments are different. In $P(a/b=0)$ [Eq. (2.1.11)], we follow the average occupation of site *a* while maintaining site *b* empty. On the other hand, in $P(a/b=1)$ [Eq. (2.1.12)], we follow the average occupation of site *a* while we secure the occupation of site *b*. In θ_a [Eq. (2.2.9)], we follow the occupation of site *a* while leaving site *b* unrestricted to bind ligands under the same conditions as if we were to measure θ, or $\bar{n}$, but monitor the binding on *a* only.

The binding isotherm is a monotonous increasing function of λ. This follows from the thermodynamic stability of the macroscopic system. Thus

$$\frac{\partial \bar{n}}{\partial \lambda}=\frac{1}{M}\left(\frac{\partial \bar{N}}{\partial \lambda}\right)_T>0 \tag{2.1.13}$$

The thermodynamic stability requires that $\bar{N}$ be a monotonous increasing function of λ. Due to the independence of the adsorbent molecules, the same is true also for $\bar{n}$ as a function of λ.

The statistical mechanical interpretation of the stability condition is quite simple. From Eq. (2.1.3) we obtain by differentiation

$$\lambda\frac{\partial \bar{n}}{\partial \lambda}=\Sigma k^2 P(k)-[\Sigma k P(k)]^2=\langle k^2\rangle-\langle k\rangle^2 \tag{2.1.14}$$

This is the fluctuation, or the variance, of *k* for a single adsorbent molecule. This is always positive since

$$\langle (k-\langle k\rangle)^2\rangle=\langle k^2\rangle-2\langle k\rangle\langle k\rangle+\langle k\rangle^2=\langle k^2\rangle-\langle k\rangle^2\geq 0 \tag{2.1.15}$$

The slope of the BI is related to the slope of the titration curve (see Section 2.6).

The latter is sometimes referred to as the binding capacity. It measures the amount of acid (or base) that should be added to the solution so that the pH will increase by one unit.

2.2. THE INTRINSIC BINDING CONSTANTS

We introduce here the quantities referred to as *intrinsic* binding constants. They are *intrinsic* in the sense that they pertain to a specific site or sites. They are also related to the free-energy change for bringing a ligand from some specified state in the reservoir onto a *specific* site. We shall also introduce the conditional binding constants, and binding constants for a group of specific sites. In all cases, whenever we *specify* the binding site (or sites), there are no combinatorial factors such as the ones that appear when the site (or sites) are not specified. The latter appear in the *thermodynamic* binding constants discussed in the next section.

To maintain brevity and keep the notation simple, we shall introduce all the relevant quantities for a three-site system. The generalization to an m-site system is quite straightforward. We start with three *different* sites, denoted by a, b, and c. The corresponding GPF is

$$\xi = \sum_{k=0}^{3} Q(k)\lambda^k \tag{2.2.1}$$

In this section we characterize our system only by the pattern of occupations of the sites and overlook the conformational states. Hence, we write

$$Q(0) = Q(0, 0, 0) \tag{2.2.2}$$

$$Q(1) = Q(a, 0, 0) + Q(0, b, 0) + Q(0, 0, c) \tag{2.2.3}$$

$$Q(2) = Q(a, b, 0) + Q(0, b, c) + Q(a, 0, c) \tag{2.2.4}$$

$$Q(3) = Q(a, b, c) \tag{2.2.5}$$

Here $Q(k)$ is the canonical PF of the system with k sites occupied, while on the rhs of Eqs. (2.2.2)–(2.2.5) we *specify* the sites that are occupied. For instance, $Q(a, 0, 0)$ is the PF for the system occupied at site a but empty at b and c.

In what follows we shall always write $\lambda = \lambda_0 C$. We assume that the ligand is provided from either an ideal gas phase or an ideal dilute solution. Hence, λ_0 is related to the standard chemical potential and is independent of the concentration C. On the other hand, for the nonideal phase, λ_0 will in general depend on concentration C. A first-order dependence on C is discussed in Appendix D. Note also that λ is a dimensionless quantity. Therefore, any units used for concentration C must be the same as for $(\lambda_0)^{-1}$.

We define the following seven *intrinsic* binding constants,

$$\left.\begin{aligned} k_a &= \frac{Q(a,0,0)}{Q(0,0,0)}\lambda_0, & k_b &= \frac{Q(0,b,0)}{Q(0,0,0)}\lambda_0, & k_c &= \frac{Q(0,0,c)}{Q(0,0,0)}\lambda_0 \\ k_{ab} &= \frac{Q(a,b,0)}{Q(0,0,0)}\lambda_0^2, & k_{bc} &= \frac{Q(0,b,c)}{Q(0,0,0)}\lambda_0^2, & k_{ac} &= \frac{Q(a,0,c)}{Q(0,0,0)}\lambda_0^2 \\ k_{abc} &= \frac{Q(a,b,c)}{Q(0,0,0)}\lambda_0^3 \end{aligned}\right\} \qquad (2.2.6)$$

We stress again that these binding constants are *intrinsic* only in the sense that they refer to a *specific* set of sites. In terms of these constants, the GPF of the adsorbent molecule is written as

$$\xi = Q(0,0,0)[1 + (k_a + k_b + k_c)C + (k_{ab} + k_{bc} + k_{ac})C^2 + k_{abc}C^3] \qquad (2.2.7)$$

and the BI per molecule now has the more familiar form

$$\bar{n} = \frac{(k_a + k_b + k_c)C + 2(k_{ab} + k_{bc} + k_{ac})C^2 + 3k_{abc}C^3}{1 + (k_a + k_b + k_c)C + (k_{ab} + k_{bc} + k_{ac})C^2 + k_{abc}C^3} \qquad (2.2.8)$$

Since λ_0 is presumed to be independent of C, all the intrinsic constants are also independent of C. Note also that the factor $Q(0, 0, 0)$ does not appear explicitly in $\bar{n}$, although it is contained in the definitions of all the binding constants.

The individual BI can be obtained either from the probabilities read from the GPF, as we have done in Section 2.1, or from Eq. (2.2.8). For instance, the individual BI of site a, expressed in terms of the intrinsic constants, is

$$\theta_a = \frac{k_aC + (k_{ab} + k_{ac})C^2 + k_{abc}C^3}{\xi'} \qquad (2.2.9)$$

with similar expressions for θ_b and θ_a. The sum of these gives

$$n = \theta_a + \theta_b + \theta_c \qquad (2.2.10)$$

Note that ξ' is simply $\xi/Q(0, 0, 0)$, i.e., the denominator of Eq. (2.2.8).

Each of the intrinsic binding constants may be interpreted either as a probability ratio or as a free-energy change for a specific binding process. Both meanings are derived directly from definitions (2.2.6). For instance,

$$k_aC = \frac{Q(a,0,0)}{Q(0,0,0)}\lambda = \frac{P(a,0,0)}{P(0,0,0)} \qquad (2.2.11)$$

and similarly

$$k_{ab}C^2 = \frac{Q(a, b, 0)}{Q(0, 0, 0)}\lambda^2 = \frac{P(a, b, 0)}{P(0, 0, 0)} \tag{2.2.12}$$

The thermodynamic interpretation of k_aC follows also from definitions (2.2.6),

$$k_aC = \exp\{-\beta[A(a, 0, 0) - A(0, 0, 0) - \mu]\} \tag{2.2.13}$$

The quantity in square brackets is the Helmholtz energy change for the process of bringing a ligand from the reservoir at a given chemical potential μ onto a specific site, here a, of an empty molecule. The process is carried out at constant temperature T.

One could also interpret k_a itself in terms of a similar process where the ligand is brought from a hypothetical "standard state" corresponding to $C = 1$. This interpretation is somewhat risky and should be avoided.* Symbolically, we write the process as

$$\mathbf{L} + (0, 0, 0) \to (a, 0, 0) \tag{2.2.14}$$

Similarly, the quantity $k_{ab}C^2$ is related to the Helmholtz energy change for the process

$$2\mathbf{L} + (0, 0, 0) \to (a, b, 0) \tag{2.2.15}$$

i.e., the process of bringing two ligands from the reservoir at μ, onto two specific sites, here a and b, of an empty molecule. Although the intrinsic constants are commonly defined for a single site, there is no reason for not defining them for any group of sites. The important point in defining an intrinsic binding constant is that we refer to a specific site or group of sites. This requirement applies both when the sites are different and when they are identical.

It is sometimes convenient to introduce *conditional* intrinsic binding constants. These are defined in the same way as the conditional probabilities, namely,

$$k_{a/b} = \frac{k_{ab}}{k_b} = \frac{Q(a, b, 0)}{Q(0, b, 0)}\lambda_0 \tag{2.2.16}$$

The probabilistic meaning of $k_{a/b}C$ [compare with Eq. (2.2.11)] is

$$k_{a/b}C = \frac{P(a, b, 0)}{P(0, b, 0)} \tag{2.2.17}$$

which is the probability ratio for the events $(a, b, 0)$ and $(0, b, 0)$.† The corresponding

*For a discussion of the meaning of such hypothetical states of $C = 1$, see Ben-Naim (1972).

†Note that this is slightly different from the conditional probability of finding sites a and b occupied and c empty, given that b is occupied and c is empty. Here, the denominator is the probability of the event $(0, b, 0)$ and not the event "site b is occupied."

process is

$$\mathbf{L} + (0, b, 0) \rightarrow (a, b, 0) \tag{2.2.18}$$

Similarly, the conditional intrinsic constant $k_{a/bc}$ is defined by

$$k_{a/bc} = \frac{k_{abc}}{k_{bc}} = \frac{Q(a, b, c)}{Q(0, b, c)} \lambda_0 \tag{2.2.19}$$

The probabilistic and thermodynamic interpretations of $k_{a/bc}$ follow directly from the definition. By substituting the various binding constants, one can rewrite the GPF as well as the BI in terms of the conditional constant. Particularly simple are the expressions for the individual BIs. For instance,

$$\theta_a = \frac{k_a C + k_a(k_{b/a} + k_{c/a})C^2 + k_a k_{b/a} k_{c/ab} C^3}{\xi'} \tag{2.2.20}$$

with similar expressions for θ_b and θ_c. Note the pattern of the conditions in each term in the numerator of Eq. (2.2.20).

We now turn to the case of identical sites. Actually, we require that the sites be identical in a strict sense, as we explain below. We use again the three-site case, but instead of three different sites a, b, and c we assume that the sites are identical. Since we are dealing with localized molecules, the sites are still distinguishable.*

The canonical PFs listed in Eqs. (2.2.2)–(2.2.5) now reduce to

$$Q(0) = Q(0, 0, 0) \tag{2.2.21}$$

$$Q(1) = Q(1, 0, 0) + Q(0, 1, 0) + Q(0, 0, 1) = 3Q(1, 0, 0) \tag{2.2.22}$$

$$Q(2) = Q(1, 1, 0) + Q(0, 1, 1) + Q(1, 0, 1) = 3Q(1, 1, 0) \tag{2.2.23}$$

$$Q(3) = Q(1, 1, 1) \tag{2.2.24}$$

We say that the sites are identical in a *weak* sense whenever the three PFs $Q(1, 0, 0)$, $Q(0, 1, 0)$, and $Q(0, 0, 1)$ have the same value. This is identical to the requirement that the single-site intrinsic constant is the same for any specific site. In this case we can replace these three PFs by three times one representative PF, as is done on the rhs of Eq. (2.2.22). We shall say that the sites are identical in a *strict* sense whenever the PF of any given occupation number is independent of the specific group of occupied sites. For instance, in an equilateral triangle all PFs with two sites occupied are equal. Hence we can replace the sum on the rhs of Eq. (2.2.23) by three times one representative PF. This cannot be done, in general, for a linear arrangement of the three sites, in which case $Q(1, 1, 0)$ is different from $Q(1, 0, 1)$, even when the sites are identical in the weak sense (see Chapter 5). Similarly, for

*They become indistinguishable when the molecule gains rotational freedom. See Appendix B.

a tetrahedral arrangement of identical sites, all pairs and all triplet sites give the same PF. Clearly, identical sites in a weak sense do not imply identity in the strict sense.* A simple example is three identical subunits arranged linearly. The PFs for nearest-neighbor pairs might not be the same as the PF for next-nearest-neighbor pairs. (This case is studied in Section 5.10). Similarly, four identical subunits arranged in a square might have two different PFs for a pair of occupied sites (nearest and next-nearest neighbors).

We proceed with the case of three identical sites in the strict sense and define the corresponding intrinsic constants

$$k_1 = \frac{Q(1,0,0)}{Q(0,0,0)}\lambda_0, \qquad k_{11} = \frac{Q(1,1,0)}{Q(0,0,0)}\lambda_0^2, \qquad k_{111} = \frac{Q(1,1,1)}{Q(0,0,0)}\lambda_0^3 \tag{2.2.25}$$

compared with the seven constants in Eq. (2.2.6). The probabilistic and thermodynamic interpretation of these constants is the same as for the different sites. One should note, however, that all of the constants in Eq. (2.2.25) refer to *specific* sites. These should be distinguished from nonintrinsic constants defined below and in the next section.

One can also define conditional constants, similar to those defined in Eqs. (2.2.17) and (2.2.19). For instance,

$$k_{1/1} = \frac{Q(1,1,0)}{Q(1,0,0)}\lambda_0 = \frac{k_{11}}{k_1} \tag{2.2.26}$$

and similarly

$$k_{1/11} = \frac{Q(1,1,1)}{Q(1,1,0)}\lambda_0 = \frac{k_{111}}{k_{11}} \tag{2.2.27}$$

Again, we stress that these conditional constants always refer to a specific configuration *before* and *after* the addition of the ligand; k_1, $k_{1/1}$, and $k_{1/11}$ may be referred to as the intrinsic constants for the *first, second*, and *third* ligands. These should be distinguished from the normally used first, second, and third *thermodynamic* constants, defined in Section 2.3. In the latter, the specification of the sites is not required.

In terms of these intrinsic constants the GPF is written as

$$\begin{aligned} \xi &= Q(0,0,0)(1 + 3k_1C + 3k_{11}C^2 + k_{111}C^3) \\ &= Q(0,0,0)(1 + 3k_1C + 3k_1k_{1/1}C^2 + k_1k_{1/1}k_{1/11}C^3) \end{aligned} \tag{2.2.28}$$

*Sometimes, the term "equivalent" is used instead of "strict sense." This can be confusing. For instance, in an equilateral triangle the three sites are equivalent, but in a linear case they are not equivalent. However, the term equivalent might not be suitable to distinguish between square and tetrahedral models. In both cases, identical sites are also equivalent because of symmetry. Yet, one has strict identical sites and the other weak identical sites in the sense defined here. For more details, see Chapter 6.

Although we shall not use nonintrinsic constants in this book, we mention them here since they are sometimes used in the literature. The nonintrinsic constants are obtained from the intrinsic constants by simply removing the requirement of a *specific* set of sites. For the three-site case, these are defined by

$$\bar{k}_1 = 3k_1, \quad \bar{k}_{11} = 3k_{11}, \quad \bar{k}_{111} = k_{111} \tag{2.2.29}$$

Thus $\bar{k}_1$ is the first binding constant to *any* site. The general relation between the two sets of binding constants is, for any l,

$$\bar{k} = \binom{m}{l} k \tag{2.2.30}$$

where $\bar{k}$ on the lhs refers to *any* l sites while k on the rhs refers to a specific set of l sites.

In summary, the intrinsic binding constant to be used throughout this book always refers to a *specific* set of sites. They are defined in terms of the molecular properties of the system through the corresponding canonical PFs. They are also interpreted as probability ratios or as free energies of binding processes. In subsequent chapters we shall see how to extract from these quantities various correlation functions or, equivalently, cooperativities.

2.3. THE THERMODYNAMIC BINDING CONSTANTS

The thermodynamic approach starts from a sequence of binding processes at equilibrium that define the corresponding equilibrium constants, after which the BI is obtained.

Consider again, for simplicity, the case $m = 3$. The adsorbent molecule **P** can form one of the three complexes which we denote by $\mathbf{P}_1$, $\mathbf{P}_2$, and $\mathbf{P}_3$, respectively, while $\mathbf{P}_0$ denotes the empty adsorbent molecule. The three binding processes and the corresponding equilibrium constants are

$$\mathbf{P}_0 + \mathbf{L} \rightarrow \mathbf{P}_1, \quad K_1 = \frac{[\mathbf{P}_1]}{[\mathbf{P}_0]C} \tag{2.3.1}$$

$$\mathbf{P}_1 + \mathbf{L} \rightarrow \mathbf{P}_2, \quad K_2 = \frac{[\mathbf{P}_2]}{[\mathbf{P}_1]C} \tag{2.3.2}$$

$$\mathbf{P}_2 + \mathbf{L} \rightarrow \mathbf{P}_3, \quad K_3 = \frac{[\mathbf{P}_3]}{[\mathbf{P}_2]C} \tag{2.3.3}$$

We use capital Ks to denote thermodynamic equilibrium constants, and we use square brackets to denote concentrations, say, in moles per unit volume.

Note that K_1, K_2, and K_3 are *sequential* equilibrium constants for the *first*, *second*, and *third* binding of ligands. The configurations of the ligands on the sites before and after the binding processes are not specified. This is the fundamental difference between the thermodynamic and the intrinsic binding constants.

Since $[\mathbf{P}_i]$ is the concentration of adsorbent molecules having i bound ligands, the total average number of bound ligands in the macroscopic system containing M molecules in volume V is

$$\overline{N} = V\sum_{i=1}^{3} i[\mathbf{P}_i] = V(K_1[\mathbf{P}_0]C + 2K_1K_2[\mathbf{P}_0]C^2 + 3K_1K_2K_3[\mathbf{P}_0]C^3) \qquad (2.3.4)$$

The total (fixed) number of adsorbent molecules is

$$M = V\sum_{i=0}^{3} [\mathbf{P}_i] = V[\mathbf{P}_0](1 + K_1C + K_1K_2C^2 + K_1K_2K_3C^3) \qquad (2.3.5)$$

Hence the average number of bound ligands per adsorbent molecule is

$$\bar{n} = \frac{\overline{N}}{M} = \frac{K_1C + 2K_1K_2C^2 + 3K_1K_2K_3C^3}{1 + K_1C + K_1K_2C^2 + K_1K_2K_3C^3} \qquad (2.3.6)$$

This is the BI expressed in terms of the measurable equilibrium constants K_i. Since we require that the BI as represented by the thermodynamic constants be the same as that represented by the intrinsic constant, we can make the identifications*

$$\left.\begin{aligned} K_1 &= (k_a + k_b + k_c) = 3k_1 \\ K_1K_2 &= (k_{ab} + k_{ac} + k_{bc}) = 3k_{11} \\ K_1K_2K_3 &= k_{abc} = k_{111} \end{aligned}\right\} \qquad (2.3.7)$$

where for each line, the first equality holds for *different* sites and the second for *identical* sites (in the strict sense).

From Eqs. (2.3.7) we also obtain

$$K_1 = 3k_1, \qquad K_2 = k_{1/1}, \qquad K_3 = \tfrac{1}{3}k_{1/11} \qquad (2.3.8)$$

These equations relate the sequential thermodynamic constants (first, second, and third) to the sequential intrinsic constants. The difference between the two sets arises from the requirement to specify the sites in the latter but not in the former. The generalization to m identical sites (in the strict sense) is quite straightforward.

*Note that in this section we have a thermodynamic system of molecules possessing translational and rotational degrees of freedom. In the previous sections we treated a system of localized molecules. Therefore, the GPFs of the two systems are different but their BI is the same, provided the approximations made in Appendix B are valid.

The generalization of Eqs. (2.3.7) is

$$K_1K_2\cdots K_j = \binom{m}{j} k_1 k_{1/1} k_{1/11} \cdots k_{1/1\cdots1} \tag{2.3.9}$$

where $k_{1/1\cdots1}$ denotes the conditional intrinsic binding constant to the jth *specific* site, given that $j-1$ *specific* sites are already occupied. Solving for the thermodynamic constants we obtain

$$K_1 = \binom{m}{1} k_1 = mk_1$$

$$K_2 = \frac{\binom{m}{2}}{\binom{m}{1}} k_{1/1} = \frac{m-1}{2} k_{1/1}$$

$$\vdots$$

$$K_j = \frac{\binom{m}{j}}{\binom{m}{j-1}} k_{1/1\cdots1} = \frac{m-j+1}{j} k_{1/1\cdots1} \tag{2.3.10}$$

The ratio between the jth thermodynamic constant and the corresponding intrinsic constant is the ratio between the number of specific configurations of the ligands before and after the addition of the jth ligand.

The overall equilibrium constant for binding j ligands to an empty molecule is defined for the reaction

$$\mathbf{P}_0 + j\mathbf{L} \rightarrow \mathbf{P}_j \tag{2.3.11}$$

by

$$\bar{K}_j = \frac{[\mathbf{P}_j]}{[\mathbf{P}_0]C^j} = K_1K_2\cdots K_j = \binom{m}{j} k_{1\cdots1} = \bar{k}_{1\cdots1} \tag{2.3.12}$$

where $\bar{k}_{1\cdots1}$ is the nonintrinsic constant for j ligands [see Eqs. (2.2.29)]. Clearly, when we remove the requirement of specific sets of sites, the overall thermodynamic constants $\bar{K}_j$ become identical to the quantities $\bar{k}_{1\cdots1}$ ($1\cdots1$ stands for j unities).

Throughout the remainder of this book we shall use only the intrinsic binding constant as defined by the corresponding canonical PFs in Section 2.2. The K_is are the quantities that are obtained directly from experiments. However, if we wish to interpret these quantities, say in terms of cooperativity, we must convert to the language of intrinsic binding constants. This was recognized implicitly by Bjerrum (1923), and later by Kirkwood and Westheimer (1938), who interpreted $4K_2/K_1$ as a quantity that measures the "interaction" between two protons in dibasic acids (this is discussed further in Chapter 4).* Since $K_1 = 2k_1$ and $K_2 = \frac{1}{2}k_{1/1}$, $4K_2/K_1 = k_{11}/(k_1)^2$, which is the pair correlation function. Thus, the inclusion of the factor 4 in $4K_2/K_1$ effectively converts from the thermodynamic into the intrinsic language.

Furthermore, K_is as phenomenological equilibrium constants cannot tell us much about the way they depend on the molecular parameters. This is possible only when our starting point is a molecular approach based on the PF of the system. This is particularly true if we are interested in studying the molecular origin of cooperativity. If one *defines* "interaction energy" (or free energy of interaction), in terms of the thermodynamic constant, one could easily be misled in interpreting these interactions. A typical example of such an erroneous interpretation is given in Chapter 5 (Section 5.10). The reason for such potential misinterpretations is that, although the thermodynamic constants are certainly *determined* by the molecular parameters, they do not reveal the way in which they depend on these parameters. Specifically, the *interaction energy* between two ligands is defined in this book independently of the adsorbent molecule. The molecular approach shows how the correlation between the two ligands depends on this interaction energy. Sometimes, the whole correlation is due exclusively to this interaction energy. At other times, it consists of the product of direct and indirect factors. Yet in other cases, the direct and indirect parts are so intertwined that no such factorization is possible. Examples of these cases will be discussed in Chapter 4.

The polynomial within parentheses in Eq. (2.3.5) [or in the denominator of Eq. (2.3.6)] is called the *binding polynomial* and is used extensively in the phenomenological approach to binding systems. It is often identified (up to a multiplicative constant) with the partition function of the system. This identification is erroneous and misleading for two fundamental reasons. First, the coefficients of the binding polynomial are determined by the equilibrium constants of the binding processes, Eqs. (2.3.1)–(2.3.3). As such, they are phenomenological constants that do not reveal their dependence on the molecular properties of the system. On the other hand, the coefficients in the partition function are *defined* in terms of molecular parameters and therefore, in principle, are calculable from molecular properties of the binding system. Specifically, all possible sources of cooperativity may be studied from the coefficients of the partition function, but none can be extracted from the phenomenological binding constants. Second, the partition-function approach is also more general in the sense that it provides, in principle, the form of the BI for systems that are not infinitely dilute with respect to either or both of the adsorbent molecules and the ligand. On the other hand, the phenom-

*Here, we employ the term "interaction" as used in the literature. The quantity $4K_2/K_1$ is actually a "correlation" and not interaction in the sense of Section 2.1.

enological approach is limited to the infinite dilute region only. It offers no means for estimating the form of the BI when the system deviates from this limit. We shall discuss in Appendix D the first-order deviation of the BI from this ideal-dilute limit.

2.4. THE SIMPLEST MOLECULAR MODEL FOR THE LANGMUIR ISOTHERM

In this section we derive the simplest adsorbing isotherm from a molecular model. It can also be derived by thermodynamic considerations, as discussed in Section 2.3. Originally, it was derived by Langmuir in 1918, applying kinetic arguments on the rate of evaporation and deposition of ligands. We describe here the "minimal model"* that gives rise to the characteristic Langmuir isotherm. Our system consists of M particles or adsorbing molecules, each having one site for binding a ligand **L**. We assume that the molecules are identical and independent. For simplicity, we also assume that they are localized so that the GPF of the entire system is $\Xi = \xi^M$. If they are not localized the GPF is $\Xi = \xi^M/M!$, but this does not affect the BI (see Appendix B). We assume that the internal degrees of freedom of both the adsorbent molecules and the ligands are unaffected by the binding process. No degrees of freedom are ascribed to the ligand occupying the binding site. When a ligand binds to the site, the energy change in the process is described by one parameter U, referred to as the binding energy.

By virtue of the assumption of localization, the adsorbent molecules are distinguishable, but the ligands are indistinguishable, i.e., interchanging two ligands occupying different sites does not produce a new configuration. The ligands are supplied from an ideal gas reservoir; the only degree of freedom that changes upon binding is its translation. Other degrees of freedom, if any, are presumed unchanged upon binding.

With these assumptions we write the GPF of the entire system and of a single molecule as

$$\Xi = \xi^M, \qquad \xi = Q(0) + Q(0)\, q_{int} q\, \lambda \tag{2.4.1}$$

where $q = \exp(-\beta U)$, q_{int} is the internal PF of the ligand, and λ is given by

$$\lambda = \exp(\beta\mu) = \frac{\Lambda^3 C}{q_{int}} = \lambda_0 C \tag{2.4.2}$$

Since we have assumed that q_{int} is the same in the ideal gas phase and at the site, it

*The minimal models used here and elsewhere in the book are models that can be described and solved to demonstrate a phenomenon with a minimal number of molecular parameters—sometimes, this is the same as the maximal number of simplifying assumptions.

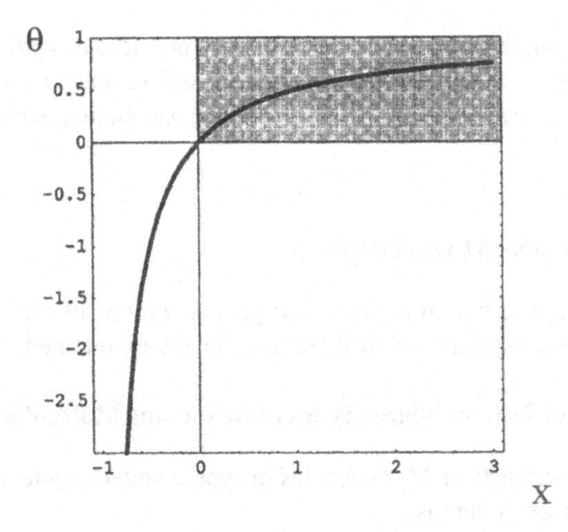

Figure 2.2. The Langmuir isotherm θ as a function of $x = kC$. The experimental range of x and θ is the shaded area $0 \le x \le \infty$ and $0 \le \theta \le 1$. The function $\theta(x)$ has been plotted here in the range $-1 \le x \le 3$ to stress the hyperbolic character of the curve. The center of the hyperbola is $x = -1$ and $\theta = 1$.

cancels out in the product $q_{int}\lambda$. The BI is thus

$$\theta = \bar{n} = \lambda \frac{\partial \ln \xi}{\partial \lambda} = \frac{kC}{1 + kC} = \frac{x}{1 + x} \tag{2.4.3}$$

where

$$k = q\Lambda^3 = \exp(-\beta U)\Lambda^3, \qquad \Lambda = \frac{h}{\sqrt{2\pi m_L k_B T}} \tag{2.4.4}$$

We see that this model produces the typical hyperbolic function depicted in Fig. 2.2.* This curve was first derived by Langmuir and it describes many experiments of binding ligands to surfaces or to independent molecules. In the thermodynamic derivation of Eq. (2.4.3), k is an equilibrium constant. Nothing can be said about its dependence on the molecular parameters of the system. On the other hand, in the molecular approach we have an explicit expression for the binding constant k in terms of the molecular parameters in Eqs. (2.4.4). In this particular model, there are only two: the binding energy U and the mass of the ligand m_L. This allows us to predict that replacing a ligand by its heavier isotope will decrease the binding constant. Likewise, if we take two ligands with identical mass but different electronic charge or dipole moment, then we can predict the direction of change of the binding constant. Note that in the Langmuir model the thermodynamic constant is equal to the intrinsic constant.

*Equation (2.4.3) can be rearranged to the form $(\theta - 1)(kC + 1) = -1$. A plot of θ as a function of $x = kC$ gives a hyperbola centered at $x = -1$ and $\theta = 1$, with the two asymptotes $y = 1$ and $x = -1$ (Fig. 2.2).

Thus, in general, the thermodynamic approach provides us with an equation of state of the form $f(\theta, T, \lambda) = 0$. The molecular approach provides us with an explicit equation of state for each specific set of molecular parameters, which in this case is $f(\theta, T, \lambda; m_L, U) = 0$.

2.5. A FEW GENERALIZATIONS

We present here a few straightforward generalizations of the Langmuir isotherm. The essential assumptions made in Section 2.4 are retained.

2.5.1. Mixture of Two (or More) Types of Adsorbing Molecules

If the system consists of M_a molecules of type a and M_b molecules of type b, the GPF of the entire system is

$$\Xi(T, M_a, M_b, \lambda) = \xi_a^{M_a}\xi_b^{M_b} = [Q_a(0) + Q_a(1)\lambda]^{M_a}[Q_b(0) + Q_b(1)\lambda]^{M_b}$$

$$= Q_a(0)^{M_a}Q_b(0)^{M_b}[1 + k_aC]^{M_a}[1 + k_bC]^{M_b} \tag{2.5.1}$$

where we set $k_aC = Q_a(1)\lambda/Q_a(0)$ and $k_bC = Q_b(1)\lambda/Q_b(0)$. The corresponding BI is

$$\bar{n} = \theta = \frac{\lambda}{M}\frac{\partial \ln \Xi}{\partial \lambda} = \frac{M_a}{M}\frac{k_aC}{1 + k_aC} + \frac{M_b}{M}\frac{k_bC}{1 + k_bC} \tag{2.5.2}$$

where $M = M_a + M_b$; k_a and k_b are the binding constants for sites of type a and b, respectively.

It should be stressed here that although each type of molecule by itself produces a Langmuir isotherm, the combination in Eq. (2.5.2) has a form different from the typical hyperbolic form of the Langmuir isotherm. Figure 2.3 shows the BI for a mixture of two types of molecules with $k_a \neq k_b$. The BIs are quite different from the Langmuir isotherm. We shall see in Chapter 4 that this is a typical BI for a two-site system with *negative* cooperativity. We shall also discuss in Sections 3.5 and 4.6 how experimental data could be misinterpreted in such cases. At this stage we stress again that our system consists of *M independent* adsorbent molecules, and cooperativity in the sense defined in this book (see Chapter 4) is not definable in this system. The generalization to any number of types of sites is quite straightforward. The BI per site of the mixture is simply

$$\bar{n} = \sum X_\alpha \frac{k_\alpha C}{1 + k_\alpha C} \tag{2.5.3}$$

where X_α is the mole fraction of the α-component in the mixture.

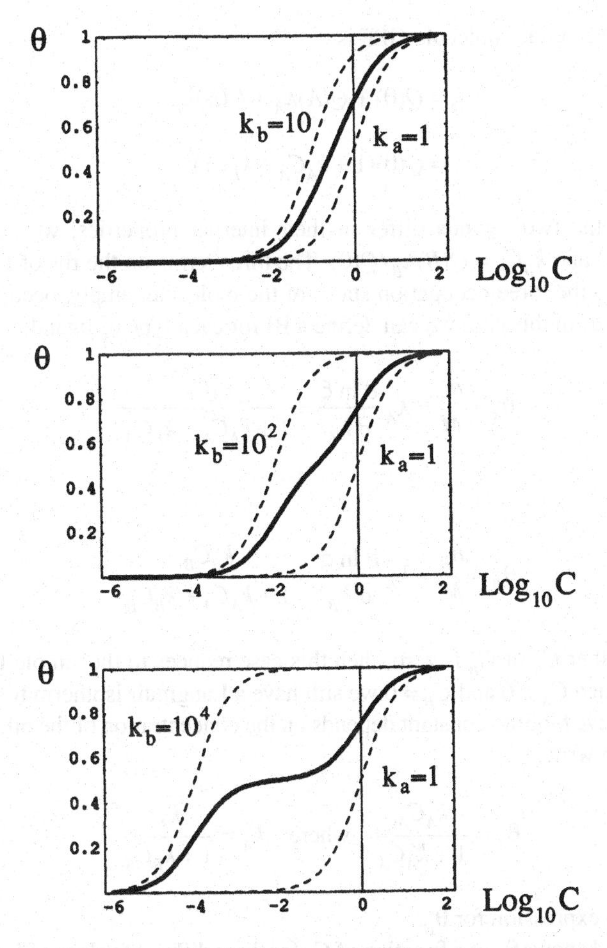

Figure 2.3. The BI (full line) of a mixture of two types of single-site molecules with $k_a = 1$ and varying k_b, as indicated next to each curve. The dashed curves are the BIs of pure a and b. [The plots are θ (log$C$); see Section 4.]

2.5.2. Mixture of Two (or More) Ligands Binding to the Same Site

Again, the system consists of M adsorbent molecules fulfilling all the simplifying assumptions listed for the Langmuir model. The sites can be occupied by either one of the two different ligands A and B which are in equilibrium with a reservoir at fixed absolute activities λ_A and λ_B, respectively. The GPF of the entire macroscopic system is

$$\Xi(T, M, \lambda_A, \lambda_B) = \xi^M \tag{2.5.4}$$

where now the single-molecule GPF is

$$\xi = Q(0) + Q(A)\lambda_A + Q(B)\lambda_B$$
$$= Q(0)(1 + k_A C_A + k_B C_B) \tag{2.5.5}$$

Here, since the two ligands differ in their internal properties, we set $k_A C_A = Q(A)\lambda_A / Q(0)$ and $k_B C_B = Q(B)\lambda_B / Q(0)$. The three terms on the rhs of Eq. (2.5.5) correspond to the three occupation states of the molecule: empty, occupied by A, occupied by B. In this case we can define a BI for each type of ligand,

$$\theta_A = \frac{\bar{n}_A}{M} = \lambda_A \frac{\partial \ln \xi}{\partial \lambda_A} = \frac{k_A C_A}{1 + k_A C_A + k_B C_B} \tag{2.5.6}$$

and

$$\theta_B = \frac{\bar{n}_B}{M} = \lambda_B \frac{\partial \ln \xi}{\partial \lambda_B} = \frac{k_B C_B}{1 + k_A C_A + k_B C_B} \tag{2.5.7}$$

Clearly, if either C_A or C_B is zero, then this case reduces to the simple Langmuir isotherm. When $C_A \neq 0$ and $C_B \neq 0$ we still have a Langmuir isotherm for A and B, but the effective binding constant depends on the concentration of the other ligand. Thus, we can write

$$\theta_A = \frac{k'_A C_A}{1 + k'_A C_A} \quad \text{where} \quad k'_A = \frac{k_A}{1 + k_B C_B} \tag{2.5.8}$$

and a similar expression for θ_B.

Figure 2.4 shows θ_A as a function of C_A for three different values of $k_B C_B$. Since B competes with A for binding on the same site, the larger the value of $k_B C_B$, the smaller the effective binding constant k'_A, and the corresponding Langmuir isotherm becomes less steep. The rate of change of θ_A with C_B for a system with given values of k_A and k_B is

$$\left(\frac{\partial \theta_A}{\partial C_B}\right)_{C_A} = \frac{-k_A k_B C_A}{(1 + k_A C_A + k_B C_B)^2} \tag{2.5.9}$$

which is always negative—this is clearly a result of the competition for the same site. Since ξ is a well-behaved function of C_A and C_B, the rate of change of θ_B with respect to C_A is the same as the rate of change of θ_A with respect to C_B, i.e., we have

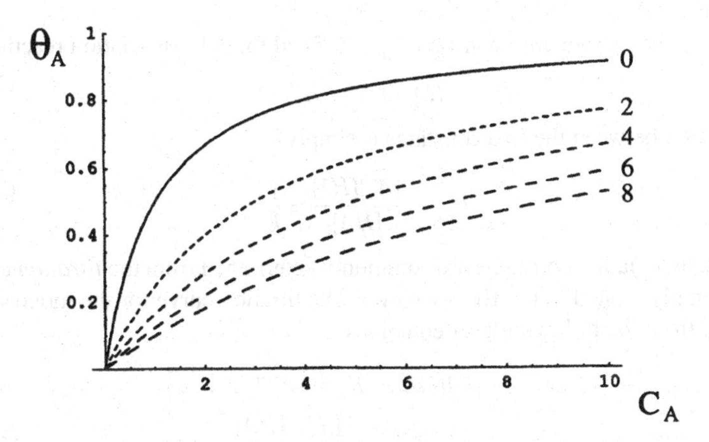

Figure 2.4. The BI of A as a function of C_A for different values of C_B, as indicated next to each curve ($k_A = k_B = 1$).

the relation

$$\left(\frac{\partial \theta_A}{\partial C_B}\right)_{C_A} = \left(\frac{\partial \theta_B}{\partial C_A}\right)_{C_B} \tag{2.5.10}$$

The generalization for any number of ligands is straightforward. The BI of ligand A for any number of ligands has the form

$$\theta_A = \frac{k_A C_A}{1 + \Sigma_\alpha k_\alpha C_\alpha} \tag{2.5.11}$$

where the sum is over all the different ligands. Again, all the BIs are of simple Langmuir form, with an effective binding constant that depends on all products $k_\alpha C_\alpha$ for all $\alpha \neq A$.

2.6. EXAMPLES

2.6.1. Normal Carboxylic Acids

The simplest binding molecule is a weak acid that releases a proton into the solution. The anionic group A^- may be viewed as an adsorbent system for a proton

$$A^- + H^+ \rightarrow HA \tag{2.6.1}$$

Normally, instead of the *binding* constant k, data reported in the literature are given in terms of the *dissociation* constant, k_{diss}, defined for the dissociation reaction

$$HA \rightarrow A^- + H^+ \tag{2.6.2}$$

The relation between the two constants is simply

$$k_{diss} = \frac{[A^-][H^+]}{[HA]} = \frac{1}{k} \tag{2.6.3}$$

The dissociation constants are commonly determined from the *titration curve*, which simply related to the BI as follows: The titration curve for the monoacid is obtained from the following four equations:

$$\left.\begin{aligned} [H][OH] &= K_w = 10^{-14} \\ k_{diss} &= [A][H]/[HA] \\ N_T &= [A] + [HA] \\ [A] + [OH] &= [H] + [N_B] \end{aligned}\right\} \tag{2.6.4}$$

The first equation defines the ionization constant of water at 25 °C (we omit the sign of the charges to simplify notation). The second is the same as Eq. (2.6.3), while the third is the conservation of the total initial concentration of the (weak) acid N_T (we assume that there is no change in volume during the titration, hence this is the same as the conservation of the total number of acid molecules). The fourth equation is the electroneutrality condition, where $[N_B]$ is the concentration of the added (strong) base.

Solving for N_B as a function of $[H]$, and letting $N_T = 1$ [say, one mole per liter, or any other units chosen consistently in Eq. (2.6.4)], we obtain the required function

$$N_B = N_B([H]) = \frac{-[H]^3 - k_{diss}[H]^2 + K_w[H] + k_{diss}K_w + k_{diss}[H]}{[H]^2 + k_{diss}[H]} \tag{2.6.5}$$

The corresponding BI expressed in terms of the dissociation constant k_{diss} is simply obtained from the Langmuir isotherm, Eq. (2.4.3), by substituting $k = 1/k_{diss}$, namely,

$$\theta = \theta([H]) = \frac{[H]}{[H] + k_{diss}} \tag{2.6.6}$$

Clearly the two functions (2.6.5) and (2.6.6) are different*; the relation between the two is

$$N_B([H]) = 1 - \theta([H]) - [H] + K_w[H]^{-1} \tag{2.6.7}$$

*In many publications, these are referred to as equivalent functions.

Figure 2.5 shows N_B and the function $1 - \theta$, plotted as a function of $pH = -\log_{10}[H]$*
for acetic acid with $pk_{diss} = 4.754$. In order to superimpose the two curves we must transform $\theta([H])$ into $1 - \theta([H])$. Since K_w is very small, the two curves $N_B([H])$ and $1 - \theta([H])$ are almost identical in the range $3 \lesssim pH \lesssim 11$. The dissociation constant k_{diss} may be obtained either from the BI at $\theta = 1/2$, or from the titration curve at $N_B = 1/2$. Table 2.1 presents some values of $pk_{diss} = -\log_{10} k_{diss}$ for linear (or normal) carboxylic acids, in water at 25 °C. It is clear that, except for formic acid, the values of pk_{diss} converge to an average value of about 4.88.

We now show that as the number n of carbon atoms in the acid increases, k_{diss} must converge to a constant value. The qualitative argument is simple. Adding a methylene group far from the carboxylic group is not expected to affect the binding properties of the carboxylic group.

A more precise argument is the following: We write the statistical mechanical expression for the dissociation constant as

$$k_{diss} = \exp\{-\beta[\mu^0(A_n) + \mu^0(H) - \mu^0[(HA_n)]\} \tag{2.6.8}$$

where $\mu^0(\alpha)$ is the standard chemical potential of the species α and A_n is the anion of length n. Assuming that the hydrocarbon chains are linear and rigid, we can write each standard chemical potential as†

$$\mu^0(\alpha) = W(\alpha) + k_B T \ln \Lambda_\alpha^3 q_{\alpha,int}^{-1} \tag{2.6.9}$$

where $W(\alpha)$ is the coupling work of the species α to the solvent, Λ_α^3 is the momentum partition functions [see Eq. (1.2.2)], and $q_{\alpha,int}$ is the internal partition function of α. Here, $q_{\alpha,\,int}$ is essentially the rotational–vibrational PF. If α is not rigid, one must take the appropriate average over all possible conformations of α; but we do not need to consider this case in the present argument. We now write the coupling works $W(A_n)$ and $W(HA_n)$ as

$$W(A_n) = W(R_n) + W(COO^-/R_n) \tag{2.6.10}$$

and

$$W(HA_n) = W(R_n) + W(COOH/R_n) \tag{2.6.11}$$

where $W(R_n)$ is the coupling work for the hydrocarbon chain of length n, and $W(COO^-/R_n)$ is the *conditional* coupling work of the carboxylate group given that R_n has already been coupled. Also, since the mass of the hydrogen ion is much smaller than the mass of the entire molecule, we can assume that adding a proton

*The experimental titration curve is normally the pH as a function of the added number of moles of base. This is the inverse function of Eq. (2.6.5).

†For details, see Chapters 1 and 2 in Ben-Naim (1987).

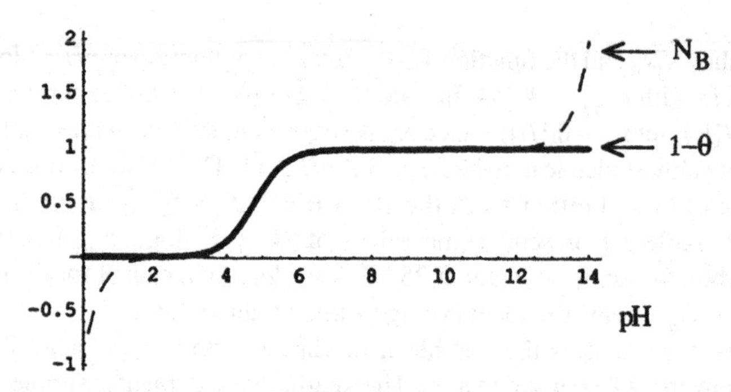

Figure 2.5. The titration curve N_B (dashed line), and the BI, plotted as $1 - \theta$ (full line) as a function of $pH = -\log_{10}[H]$ for acetic acid with $pK_w = 14$, $pk_{diss} = 4.754$. (Note that higher pH corresponds to lower concentration $[H]$.)

does not change much the translational and the rotational PF of the molecule; hence we set (see also Appendix B)

$$\Lambda^3_{HA_n} \approx \Lambda^3_{A_n}, \qquad q_{HA_{n,int}} \approx q_{A_{n,int}} \tag{2.6.12}$$

With these approximations we rewrite k_{diss} as

$$k_{diss} = \exp\{-\beta[-U + W(COO^-/R_n) - W(COOH/R_n)]\} \frac{\exp[-\beta W(H)]}{\Lambda^3_H} \tag{2.6.13}$$

The factor on the rhs of Eq. (2.6.13) is simply λ_0 defined in Section 2.4 and in Appendix *B*; *U* is the binding energy, i.e., the interaction *energy* between the proton

Table 2.1
Values of $pk_{diss} = -\log_{10} k_{diss}$ for Normal Carboxylic Acids in Water at 25 °C ($pk = -pk_{diss}$)[a]

Carboxylic acid	pk_{diss}
H–COOH	3.752
CH_3–COOH	4.756
CH_3–CH_2–COOH	4.874
CH_3–$(CH_3)_2$–COOH	4.820
CH_3–$(CH_2)_3$–COOH	4.842
CH_3–$(CH_2)_4$–COOH	4.857
CH_3–$(CH_2)_5$–COOH	4.893
CH_3–$(CH_2)_6$–COOH	4.894
CH_3–$(CH_2)_7$–COOH	4.955

[a]Data from Robinson and Stokes (1959).

and the molecule at the binding site. Clearly, as n increases all the factors in Eq. (2.6.13) become independent of n.*

If there is no solvent, then k_{diss} reduces to

$$k_{diss} = e^{\beta U}/\Lambda_H^3 \tag{2.6.14}$$

which is the inverse of the (binding) constant k, given in Section 2.4.

2.6.2. Normal Amines

Table 2.2 presents the dissociation constants for linear amines. Here, the values of pk_{diss} converge to about 10.64 for large n.

The statistical mechanical expression for the dissociation constant

$$k_{diss} = \frac{[B][H]}{[HB]} \tag{2.6.15}$$

is similar to Eq. (2.6.13) and has the form

$$k_{diss} = \exp\{-\beta[-U + W(NH_2/R_n) - W(NH_3^+/R_n)]\} \frac{\exp[-\beta W(H)]}{\Lambda_H^3} \tag{2.6.16}$$

The factor on the rhs is the same as in Eq. (2.6.13).

We now take the ratio of the average values of k_{diss} for large n for the acids and bases. From Eqs. (2 6.13) and (2.6.16) we have

$$\frac{k_{diss}(acid)}{k_{diss}(base)} = \exp[\beta(U(acid) - \beta U(base)]$$

$$\times \exp\{\beta[W(COOH/R_n) + W(NH_2/R_n) - W(COO^-/R_n) - W(NH_3^+/R_n)]\}$$

$$= (binding\ energies)\ (solvation) \tag{2.6.17}$$

The experimental value of this ratio is

$$\frac{k_{diss}(acid)}{k_{diss}(base)} = \frac{10^{-4.88}}{10^{-10.64}} = 5.7 \times 10^5 \tag{2.6.18}$$

*The conditional coupling works $W(COO^-/R_n)$ and $W(COOH/R_n)$ depend on n for small n. The range of this dependence is the same as the range of the pair correlation function between two particles in a solvent. For more details, see Ben-Naim (1987) and Ben-Naim and Mazo (1993).

Table 2.2
Values of $pk_{diss} = -\log_{10} k_{diss}$ for Normal Amines at 25 °C ($pk = -pk_{diss}$)[a]

Amine	pk_{diss}
$CH_3-CH_2-NH_3^+$	10.631
$CH_3-(CH_2)_2-NH_3^+$	10.530
$CH_3-(CH_2)_3-NH_3^+$	10.597
$CH_3-(CH_2)_4-NH_3^+$	10.630
$CH_3-(CH_2)_5-NH_3^+$	10.640
$CH_3-(CH_2)_6-NH_3^+$	10.66
$CH_3-(CH_2)_7-NH_3^+$	10.65
$CH_3-(CH_2)_8-NH_3^+$	10.64
$CH_3-(CH_2)_9-NH_3^+$	10.64

[a]Data from Robinson and Stokes (1959).

In Eq. (2.6.17) we see that this ratio is determined by the product of two factors. One factor depends on the difference in the binding energies of the proton to the negatively charged carboxylate and to the neutral amine group. Clearly, this must be a very small number on the order of

$$\exp\left(-\beta \frac{e^2}{r_0}\right) \sim 10^{-244} \tag{2.6.19}$$

with $k_B T \approx 0.6$ kcal/mol, $e^2 = 332.8$ kcal Å/mol, and $r_0 \sim 1$ Å. On the other hand, the second factor, referred to as the solvation effect, is very large—it involves cancellation of the conditional solvation free energies of the charged groups to form neutral molecules. The reaction is shown schematically in Fig. 2.6a.

To estimate the solvation effect, consider the two "reactions" b and c in Fig. 2.6. The free energies of these "reactions" are 207 and 174 kcal/mol, respectively*; hence the second factor is on the order of

$$\exp(207\beta) \approx 5 \times 10^{151} \tag{2.6.20}$$

The two estimates (2.6.19) and (2.6.20) should not be taken too seriously. The purpose of making these estimates is only to show that the ratio (2.6.18) is determined by the product of two factors: one very small and the other very large. The experimental result (2.6.18) indicates that the *solvation* factor is the one that gains in the competition. Clearly, had we ignored solvation effects leaving only the

*Data from Ben-Naim (1987).

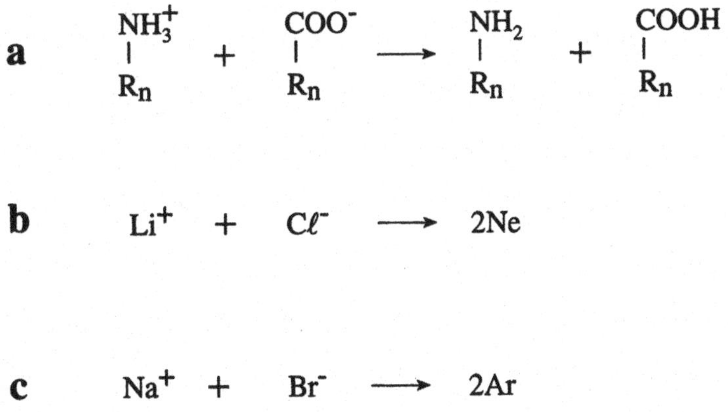

a $\underset{R_n}{\overset{NH_3^+}{|}} + \underset{R_n}{\overset{COO^-}{|}} \longrightarrow \underset{R_n}{\overset{NH_2}{|}} + \underset{R_n}{\overset{COOH}{|}}$

b $Li^+ + C\ell^- \longrightarrow 2Ne$

c $Na^+ + Br^- \longrightarrow 2Ar$

Figure 2.6. (a) The transfer of a proton from the ammonium ion to the carboxylate ion, giving two neutral molecules. (b) and (c) Approximate "reactions" where charge is transferred from one ion to another to form two neutral atoms of similar size.

Table 2.3
Values of $pk_{diss} = -\log_{10} k_{diss}$ for Some Substituted Acetic Acids at 25 °C[a]

Acid	pk_{diss}
Acetic acid	4.756
Iodoacetic acid	3.174
Bromo acetic acid	2.901
Chloro acetic acid	2.865
Fluoro acetic acid	2.584

[a]Data from Robinson and Stokes (1959).

binding energies to determine the ratio $k_{diss}(acid)/k_{diss}(base)$, we would have obtained an extremely small value. This is expected intuitively. Since $U(acid)$ is far more negative than $U(base)$, the *binding* constant to the acid must be much larger than the binding constant to the base, i.e., $k(acid) >> k(base)$ or, equivalently, $k_{diss}(acid) << k_{diss}(base)$.

Table 2.3 presents some values of pk_{diss} for various substituted acetic acids. Since the electronegative substituents on the methyl group reduce the negative charge on the carboxylate ion, the binding constant becomes smaller, or the dissociation constant becomes larger, i.e., increasing the acidity of the acid.

3

Adsorption on a Single-Site Polymer with Conformational Changes Induced by the Binding Process

3.1. INTRODUCTION

One of our main assumptions in the derivation of the Langmuir model [and implicitly made by Langmuir himself (1918)] is that the binding process does not affect the distribution of states of the adsorbent molecules. Removal of this assumption has a profound effect on the form of the BI of systems with more than a single site.

The qualitative reason is quite simple. Consider an adsorbent molecule the states of which are labeled by index j. The probability of finding an empty molecule in state j is $P_j(0)$. Suppose the molecule has m binding sites, and the binding constant to the first site is $k_1(j)$ when the molecule is in state j. If $m = 1$, then, as in Section 2.5, we shall obtain a Langmuir isotherm with binding constant k_1 which is simply an average $\Sigma_j k_1(j)P_j(0)$, the sum being over all the states of the molecule. If there are m identical sites, the intrinsic binding constant is again $k_1 = \Sigma k_1(j)P_j(0)$. When a second ligand approaches the molecule, the conditional binding constant $k_{1/1}$ might differ from k_1 for two conceptually different reasons. It might interact with the ligand on the first site, which will produce correlation between the two sites, to which we shall refer as *direct* correlation (Section 4.3). A second cause for the difference between k_1 and $k_{1/1}$ might be a change in the distribution of states induced by the first ligand, say from $P_j(0)$ to $P_j(1)$. Thus, the second ligand approaching the molecule with a new distribution of states $P_j(1)$ will have a binding constant of the form $k_{1/1} = \Sigma k_{1/1}(j)P_j(1)$, where $k_{1/1}(j)$ is the conditional binding constant for the state j. Clearly, even when there are no direct interactions between the ligands (e.g., between two oxygens in hemoglobin), in which case $k_{1/1}(j) = k_1(j)$ for each state j, the average conditional binding constant $k_{1/1}$ will differ from k_1 simply because the distribution has changed from $P_j(0)$ to $P_j(1)$. We shall explore this effect on the cooperativity of the system beginning in Chapter 4. In this chapter we examine only the change in the distribution of states induced by a single ligand. We shall do so in a "minimal model" of two states, denoted by L and H, with energies E_L and E_H, respectively.

If the study of cooperativity is likened to the study of chemical bonds, then the model studied in this chapter would be the analogue of the hydrogen atom model. The hydrogen atom does not have a chemical bond, yet its thorough understanding is crucial for understanding molecules.

Likewise, this book is devoted to the study of cooperativity arising from correlation between at least two ligands occupying different sites. The model of this chapter, like the Langmuir model, has by definition no cooperativity.* However, its thorough understanding is crucial for understanding the mechanism by which ligands communicate in a multisubunit system.

3.2. THE MODEL AND ITS PARTITION FUNCTION

As in the Langmuir model we focus on a single adsorbent molecule, **P**, having a single site for accommodating a ligand **L**. The new feature of this system is that each **P** molecule can be in either one of the two conformational states. We shall refer to these two states as the *energy levels* of our system—although in reality these are *free-energy levels* (see also Appendix C).† The two states are denoted by L and H for low and high, respectively. The corresponding energies are denoted by E_L and E_H, and we shall assume that

$$E_L < E_H \tag{3.2.1}$$

The PF of the empty system is simply

$$\xi(0) = Q_L + Q_H \tag{3.2.2}$$

where

$$Q_L = \exp(-\beta E_L), \qquad Q_H = \exp(-\beta E_H) \tag{3.2.3}$$

At equilibrium, the probability of finding the empty system in one of the two states is the same as the equilibrium mole fractions, and these can be read from the PF,

*Wyman and Gill (1990) express an apparently different view, referring to single-site systems as cooperative. We shall see in Section 3.5 and in Section 4.6 that this is not a genuine cooperativity, and we shall refer to it as spurious cooperativity.

†We suppress the free-energy character of these two states in order to highlight the emergence of new free energies whenever we average either over conformational states (e.g., Section 3.3) or over solvent configurations (see Chapter 9).

$\xi(0)$, in Eq. (3.2.2) as follows:

$$X_L^0 = \frac{Q_L}{Q_L + Q_H}, \quad X_H^0 = \frac{Q_H}{Q_L + Q_H} \tag{3.2.4}$$

where we denote by X_L^0 and X_H^0 the mole fractions of L and H in an ensemble of a large number of empty systems at equilibrium. The equilibrium constant for the "reaction"

$$L \rightleftharpoons H \tag{3.2.5}$$

is thus

$$K = \frac{Q_H}{Q_L} = \exp[-\beta(E_H - E_L)] = \frac{X_H^0}{X_L^0} \tag{3.2.6}$$

In terms of K, we write the mole fractions X_L^0 and X_H^0 as

$$X_L^0 = \frac{1}{1+K}, \quad X_H^0 = \frac{K}{1+K} \tag{3.2.7}$$

since we have chosen $E_L - E_H < 0$, i.e., L is more stable than H, $0 \leq K \leq 1$ and we always have $X_H^0 < X_L^0$.

We note that in Eq. (3.2.6) we have *defined* K in terms of the molecular parameters of the system (E_L and E_H). By virtue of the equilibrium condition, K is also equal to the ratio of the two mole fractions X_H^0 and X_L^0. In the subsequent sections we shall follow the changes of X_H and X_L upon binding of ligands, and we shall continue to use the parameter K as *defined* in terms of Q_H and Q_L. The *equilibrium constant* for the reaction $L \rightleftharpoons H$ will, however, be changed upon binding.

Next, we turn to the PF of a system with ligands. There are altogether four states of our system, denoted symbolically by

$$(L, 0), (H, 0), (L, 1), (H, 1) \tag{3.2.8}$$

i.e., empty L, empty H, filled L, and filled H. We denote by U_L and U_H the binding energies of **L** on the two conformations L and H, respectively. The PF of the system is now

$$\xi(T, \lambda) = Q_L + Q_H + Q_L q_L \lambda + Q_H q_H \lambda \tag{3.2.9}$$

where $\lambda = \exp(\beta\mu)$ is the absolute activity of the ligand and we have introduced the notation

$$q_L = \exp(-\beta U_L), \quad q_H = \exp(-\beta U_H) \tag{3.2.10}$$

The four terms in the PF correspond to the four states of the system, as described symbolically in (3.2.8).

The probabilities of these four states are

$$P(L, 0) = Q_L/\xi, \quad P(H, 0) = Q_H/\xi$$
$$P(L, 1) = Q_L q_L \lambda/\xi, \quad P(H, 1) = Q_H q_H \lambda/\xi \tag{3.2.11}$$

These are the probabilities of the fundamental states of our system, as described in (3.2.8).

The *marginal* probabilities are obtained by summing over all possible states that we do not care to specify. Thus, the probability of finding the system empty, irrespective of its conformational state, is

$$P(0) = P(L, 0) + P(H, 0) = \frac{Q_L + Q_H}{\xi} \tag{3.2.12}$$

Similarly,

$$P(1) = P(L, 1) + P(H, 1) = \frac{(Q_L q_L + Q_H q_H)\lambda}{\xi} \tag{3.2.13}$$

The probability of finding the system at conformational state L, irrespective of its occupational state, is

$$P(L) = \frac{Q_L + Q_L q_L \lambda}{\xi} \tag{3.2.14}$$

and similarly

$$P(H) = \frac{Q_H + Q_H q_H \lambda}{\xi} \tag{3.2.15}$$

Some *conditional* probabilities are

$$P(L/0) = \frac{P(L, 0)}{P(0)} = \frac{Q_L}{Q_L + Q_H} \tag{3.2.16}$$

which is clearly the same as X_L^0 in Eq. (3.2.4), only here we have explicitly used the condition "0" (empty) in the notation. Note that $P(L, 0)$ in Eqs. (3.2.11) refers to the probability of finding "state L *and* empty," whereas $P(L/0)$ in Eq. (3.2.16) is the *conditional* probability of finding state L *given* that it is empty. Similarly,

$$P(L/1) = \frac{P(L,1)}{P(1)} = \frac{Q_L q_L}{Q_L q_L + Q_H q_H} \tag{3.2.17}$$

Note that the probability of finding state L changes with λ according to Eq. (3.2.14), which is also the mole fraction of the L state as a function of λ. We shall denote this function by $X_L(\lambda)$, which changes from X_L^0 [i.e., X_L $(\lambda = 0)$], when $\lambda = 0$ to $X_L^{(1)}$ for the fully occupied system at $\lambda \to \infty$. Thus,

$$X_L^0 = X_L(\lambda = 0) = P(L/0) = \frac{Q_L}{Q_L + Q_H} \tag{3.2.18}$$

and

$$X_L^{(1)} = X_L(\lambda = \infty) = P(L/1) = \frac{Q_L q_L}{Q_L q_L + Q_H q_H} \tag{3.2.19}$$

The equilibrium constant for the reaction $L \rightleftharpoons H$ is also a function of λ,

$$K(\lambda) = \frac{X_H(\lambda)}{X_L(\lambda)} = \frac{Q_H + Q_H q_H \lambda}{Q_L + Q_L q_L \lambda} \tag{3.2.20}$$

The equilibrium constant defined in the empty system, Eq. (3.2.6), is now identified as $K = K(\lambda = 0)$. For $\lambda \to \infty$ we have the fully occupied system, for which the equilibrium constant is

$$K^{(1)} = K(\lambda = \infty) = \frac{Q_H q_H}{Q_L q_L} \tag{3.2.21}$$

Compared with K [i.e., $K(\lambda = 0)$], the new equilibrium constant $K^{(1)}$ corresponds to a new system of two levels with energies $E_H + U_H$ and $E_L + U_L$, as depicted in Fig. 3.1.

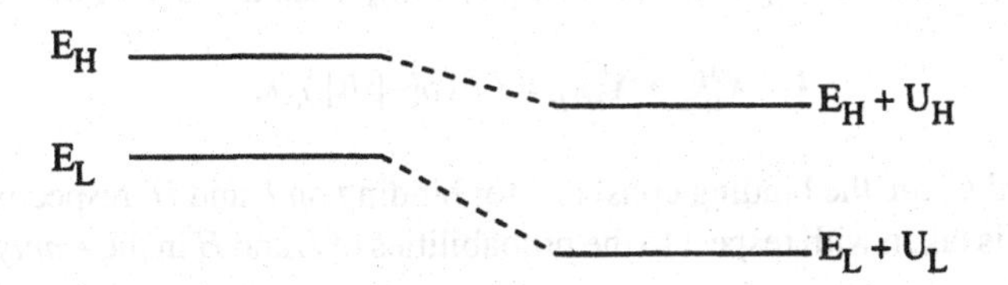

Figure 3.1. Schematic energy-level diagram of a system before (left) and after (right) the binding of a ligand.

3.3. THE BINDING ISOTHERM

From the GPF in Eq. (3.2.9) we derive the average number of ligands per system,*

$$\theta = n = \lambda \frac{\partial \ln \xi}{\partial \lambda} = \frac{(Q_L q_L + Q_H q_H)\lambda}{\xi} \tag{3.3.1}$$

Incidentally, $\bar{n}$ is the same as $P(1)$. The reason is that $\bar{n}$, as an average over two possible occupation states, can also be written as

$$n = \sum_{i=0}^{1} iP(i) = 0 \cdot P(0) + 1 \cdot P(1) = P(1) \tag{3.3.2}$$

We now define the *intrinsic* binding constant k as the ratio $Q(1)\lambda_0/Q(0)$ (Section 2.2),

$$k = \frac{Q(1)\lambda_0}{Q(0)} = \frac{Q_L q_L + Q_H q_H}{Q_L + Q_H}\lambda_0 \tag{3.3.3}$$

and rewrite the BI as

$$\theta = n = \frac{kC}{1 + kC} \tag{3.3.4}$$

which is the same as the BI for the simple Langmuir model. It is easy to show that the form of the BI is independent of the number of states chosen for **P** (see Appendix C).

If we define the binding energy function† by

$$B(\alpha) = \begin{cases} U_L & \text{if} \quad \alpha = L \\ U_H & \text{if} \quad \alpha = H \end{cases} \tag{3.3.5}$$

we can express the binding constant k [Eq. (3.3.3)] as an average of the form

$$k = X_L^0 k_L + X_H^0 k_H = \langle \exp[-\beta B] \rangle_0 \lambda_0 \tag{3.3.6}$$

where k_L and k_H are the binding constants for binding on L and H, respectively, and the average is taken with respect to the probabilities of L and H in the *empty* system.

*We shall use $\bar{n}$ for the average number of ligands per system or per adsorbent molecule, and θ for the average per site. Here, they are identical.

†This is a random variable defined over the space of the two conformational states L and H.

In the Langmuir model k is related to the *energy change* for the process of binding. More precisely,

$$kC = \frac{Q(1)}{Q(0)}\lambda = \exp\{-\beta[A(1) - A(0) - \mu]\} \tag{3.3.7}$$

$A(1) - A(0)$ is simply U in the Langmuir model. Here, on the other hand, $A(1) - A(0)$ is a free-energy change, not an energy change. Recall that in defining U we emphasize that U is chosen as an *energy* of binding, although in reality it is always a free energy (see Appendix C for more details). This was done to emphasize the emergence of free energy as a result of the averaging over the states of the molecule in Eq. (3.3.6). In the simple Langmuir model, U is presumed to be independent of temperature. On the other hand, the quantity $A(1) - A(0)$, which replaces U in the present model, is temperature-dependent,

$$A(1) - A(0) = -k_BT \ln \frac{Q(1)}{Q(0)} = -k_BT \ln \frac{Q_L q_L + Q_H q_H}{Q_L + Q_H}$$

$$= -k_BT \ln (X_L^0 q_L + X_H^0 q_H) \tag{3.3.8}$$

If there is only one state, or when $q_L = q_H$, then $A(1) - A(0) = U$. However, when there are more than one state and $q_L \neq q_H$, the change in temperature would change the equilibrium concentrations of L and H, and hence $A(1) - A(0)$ becomes temperature-dependent. This has important consequences for the thermodynamics of the binding process, such as the entropy and the energy of the binding process.*

3.4. INDUCED CONFORMATIONAL CHANGES

We now examine the extent of conformational changes induced by the binding process. Since for any λ we must have

$$X_L(\lambda) + X_H(\lambda) = 1 \tag{3.4.1}$$

it is sufficient to study only one of the mole fractions. We already had

$$X_L(\lambda) = P(L) = \frac{Q_L + Q_L q_L \lambda}{\xi} \tag{3.4.2}$$

where $P(L)$ is the probability of finding the state L, here expressed as a function of λ. We have earlier seen two limiting cases of $X_L(\lambda)$ in Eqs. (3.2.18) and (3.2.19).

Instead of examining the nonlinear function $X_L(\lambda)$ in the range $0 \leq \lambda \leq \infty$, it is more convenient to study the function $X_L(\theta)$ in the range $1 \leq \theta \leq 1$. By eliminating

*More details on this topic can be found in Chapter 3 of Ben-Naim (1992).

λ from Eqs. (3.4.2) and (3.3.1) and using the two dimensionless quantities

$$K = \frac{Q_H}{Q_L}, \quad h = \frac{q_H}{q_L} \tag{3.4.3}$$

we obtain the explicit dependence of X_L on θ,

$$K_L(\theta) = X_L^0 + \theta \frac{K(1-h)}{(1+K)(1+hK)} \tag{3.4.4}$$

This is a linear function in θ (Fig. 3.2a), the slope of which is

$$d_L = \left(\frac{\partial X_L}{\partial \theta}\right)_T = \frac{K(1-h)}{(1+K)(1+hK)} \tag{3.4.5}$$

Note that the slope is determined by the parameter h. When $h = 1$, $d_L = 0$. The sign of the slope d_L depends on whether $h > 1$ or $h < 1$. Figure 3.3 shows d_L as a function of X_L^0 for various values of h. This function is obtained from Eq. (3.4.5) by substituting $K = (1 - X_L^0)/X_L^0$, i.e.,

$$d_L = \frac{(1-h)(1-X_L^0)}{(1-h+h/X_L^0)} \tag{3.4.6}$$

The quantity d_L will be referred to as the extent of the conformational change induced by the ligand. Whenever $d_L = 0$, we shall say that the binding of the ligand does not induce conformational changes. This can occur either because $K = 0$ (or $K = \infty$), in which case we say that the adsorbent molecule is not *responsive* to the binding process, or because $h = 1$, in which case we say that the ligand *cannot induce* conformational changes in the molecule **P**. Clearly, "either $K = 0$ or $h = 1$" is a necessary and sufficient condition for $d_L = 0$. Because of its fundamental role in the transmission of information between ligands—hence the emergence of indirect cooperativity—it is worthwhile pausing to examine the behavior of d_L in some limiting cases:

1. When $K \to 0$, the energy difference $E_H - E_L \to \infty$. The L conformer is infinitely more stable relative to H. Thus, for any finite value of h the system is *infinitely resistant* to conformational changes. (Clearly, the same is true for $K \to \infty$, but we have chosen the energy levels such that $E_H - E_L > 0$.)
2. When $h = 1$, i.e., $U_L = U_H$, the ligand has no preference for binding on L or on H. Therefore, for any finite value of K the ligand cannot induce conformational change.
3. When $h \to 0$, $U_H - U_L \to \infty$, in which case binding on L is overwhelm-

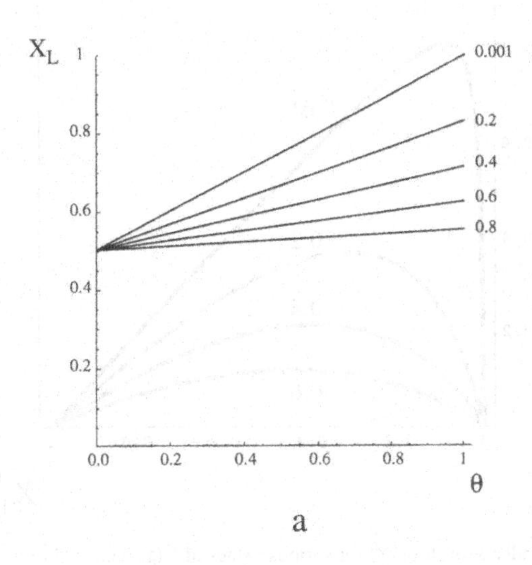

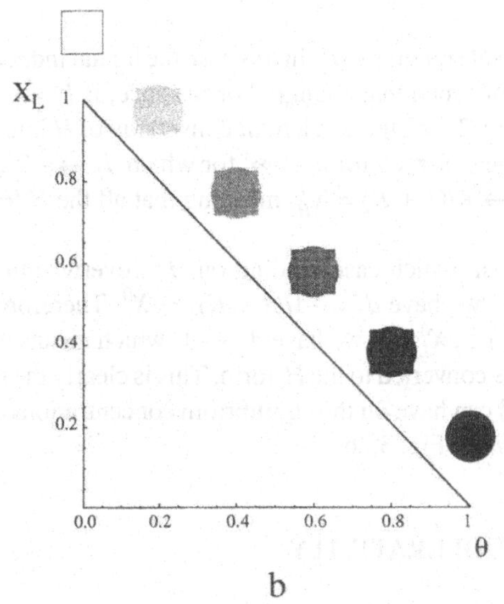

Figure 3.2. X_L as a function of θ [Eq. (3.4.4)] for various values of h. (a) For $K = 1$, at $\theta = 0$, the initial value of X_L is $X_L^0 = 1/2$. (b) For $K = 0.001$, the initial value of X_L is $X_L^0 = (1.001)^{-1} \approx 1$. With $h = 1000$, X_L drops to almost zero at $\theta = 1$. Pure L and pure H are shown as a square and a circle, respectively. Intermediate equilibrium values of L and H are indicated by the intermediate shades of gray.

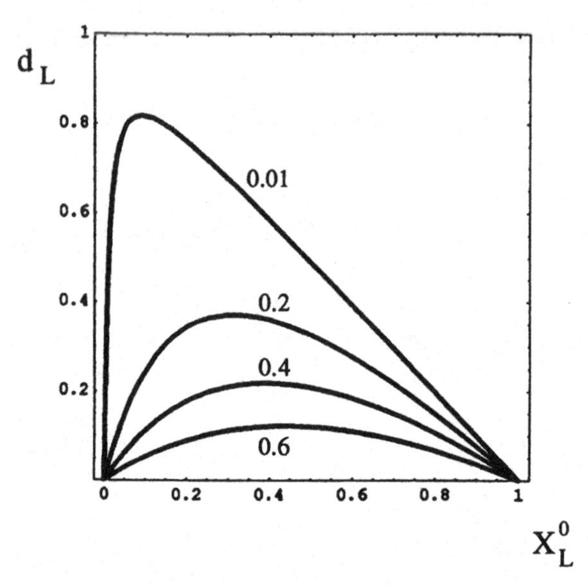

Figure 3.3. Dependence of d_L on X_L^0 for various values of h (as indicated next to each curve).

ingly preferable relative to H. In this case the ligand induces the maximum possible conformational change. For instance, if $K = 1$, then $X_L^0 = X_H^0 = 1/2$ and $d_L = 1/2$, i.e., there is a *total* conversion of H into L. The opposite effect will be observed for $h \to \infty$, for which $d_L \to -1/2$. In general, for $h \to 0$, $d_L \to K/(1+K) = X_H^0$, meaning that all the H form is converted to L.

4. For $h \to \infty$, in which case binding on H is overwhelmingly preferable relative to L, we have $d_L = -1/(1+K) = -X_L^0$. Therefore, as K becomes very small,* i.e., $X_L^0 \sim 1$, we have $d_L \sim -1$, which means that almost all of the L-form is converted to the H-form. This is clearly the maximum effect that a ligand can have on the equilibrium concentrations of H and L. This case is shown in Fig. 3.2b.

3.5. SPURIOUS COOPERATIVITY

The term cooperativity, as defined qualitatively in Section 2.1 and in more detail in Chapter 4, requires at least two ligands that can "communicate" on the same adsorbent molecule. Clearly since this chapter is devoted to single-site molecules, the term cooperativity is not even definable in such systems. Nevertheless, we discuss in this section a phenomenon referred to as *spurious* cooperativity, which

*One should be careful about the order in which we let $h \to \infty$ and $K \to 0$. Here, we first let h be very large so that $d_L \sim -1/(1+K)$, and then let K be small enough so that $d_L \sim -1$.

is not a genuine cooperativity.[†] Spurious cooperativity can occur in any system consisting of a mixture of different adsorbing molecules.

In multiple-site systems, spurious cooperativity can occur along with genuine cooperativity (as defined in subsequent chapters). It is only in the single-site system that any apparent cooperativity is necessarily *spurious*, and therefore we place the discussion of this phenomena in this section. We shall return to spurious cooperativity in two-site systems in Section 4.6. The reader should keep in mind the possibility of spurious cooperativity whenever processing and interpreting experimental data, especially when one has reason to suspect that the two or more conformations might not be in equilibrium.

In this section we find it more convenient to start with an ensemble of M independent and indistinguishable systems (i.e., the systems are identical but not localized, as assumed in Section 2.4), each of which has a single binding site. We stress from the outset that the concept of cooperativity, as defined in Section 4.2, does not apply to such systems. What we shall show is that under certain conditions a single-site system can exhibit behavior that is similar to the behavior of a cooperative system.

The GPF of the ensemble is

$$\Xi(T, M, \lambda) = \xi^M / M! \tag{3.5.1}$$

If each system can be in one of two conformational states, L or H, we write[‡]

$$\Xi(T, M, \lambda) = (\xi_L + \xi_H)^M / M! = \sum_{M_L + M_H = M} \frac{\xi_L^{M_L}}{M_L!} \frac{\xi_H^{M_H}}{M_H!} = \sum_{M_L + M_H = M} \Xi^*(T, M_L, M_H, \lambda) \tag{3.5.2}$$

where $M_H = M - M_L$. The quantity Ξ^* is the GPF of the same system but having a fixed number M_L of L-systems, and the rest having M_H of H-systems. Since $\Xi(T, M, \lambda)$ is a sum over many positive numbers $\Xi^*(T, M_L, M_H, \lambda)$, we must have the inequality

$$\Xi(T, M, \lambda) > \Xi^*(T, M_L, M_H, \lambda) \tag{3.5.3}$$

Thus, the GPF of the equilibrated system, defined by the variables (T, M, λ), is always *larger* than the GPF of a system (T, M_L, M_H, λ) with any arbitrary but fixed

[†]A minimum of two ligands must cooperate to observe a cooperative system. A different view is expressed in Wyman and Gill (1990). These authors correctly point out that in a single binding site there can be *no cooperativity* (page 51). However, three pages later they refer to a mixture of single-site molecules as being negative cooperative. No less confusing is the usage of two different definitions of "macroscopic" and "microscopic" cooperativities [Bradsley and Waight (1978), Whitehead (1980), Briggs 1984), DiCera (1996)]. These authors rely on the *shape* of the BI to define cooperativity. As we shall see in this and in the next chapter, definitions of cooperativity based on the shape of the BI could be quite misleading.

[‡]The GPF of this system depends also on the volume V, but we suppress this dependence in our notation.

values of M_L and $M_H = M - M_L$. We shall refer to the system (T, M_L, M_H, λ) as the "frozen-in" system.

There is a particular value of M_L (and hence of $M_H = M - M_L$) for which the inequality (3.5.3) turns into an almost equality. This is the equilibrium value of M_L, denoted by M_L (hence $M_H = M - M_L$), given by

$$M_L = M\frac{\xi_L}{\xi}, \quad M_H = M\frac{\xi_H}{\xi} \tag{3.5.4}$$

for which

$$\Xi(T, M, \lambda) = \Xi^*(T, M_L, M_H, \lambda) \tag{3.5.5}$$

Although we have used the equality sign in Eq. (3.5.5), the reader should be aware of the fact that this is not an equality in the strict mathematical sense. It is, however, an equality only in the thermodynamic sense; i.e., when we take the logarithm of the two quantities, the difference in the resulting quantities is negligibly small, hence we can view them as practically equal.† A qualitative argument is the following: Suppose we start with a fully equilibrated ensemble of systems, for which the PF is $\Xi(T, M, \lambda)$. We then "freeze-in" the conversion $L \rightleftharpoons H$ (say, by introducing an inhibitor). We convert our ensemble into a mixture of *fixed* values of M_L and M_H. The thermodynamic quantities obtained directly from Ξ, or by its first derivatives (such as the entropy, energy, free energy, etc.), are unaffected by this operation. Note, however, that second and higher derivatives are affected.‡

We next write the BI of the two systems. The first is the "equilibrated system," the GPF of which is Ξ on the lhs of Eq. (3.5.5),

$$\begin{aligned}\theta^{eq} &= \frac{N}{M} = \frac{\lambda}{M}\frac{\partial \ln \Xi}{\partial \lambda} = \frac{\lambda}{\xi}\left(\frac{\partial \xi_L}{\partial \lambda} + \frac{\partial \xi_H}{\partial \lambda}\right) \\ &= \lambda X_L^{eq}\frac{\partial \ln \xi_L}{\partial \lambda} + \lambda X_H^{eq}\frac{\partial \ln \xi_H}{\partial \lambda} = X_L^{eq}\theta_L + X_H^{eq}\theta_H\end{aligned} \tag{3.5.6}$$

Note that X_L^{eq} and X_H^{eq} are the mole fractions of L and H in the *equilibrated* system.

Similarly, the BI for the "frozen-in" system is obtained from Ξ^*, defined in Eq. (3.5.2), so

$$\theta^f = \frac{N}{M} = \frac{\lambda}{M}\frac{\partial \ln \Xi^*}{\partial \lambda} = X_L^f\theta_L + X_H^f\theta_H \tag{3.5.7}$$

†A more detailed discussion of this point can be found in Chapter 2 of Ben-Naim (1992).
‡Examples are the heat capacity, the compressibility, etc. For more details, see Chapter 7 in Ben-Naim (1992).

In both cases θ^{eq} and θ^f are averages of the BIs of the pure individual conformers L and H. However, the weights used in θ^{eq} and in θ^f are different. The values of the weights X_L^{eq} and X_H^{eq} are *determined* by the equilibrium conditions, given by Eqs. (3.5.4), while X_L^f and X_H^f are determined by the arbitrary values of M_L and M_H *chosen* in the preparation of the mixture of L and H. Since both θ^{eq} and θ^f are averages of θ_L and θ_H, the two curves $\theta^{eq}(\lambda)$ and $\theta^f(\lambda)$ will always fall between the curves θ_L and θ_H. Specifically, for a single-site system $\theta^{eq}(\lambda) > \theta^f(\lambda)$ for any λ, i.e., the equilibrated BI is everywhere "above" the frozen-in curve. Details are given in Appendix E.

The reason for the different behavior of θ^{eq} and θ^f is quite simple. Suppose we start with an equilibrated system and follow the BI, $\theta^{eq}(\lambda)$. If, at some point, we "freeze-in" the equilibrium $L \rightleftharpoons H$, then at that point $X_L^{eq} = X_L^f$ (hence also $X_H^{eq} = X_H^f$), and also $\theta^{eq} = \theta^f$. It is only when we continue the binding process that the two curves diverge. It can be shown (see Appendix E) that at that point the slope of θ^{eq} is always larger than that of θ^f.

We next apply Eqs. (3.5.6) and (3.5.7) to the special case of a single-site system, for which we have

$$\xi_L = Q_L + Q_L q_L \lambda, \quad \xi_H = Q_H + Q_H q_H \lambda \tag{3.5.8}$$

The equilibrated BI is [see Eq. (3.5.6)]

$$\theta^{eq} = \frac{\langle q \rangle \lambda}{1 + \langle q \rangle \lambda} = \frac{kC}{1 + kC} \tag{3.5.9}$$

where $k = (X_L^0 q_L + X_H^0 q_H)\lambda_0$, as obtained in Section 3.3.

Next, we derive the BI θ^f for the *particular* choice of the mole fractions X_L^0 and X_H^0 (i.e., we freeze the equilibrium of the empty system, $\lambda = 0$). Equation (3.5.7) yields

$$\theta^f = X_L^0 \frac{q_L \lambda}{1 + q_L \lambda} + X_H^0 \frac{q_H \lambda}{1 + q_H \lambda} = X_L^0 \theta_L + X_H^0 \theta_H \tag{3.5.10}$$

In the general equations (3.5.6) and (3.5.7), we viewed both θ^{eq} and θ^f as averages of θ_L and θ_H, but with *different* weighting mole fractions X_L^{eq}, X_H^{eq} in the first and X_L^f, X_H^f in the second. In the special case derived above θ^{eq} is viewed as a simple Langmuir isotherm, the binding constant k of which is an average of the binding constants for L and H. Thus, in θ^{eq} we *first* average over k_L and k_H, and then form the Langmuir isotherm. In θ^f, on the other hand, we first form the BI of L and H, and then average over θ_L and θ_H. Clearly, the resulting curves will differ. They

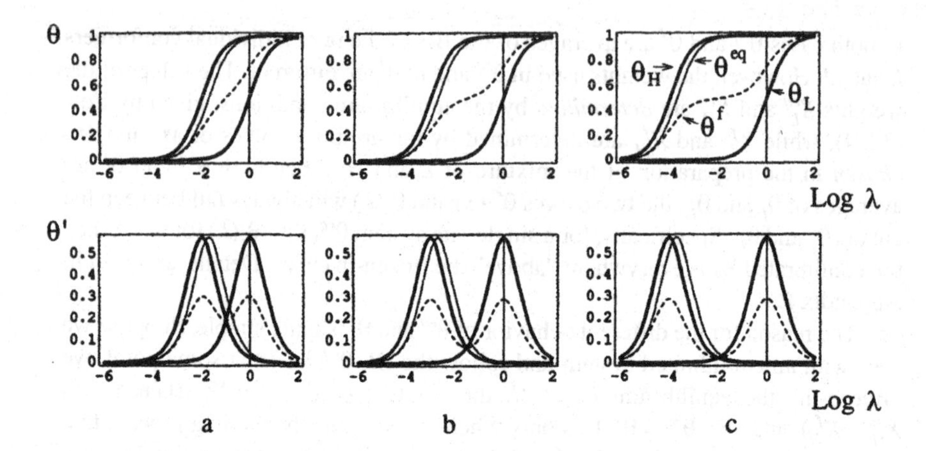

Figure 3.4. The BI (upper panel) and corresponding slopes (lower panel) for the equilibrated (dotted) and "frozen-in" (dashed) systems. The full lines are θ_L and θ_H. The parameters chosen for these illustrations are $Q_L = Q_H = 1$ and (a) $q_L = 1, q_H = 100$; (b) $q_L = 1, q_H = 10^3$; (c) $q_L = 1, q_H = 10^4$.

do have initially the same slope, i.e.,

$$\left.\frac{\partial\theta^{eq}}{\partial\lambda}\right|_{\lambda=0} = \left.\frac{\partial\theta^{f}}{\partial\lambda}\right|_{\lambda=0} = X_L^0 q_L + X_H^0 q_H \tag{3.5.11}$$

but the higher derivatives are different, e.g.,

$$\left.\frac{\partial^2\theta^{eq}}{\partial\lambda^2}\right|_{\lambda=0} = -(X_L^0 q_L + X_H^0 q_H)^2, \qquad \left.\frac{\partial^2\theta^{f}}{\partial\lambda^2}\right|_{\lambda=0} = -(X_L q_L^2 + X_H^0 q_H^2) \tag{3.5.12}$$

Figure 3.4 shows the BIs θ_L and θ_H and the derived quantities θ^{eq} and θ^f for the case $Q_L = Q_H = 1$, hence $X_L^0 = X_H^0 = 1/2$, and with (a) $q_L = 1$, $q_H = 100$; (b) $q_L = 1$, $q_H = 10^3$; (c) $q_L = 1$, $q_H = 10^4$. For reasons to be explained in Section 4.3, it is more convenient to draw θ as a function of $\log_{10}\lambda$ or $\log_{10} C$. This is also in accord with the tradition of plotting titration curves, which are similar to the BI; see Section 2.6. In Fig. 3.4 we also show the slopes of the BI, again drawn as a function of $\log_{10}\lambda$.

The locations of the maximal slope of θ_L and θ_H are clearly discernible in the maxima in the plots of the slopes of the BI. Thus, in Fig. 3.4, the maximum slope of θ_L is always at the point $\log_{10}\lambda = 0$, but as q_H increases the location of the maximum slope of θ_H moves leftward. Note that θ^{eq} is similar in shape to the Langmuir isotherm, but the form of θ^f is quite different. As can be seen from the lower panel of Fig. 3.4, in this particular case there are two points at which the curve

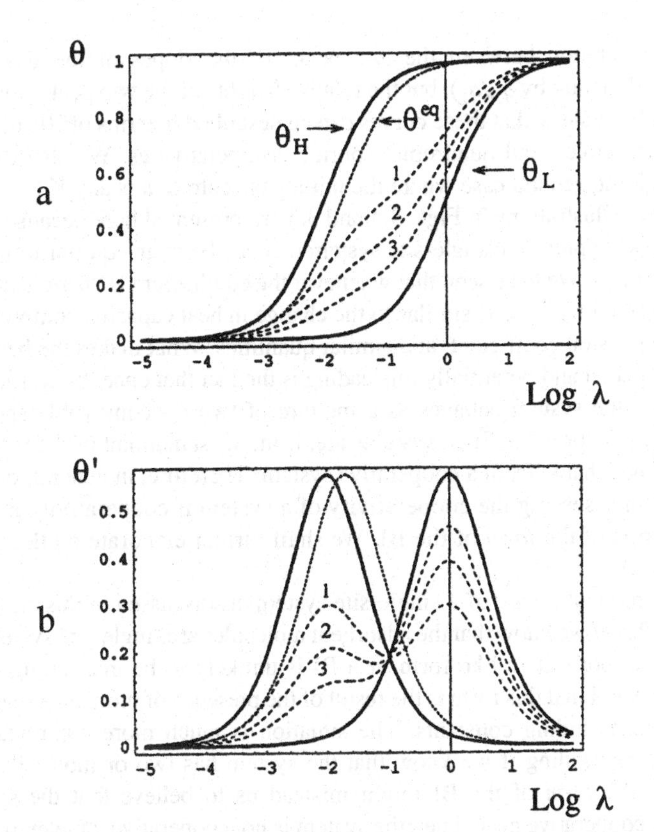

Figure 3.5. The BI and corresponding slope for the case $q_L = 1$ and $q_H = 10^3$, but different values of X_L^0 and X_H^0. Curves 1, 2, and 3 correspond to $X_L^0 = X_H^0 = 1/2$; $X_L^0 = 2/3$, $X_H^0 = 1/3$; and $X_L^0 = 3/4$, $X_H^0 = 1/4$, respectively. Note that the distance between the locations of the maximal slopes of θ^f is the same for the three cases. The relative steepness of the curves is a measure of higher-order spurious cooperativities (Appendix F).

has a maximum. The separation between the two peaks of the slopes of the BIs is determined by the difference $q_L - q_H$ or $k_L - k_H$.*

Figure 3.5 shows how the BI θ^f changes with the ratio $K = Q_H/Q_L$, i.e., the initial concentrations X_L^0 and X_H^0. For this illustration we fixed the values of $q_L = 1$, $q_H = 100$, and chose different values of K, i.e., $K = 1$, $K = 1/2$, and $K = 1/3$ or, equivalently, $X_L^0 = X_H^0 = 1/2$; $X_L^0 = 2/3$, $X_H^0 = 1/3$; and $X_L^0 = 3/4$, $X_H^0 = 1/4$. Note that

*One may wonder how a system with two conformers, having the same energies, can differ so widely in their binding energies. This is possible. Consider a binding molecule with an asymmetric (or chiral) center. In that case the two enantiomers, one being a mirror image of the other, have exactly the same energy. If the ligand is a single enantiomer of an asymmetric molecule, then it is possible that it will bind with very different binding energies to the two enantiomers of the adsorbent molecule.

here the separation between the two peaks of the slopes of the BIs is fixed (determined mainly by q_H/q_L), but the relative heights of the two peaks changes. In Appendix F we show that these curves can be described in terms of BIs of systems having pair, triplet, and quadruplet spurious cooperativities. We also discuss in Appendix F the general case where the initial concentration is any X_L^0.

The two illustrations in Figs. 3.4 and 3.5 are presented here because we have devoted this chapter to a single-site system that can be in an equilibrium between two conformers. We have seen that whenever the equilibrium is "frozen-in," the BI behaves differently. This is similar to the change in heat capacity, compressibility, thermal expansion coefficient, and similar quantities. What makes the behavior of the BI so special and potentially misleading is the fact that once the equilibrium is "frozen-in," the system behaves as a mixture of two nonconvertible species, as discussed in Section 2.5. Such systems are, from the standpoint of the form of the BI, indistinguishable from a cooperative system. Therefore, in general, one might be misled in assessing the cooperativity of a system if cooperativity is defined only in terms of the *form* of the BI. We shall further elaborate on this point in Section 4.3.

In the particular case of a single-site system, as discussed in this chapter and in Section 2.5, if we know that the adsorbent molecules are single-site systems, then observing a cooperative-like form of a BI is unlikely to be misleading. We can immediately tell that this form is the result of the presence of more than one species with different binding constants. The situation is much more complicated and potentially misleading if we know that the system has two or more sites. Here, relying on the *form* of the BI might mislead us to believe that the system is negatively cooperative even when the system is noncooperative, or even positively cooperative. We defer the discussion of these cases to Chapter 4.

In this section we derived the BIs of a system where the equilibrium conversion between the species L and H has been "frozen-in." The more general case is a mixture of two (or more) components of any concentration. This case is discussed further in Appendix F.

4

Two-Site Systems: Direct and Indirect Cooperativity

4.1. INTRODUCTION

In this chapter we begin to study cooperative systems. We start by defining the term correlation between *any* two events. The term "cooperativity" is then identified with the term "correlation" when applied to a particular event such as "sites $i_1, \ldots, i_n$ are occupied." We examine two fundamental sources of cooperativity: direct and indirect. The first is due to *direct* ligand–ligand interaction. The second can arise from various sources: (1) effect of the ligand on the translational and rotational PF of the molecule; (2) effect of the solvent; (3) effect of the ligand on the conformational state of the molecule. The first possibility is usually negligible since the ligands are very small compared with the adsorbent molecule (Appendix B). The binding of proteins to DNA, discussed in Section 5.10, is an exception. The second is important and should be considered whenever a solvent is present. We shall discuss solvent effects in Chapter 9. What remains is the effect of the ligand on the conformational state of the adsorbent molecule. This is discussed beginning in Section 4.5 and is shown to be one of the most interesting aspects of cooperativity. In Chapters 5 and 6, we shall see that the mechanism of transmitting information by the adsorbent molecule is, in some formal sense, essentially the same as through the solvent. We shall see that in order to transmit information through the adsorbent molecule, which in general consists of several subunits, three conditions must be fulfilled. First, the ligand must discriminate between the different conformations. In the case of a two-conformation model, this means that the binding energies to the two conformations must be different. Second, the conformational equilibrium denoted by $H \rightleftharpoons L$ between the states L and H must be responsive, i.e., conformational changes in the adsorbent molecule must be induced upon ligation (see Section 3.4). Finally, when there are two (or more) subunits, there should be transmission of information across the subunit–subunit boundaries. Corresponding to these three effects we shall introduce the three fundamental parameters h, K, and η. We shall see also that the same parameters are necessary to describe indirect cooperativities in more complicated binding systems.

If the model treated in Section 3.4 was likened to the "hydrogen atom" of the binding system, the models of this chapter may be referred to as the analogue of the "hydrogen molecule." The new phenomenon of the chemical bond, formed when there are two hydrogen atoms, is the analogue of the *cooperativity* between two (or more) ligands in a binding system.

4.2. THE GENERAL DEFINITION OF CORRELATION AND COOPERATIVITY IN A TWO-SITE SYSTEM

Our system is a single adsorbent molecule denoted by **P**, having two binding sites denoted by a and b. These could be identical or different, but in this section the treatment is general and applies to any case.

We shall use probabilistic language to define the terms "correlation" and "cooperativity." The probabilities pertaining to the various events are read from the appropriate terms in the GPF, which for our system is

$$\xi(T, \lambda) = Q(0, 0) + [Q(a, 0) + Q(0, b)]\lambda + Q(a, b)\lambda^2 \tag{4.2.1}$$

The four terms in Eq. (4.2.1) correspond to the four occupancy states of the system (Fig. 4.1).

The average number of ligands *per system* is

$$n = \lambda \frac{\partial \ln \xi}{\partial \lambda} = \frac{[Q(a, 0) + Q(0, b)]\lambda + 2Q(a, b)\lambda^2}{Q(0, 0) + [Q(a, 0) + Q(0, b)]\lambda + Q(a, b)\lambda^2} \tag{4.2.2}$$

and the BI *per site* is

$$\theta = \frac{1}{2} n \tag{4.2.3}$$

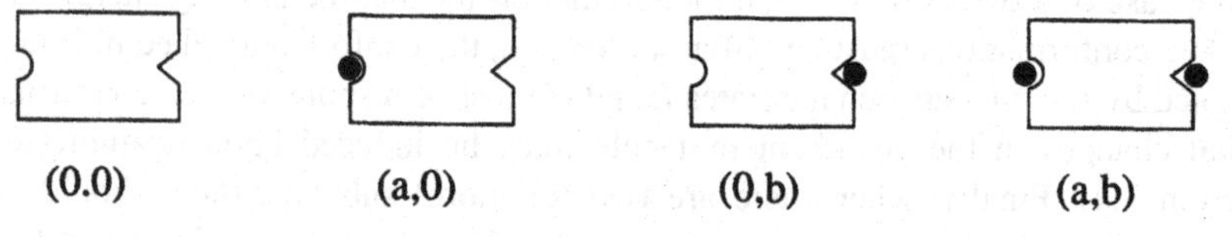

Figure 4.1. The four occupancy states of a two-site system.

In terms of the intrinsic binding constants, we transform the BI into

$$\theta = \frac{1}{2}\frac{(k_a + k_b)C + 2k_{ab}C^2}{1 + (k_a + k_b)C + k_{ab}C^2} \tag{4.2.4}$$

where the binding constants are defined in terms of the PFs as

$$k_a = \frac{Q(a,0)}{Q(0,0)}\lambda_0, \qquad k_b = \frac{Q(0,b)}{Q(0,0)}\lambda_0, \qquad \text{and } k_{ab} = \frac{Q(a,b)}{Q(0,0)}\lambda_0^2 \tag{4.2.5}$$

with $\lambda = \lambda_0 C$. The four fundamental probabilities corresponding to the four "events" depicted in Fig. 4.1 may be read directly from the PF in Eq. (4.2.1). These are

$$\begin{aligned} P(0,0) &= Q(0,0)/\xi, \qquad P(a,0) = Q(a,0)\lambda/\xi, \\ P(0,b) &= Q(0,b)\lambda/\xi, \qquad P(a,b) = Q(a,b)\lambda^2/\xi \end{aligned} \tag{4.2.6}$$

We shall also need the (marginal) probabilities

$$\begin{aligned} P(a) &= P(a,\_) = P(a,0) + P(a,b) \\ P(b) &= P(\_,b) = P(0,b) + P(a,b) \end{aligned} \tag{4.2.7}$$

A blank space in the argument of P means "unspecified" or "anything." Thus, $P(a,\_)$ [or simply $P(a)$] is the probability of the event "site a is occupied." The state of site b is unspecified and could be anything, empty or occupied.

The *correlation* between *any* two events $\mathcal{A}$ and $\mathcal{B}$ is defined by

$$g(\mathcal{A},\mathcal{B}) = \frac{\mathcal{P}(\mathcal{A}\cdot\mathcal{B})}{\mathcal{P}(\mathcal{A})\cdot\mathcal{P}(\mathcal{B})} \tag{4.2.8}$$

Clearly, this quantity measures the extent of *dependence* between the two events. The two events $\mathcal{A}$ and $\mathcal{B}$ are independent if and only if*

$$P(\mathcal{A}\cdot\mathcal{B}) = P(\mathcal{A})\cdot\mathcal{P}(\mathcal{B}) \tag{4.2.9}$$

or, equivalently,

$$g(\mathcal{A},\mathcal{B}) = 1 \tag{4.2.10}$$

In the context of this book the term *correlation* as defined in Eq. (4.2.8) applies for *any* two events. For instance, the correlation between the events "site a is empty" and "site b is empty" is

$$g(0,0) = \frac{P(0,0)}{P(0,\_)P(\_,0)} \tag{4.2.11}$$

*See also the footnote between Eqs. (1.1.8) and (1.1.9) in Section 1.1.

The same applies for the two events "site a is occupied" and "site b is occupied," for which the correlation is

$$g(a, b) = \frac{P(a, b)}{P(a)P(b)} = \frac{P(a, b)}{P(a, \_)P(\_, b)} \tag{4.2.12}$$

It is only for the latter events that we shall apply the term cooperativity. Thus, while the term *correlation* is applied, as a measure of the extent of the dependence, between *any* two (or more) events, the term *cooperativity* is applied only to a subclass of events involving ligands occupying sites. It will measure the extent of the dependence between two (or more) *ligands* bound to their sites. For the latter events, the terms cooperativity and correlation will be used synonymously. Whenever events of this type are correlated, we shall say that the ligands occupying the two sites *cooperate*, or simply the system is a *cooperative*. Sometimes we shall also say loosely that in this case there exists *communication* between the two ligands occupying the sites a and b. Communication is used here in the sense that a ligand at one site "knows" or "senses" the presence of another ligand at the second site.

Whenever $g(a, b) = 1$, we say that there is no correlation between the events "site a is occupied" and "site b is occupied," and hence the system is *noncooperative*. If $g(a, b) > 1$, we say that the two ligands cooperate positively, and when $g(a, b) < 1$, they cooperate negatively.*

In terms of conditional probabilities we have

$$P(a/b) = \frac{P(a, b)}{P(b)} = P(a)g(a, b) \quad \text{and} \quad P(b/a) = \frac{P(a, b)}{P(a)} = P(b)g(a, b) \tag{4.2.13}$$

Thus, positive cooperativity means that the conditional probability, say $P(a/b)$, is larger than the unconditional probability,† i.e., the fact that "b is occupied" enhances the probability of the event "a is occupied." Since $g(a, b)$ is defined symmetrically with respect to a and b, it follows from $g(a, b) > 0$ that both $P(a/b) > P(a)$ and $P(b/a) > P(b)$. Similarly, negative cooperativity, $g(a, b) < 1$, implies $P(a/b) < P(a)$ and $P(b/a) < P(b)$.

In general, $g(a, b)$ is dependent on λ. Two limiting cases are of interest:

$$\lim_{\lambda \to \infty} g(a, b) = 1 \tag{4.2.14}$$

and

$$g^0(a, b) \equiv \lim_{\lambda \to 0} g(a, b) = \frac{Q(a, b)Q(0, 0)}{Q(a, 0)Q(0, b)} \tag{4.2.15}$$

*In some earlier publications the term "cooperativity" is used for *positive cooperativity* and "anticooperativity" is used for *negative cooperativity*. In this book "cooperativity" is used whenever $g \neq 1$.

†Sometimes, when there is positive cooperativity one says that a ligand at a supports or favors the binding of a ligand at b, and vice versa.

When $\lambda \to \infty$, each of the probabilities $P(a)$, $P(b)$, and $P(a, b)$ tends to unity, hence also $g(a, b) \to 1$. On the other hand, when $\lambda \to 0$, all these probabilities tend to *zero*, but the ratio defining $g(a, b)$ in Eq. (4.2.12) tends to a constant.*

Figure 4.2 shows he dependence of $g(a, b)$ on λ various values of $g^0(a, b)$. Whatever the initial value of $g^0(a, b)$, the correlation function $g(a, b)$ will tend to unity when $\lambda \to \infty$.

In all our applications we shall need only $g^0(a, b)$. This is the only quantity that appears in the BI. For instance, if we wish to express k_{ab} in Eq. (4.2.5) in terms of k_a and k_b, we find that

$$k_{ab} = \frac{Q(a, b)}{Q(0, 0)} \lambda_0^2 = g^0(a, b) k_a k_b \tag{4.2.16}$$

For this reason, we shall always refer to $g^0(a, b)$ as the pair correlation in the system, and drop the superscript zero.

At this point we again stress the sequence of definitions leading to Eq. (4.2.16). First, the correlation function is *defined* as a measure of the extent of the dependence between the two events in Eq. (4.2.12) [or, equivalently, in Eq. (4.2.13)]. The probabilities used in the definition of $g(a, b)$ were read from the GPF of the system, e.g., (4.2.1). This "side" of $g(a, b)$ allows us to investigate the *molecular content* of the correlation function, which is the central issue of this book. The other "side" of $g(a, b)$ follows from the recognition that the limiting value of $g(a, b)$, denoted by $g^0(a, b)$, connects the binding constants k_{ab} and $k_a \cdot k_b$. This "side" of $g^0(a, b)$ allows us to extract information on the cooperativity of the system from the *experimental* data. In other words, these relationships may be used to *calculate* the correlation function from experimental data, on the one hand, and to *interpret* these correlation functions in terms of molecular properties, on the other.

The more traditional approach is to *define* an "interaction coefficient" [the equivalent of our $g^0(a, b)$] in terms of the experimental binding constants by

$$\alpha(a, b) \equiv \frac{k_{ab}}{k_a k_b} \tag{4.2.17}$$

Whenever $\alpha(a, b) \neq 1$, one says that the two events "site *a* occupied" and "site *b* occupied" are dependent. However, based on experimental data for k_a, k_b, and k_{ab}, one cannot trace the source of this dependence. In some cases it can be due to *direct* ligand–ligand interaction, in other cases it could be due to *indirect* communication between the ligands mediated through the adsorbent molecule or through the solvent. It could also be due to the effect of the ligand on the momentum and the

*Incidentally, we note that $g(0, 0)$ defined in Eq. (4.2.11) behaves in an opposite manner. In these limits, $g(0, 0) \to g^0(a, b)$ for $\lambda \to \infty$, but $g(0, 0) \to 1$ for $\lambda \to 0$. For other similar behavior, see Chapter 4 in Ben-Naim (1992).

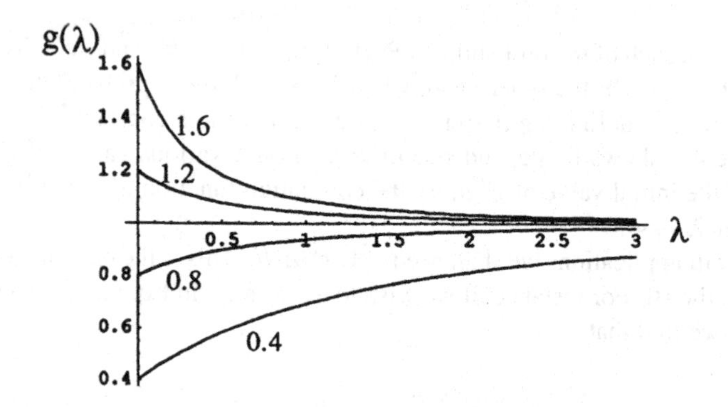

Figure 4.2. The dependence of $g(a, b)$, defined in Eq. (4.2.12), on λ for the special case $Q(a, b) = S$, $Q(a, 0) = Q(0, b) = Q(0, 0) = 1$. This is the case of *direct* correlation only, discussed in Section 4.3. The values of the initial correlation $g(\lambda = 0) = g^0(a, b) = S$ are indicated next to each curve.

rotational partition function (see Section 5.10). Therefore, care must be exercised when ascribing the meaning of "interaction parameters" to the quantity $\alpha(a, b)$.

The concept of cooperativity can also be translated into free-energy change by the relation

$$W(a, b) = -k_B T \ln g^0(a, b) \tag{4.2.18}$$

where $W(a, b)$ is the work* involved in the "reaction" that may be written symbolically as

$$(a, 0) + (0, b) \rightarrow (a, b) + (0, 0) \tag{4.2.19}$$

i.e., the formation of a fully occupied system (a, b) from two singly occupied systems.

The quantity $W(a, b)$ is the analogue of the potential of the mean force in the theory of liquids. We shall see in the following sections that this quantity sometimes behaves as an *energy*, but in most cases it is a *free energy*. In terms of $W(a, b)$ we may also define noncooperative systems whenever $W(a, b) = 0$, and positive and negative cooperativity for $W(a, b) < 0$ and $W(a, b) > 0$, respectively. The reader should note the potentially confusing statement that a *positive* cooperativity involves *negative* values of $W(a, b)$, and vice versa.

*If the system is a single adsorbent molecule, then $W(a, b)$ is either a Gibbs or a Helmholtz energy. When the system is immersed in a solvent, then there is a difference between the Gibbs and the Helmholtz energies according to the conditions under which the reaction (4.2.18) is performed (constant pressure or constant volume).

4.3. TWO IDENTICAL SITES: DIRECT CORRELATION

The simplest case of cooperativity occurs when there exists *direct* interaction between the two ligands occupying the two sites. The internal properties of the adsorbent molecule are still assumed to be unaffected by the binding process. The four possible states of the system are as in Fig. 4.1, but a and b are now identical. The corresponding coefficients in the GPF are

$$Q(0, 0), \qquad Q(1, 0) = Q(0, 1) = Q(0, 0)q, \qquad Q(1, 1) = Q(0, 0)q^2S \tag{4.3.1}$$

where

$$q = \exp(-\beta U), \qquad S = \exp[-\beta U(1, 1)] \tag{4.3.2}$$

U is the binding *energy* of the ligand to the site and $U(1, 1)$ is the *direct* interaction energy between the two ligands. By *direct* we mean the intermolecular interaction energy between the ligands in *vacuum*. Figure 4.3 shows the general form of this interaction energy for two oxygen molecules and for two protons in vacuum. Any modification to the correlation between the two ligands imposed by the presence of the adsorbent molecule (Section 4.5) or of the solvent (Chapter 9) will be referred to as *indirect*.

The BI, per site, for this system is

$$\theta = \frac{n}{2} = \frac{q\lambda + q^2S\lambda^2}{1 + 2q\lambda + q^2S\lambda^2} = \frac{kC + k^2SC^2}{1 + 2kC + k^2SC^2} \tag{4.3.3}$$

where the intrinsic binding constant is $k = q\lambda_0$. In this particular model, we identify

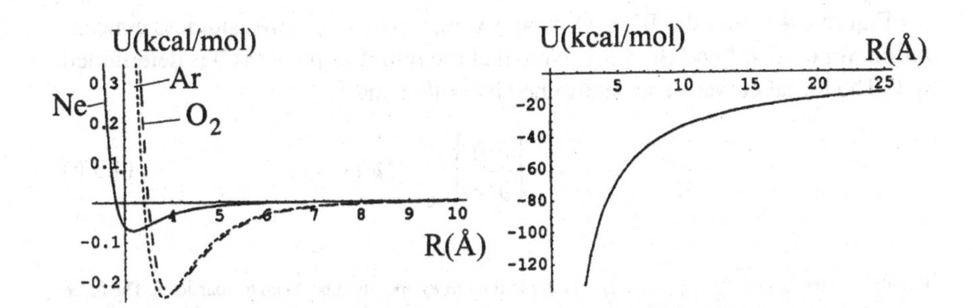

Figure 4.3. Examples of *direct* interaction between particles. (a) Lennard-Jones potential $U(1, 1) = 4\varepsilon[(\sigma/R)^{12} - (\sigma/R)^6]$ for neon ($\varepsilon = 0.071$ kcal/mol, $\sigma = 2.82$ Å), argon ($\varepsilon = 0.238$ kcal/mol, $\sigma = 3.4$ Å, and oxygen ($\varepsilon = 0.224$ kcal/mol, $\sigma = 3.5$ Å). (b) Two protons in vacuum. The Coulombic interaction is of the form e^2/R, where e is the electron charge and R is the distance; $e^2 = 332.8$ kcal Å/mol.

the correlation function

$$g(1,1)=\frac{Q(1,1)Q(0,0)}{Q(1,0)Q(0,1)}=S=\exp[-\beta U(1,1)] \qquad (4.3.4)$$

and the corresponding work [Eq. (4.2.18)]

$$W(1,1)=-k_BT\ln g(1,1)=U(1,1) \qquad (4.3.5)$$

Note that in the case of two identical sites, occupied and empty sites are denoted by 1 and 0, respectively. When the sites are different we replace the "1" by the symbol used to name the site, say "*a*" or "*b*," and insert zero when the site is empty.

We shall refer to this particular correlation as *direct* correlation, simply because it originates from the *direct* interaction between the two ligands. These interactions could be quite strong, as in the case of two protons in dicarboxylic acids (Section 4.8). In some cases of interest they are very weak and can be neglected. Such is the case for oxygen molecules occupying the sites of hemoglobin (Chapter 6).

In the present system, $S=1$, [or, equivalently, $U(1,1)=0$] means that the system is noncooperative. The system is positively, or negatively, cooperative when $S>1$, or $S<1$, respectively.

The effect of $U(1,1)$ on the conditional probabilities [see Eq. (4.2.13)] is easily understood. When there exists attraction* between the ligands, a ligand occupying one site *attracts* the second ligand, and hence increases the probability of binding the second ligand—hence positive cooperativity. The reverse holds when the two ligands repel each other.

Two properties of the direct correlation should be noted, especially when compared with indirect correlation. (1) The interaction energy $U(1,1)$, and hence $g(1,1)$, are strongly dependent on the distance between the ligands. (2) Since $U(1,1)$ is temperature-independent,† the correlation $g(1,1)$ depends on T only through $\beta=(k_BT)^{-1}$.

Figure 4.4 shows the BI, $\theta(x)$, where $x=q\lambda=kC$ is a dimensionless concentration, for (a) $S\geq 1$ and (b) $S\leq 1$. Note that the initial slope of $\theta(C)$ is determined by k. The initial curvature is determined by both k and S,

$$\left(\frac{\partial\theta}{\partial C}\right)_{C=0}=k, \qquad \left(\frac{\partial^2\theta}{\partial C^2}\right)_{C=0}=2k^2(S-2) \qquad (4.3.6)$$

*Usually, the terms attraction and repulsion refer to the *forces* operating between the particles. The force is the gradient of the potential. Therefore, it is not always true that a negative potential is attractive, or that a positive potential is repulsive. The important quantity that determines the cooperativity is the potential and, in more general cases, the work $W(1,1)$, not the force.

†Note, however, that $U(1,1)$ itself could be a free energy—see Appendix C. Here, as well as in other parts of the book, we suppress the free-energy character of $U(1,1)$ to stress the emergence of free energy that is superimposed on $U(1,1)$ whenever we average over states of **P**, or of the solvent.

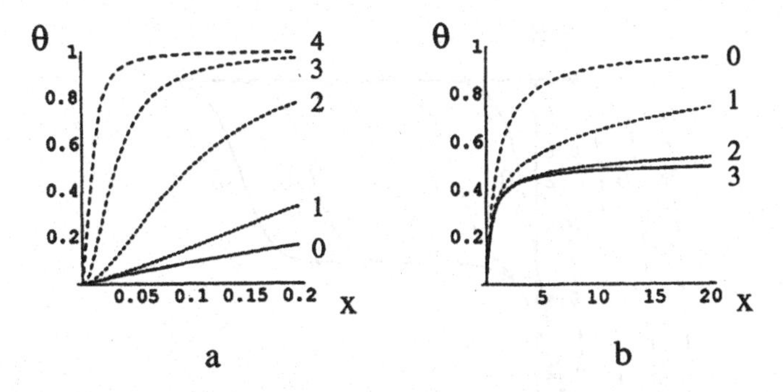

Figure 4.4. The BI for (a) positive cooperativity with $S = 10^i$ ($i = 0, 4$) and (b) negative cooperativity with $S = 10^{-i}$ ($i = 0, 3$). Values of i are indicated next to each curve.

For a given k, the BI with larger S will be everywhere "above" the BI of a system with a smaller S, i.e.,

$$\theta(x, S_2) - \theta(x, S_1) > 0 \text{ for } S_2 > S_1, \text{ for any } x \tag{4.3.7}$$

For positive cooperative systems, all the curves with $S > 2$ start with positive curvature [Eq. (4.3.6)] and then, at $x = x_{max}$, the curvature becomes negative, where x_{max}, is defined as the point for which the slope is maximal, i.e.,

$$\theta''(x = x_{max}) = 0 \tag{4.3.8}$$

For $S = 2$, the value of x_{max} is zero. As we increase $S > 2$, the value of x_{max} initially increases up to $S = 6.4$, where x_{max} attains its maximal value of 0.125. Further increase of S moves x_{max} to lower values, and at $S \to \infty$, $x_{max} \to 0$, and $\theta'(x_{max}) \to \infty$.

When $S < 1$, i.e., negative cooperativity, the BI starts with a negative curvature [see Eq. (4.3.6)]. When S is very small, we observe an apparent "saturation" at $\theta = 1/2$, but at higher values of x we reach the eventual saturation at $\theta = 1$. This is difficult to observe in a single plot of $\theta(x)$, as in Fig. 4.4. It is therefore more convenient to plot θ as a function of $y = \log_{10} x$,* where we can plot the BI for both positive and negative cooperativities, with a much wider range of values of S. Figure

*One should be careful about the double meaning of the symbol θ: once as the *name* of the function, and once as the *value* of the function, at some specified value, say, $x = x_1$. In writing $\theta(\log_{10}(x))$ we do not refer to the compound function $f(x) = \theta(\log_{10}(x))$, but to the *values* of θ plotted as a function of the new variable $y = \log_{10} x$. The actual function plotted is $g(y) = \theta(10^y)$, i.e., it is obtained from the original BI, $\theta(x)$, by substituting $x \to 10^y$. Likewise, the slopes of the plotted functions are $g'(y)$ plotted as a function of y. The choice of $y = \log_{10} x$ is convenient because it allows a large variation in x, and it also conforms to the tradition of plotting titration curves, where $pH = -\log_{10}[H]$.

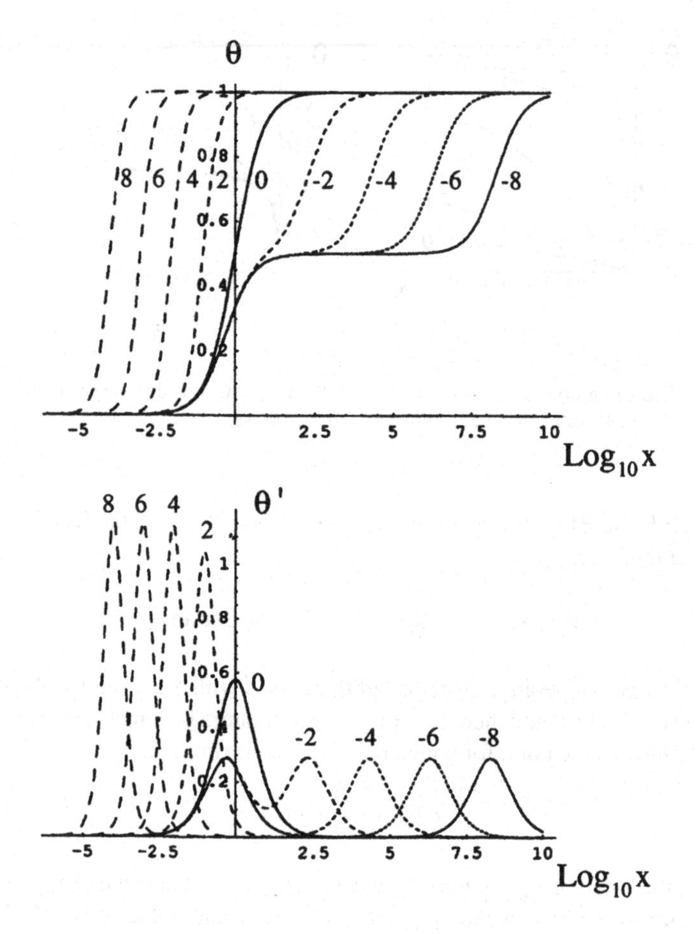

Figure 4.5. θ as a function of $\log_{10} x$ for $S = 10^i$ ($-8 \le i \le 8$). The plotted function is $f(y) = \theta(10^y)$, where we substitute $x \to 10^y$ in $\theta(x)$. Below are the derivatives, $f'(y)$, plotted as a function of $y = \log_{10} x$. Values of i are indicated next to each curve.

4.5 shows $\theta(\log_{10} x)$ for $S = 10^i$, $-8 \le i \le 8$, and the corresponding derivatives.* It is clearly seen that both positive and negative cooperativities can be exhibited in one plot. Also, the apparent saturation at $\theta = 1/2$ and the eventual saturation at $\theta = 1$ as $x \to \infty$ can be seen in the same plot. These plots are also closely related to the titration curves, where we choose the pH scale for the abscissa (see Fig. 2.5 and Appendix G). Note that the $S = 0$ curve has one point of maximum slope at $x = 0$.

*We note, however, that there is nothing more fundamental in plotting θ as a function of x (or of λ), or as a function of $\log_{10} x$ (or log λ). The chemical potential is not more fundamental than the absolute activity. (Wyman and Gill, 1990).

For $S \geq 2$ there is only one point of maximum slope, but for $S < 0$ there are two points of maximum slope.

There are several ways of obtaining the cooperativity S from the experimental data. The simplest is to take the value of x at which $\theta(x) = 1/2$; the solution for x is

$$x_{1/2} = \frac{1}{\sqrt{S}} \tag{4.3.9}$$

Another graphical method is the so-called Hill plot, which is a plot of $\log[\theta/(1-0)]$ as a function of $\log x$. The Hill coefficient is defined by

$$n_H = \frac{\partial \log[\theta/(1-\theta)]}{\partial \log x} = \frac{1}{\theta(1-\theta)} \frac{\partial\theta}{\partial \log x} \tag{4.3.10}$$

and, at the point, $x_{1/2}$, the value of n_H is

$$n_H(x = x_{1/2}) = 4\frac{\partial\theta}{\partial \log x} = 2\frac{\sqrt{S}}{1+\sqrt{S}} \tag{4.3.11}$$

Clearly, the quantity $n_H(x = x_{1/2})$ maps the region $0 \leq S \leq \infty$ into the interval $0 \leq n_H \leq 2$. The value of $n_H = 2$ is the maximum value of the Hill coefficient for the case $m = 2$. One should be careful, however, to note that these particular methods are valid *only* for the case of two sites. When $m > 2$ there are various types of cooperativities and, in general, there is no single parameter that describes the cooperativity in the system. Even for the case $m = 2$ one could be misled in estimating the cooperativity of the system if one were to rely only on the *form* or the *shape* of the BI or any of its transformed functions, as will be demonstrated in Section 4.6 and again in Section 4.8 and Appendix F.

To conclude, we emphasize that the *form* or the *shape* of the BI (or any of its transformed functions) is a manifestation of the type of cooperativity in the system. In the particular case ($m = 2$) discussed in this section, either Eq. (4.3.9) or Eq. (4.3.11) may be used to characterize the cooperativity of the system. In the general case ($m > 2$), one cannot use the *form* of the BI (or of any of its transformed functions) either to characterize or to define cooperativity. Unfortunately, the characterization of cooperativity by the *form* (especially of the Hill plot) is still very common in the biochemical literature.

4.4. TWO DIFFERENT SITES: SPURIOUS COOPERATIVITY

The partition function and the binding isotherm of a general two-site model were discussed in Section 4.2. Here, we examine a special case of *direct* correlation only. The model is essentially the same as in the previous section, except that now we have two *different* sites. The corresponding coefficients of the PF are

$$Q(0,0), \quad Q(a,0) = Q(0,0)q_a, \quad Q(0,b) = Q(0,0)q_b, \quad Q(a,b) = Q(0,0)q_a q_b S_{ab} \tag{4.4.1}$$

where

$$q_a = \exp(-\beta U_a), \qquad q_b = \exp(-\beta U_b), \qquad S_{ab} = \exp[-\beta U(a,b)] \tag{4.4.2}$$

The BI for this case is [compare with Eq. (4.3.3)]

$$\begin{aligned}\theta = \frac{n}{2} &= \frac{1}{2}\frac{(q_a + q_b)\lambda + 2q_a q_b S_{ab}\lambda^2}{1 + (q_a + q_b)\lambda + q_a q_b S_{ab}\lambda^2} \\ &= \frac{1}{2}\frac{(k_a + k_b)C + 2k_a k_b S_{ab}C^2}{1 + (k_a + k_b)C + k_a k_b S_{ab}C^2}\end{aligned} \tag{4.4.3}$$

where $k_\alpha = q_\alpha \lambda_0$ is the intrinsic binding constant for site α. Note that the BI in this model is determined by *three* constants, k_a, k_b, and S_{ab}, as compared with only two in the model of Section 4.3. Experimentally, only the two quantities K_1 and K_2 defined by

$$K_1 = k_a + k_b, \qquad K_1 K_2 = k_a k_b S_{ab} \tag{4.4.4}$$

may be determined from the BI in Eq. (4.4.3). To obtain all three parameters k_a, k_b, and S_{ab} one can either use some approximations (see the example in Section 4.8.3), or measure the *individual* BIs for the two sites. The latter are defined as the average occupation number of ligands at the specific site. Thus,

$$\begin{aligned}\theta_a = n_a &= P(a, \_) = 0 \cdot P(0, \_) + 1 \cdot P(a, \_) \\ &= P(a,0) + P(a,b) = \frac{Q(a,0)\lambda}{\xi} + \frac{Q(a,b)\lambda^2}{\xi} \\ &= \frac{k_a C + k_a k_b S_{ab} C^2}{1 + (k_a + k_b)C + k_a k_b S_{ab}C^2}\end{aligned} \tag{4.4.5}$$

and similarly

$$\theta_b = \frac{k_b C + k_a k_b S_{ab} C^2}{1 + (k_a + k_b)C + k_a k_b S_{ab}C^2} \tag{4.4.6}$$

where $\theta_a + \theta_b = 2\theta$. Clearly, having experimental data on θ_a and θ_b, one can easily resolve for the three parameters k_a, k_b, and S_{ab}.

It is important to note that θ_a is *not* the BI of an *isolated* site of type *a*. To see this we rewrite θ_a in Eq. (4.4.5) in the modified form

$$\theta_a = \frac{k_a^* C}{1 + k_a^* C} \quad \text{with} \quad k_a^* = k_a \frac{1 + k_b S_{ab} C}{1 + k_b C} \tag{4.4.7}$$

When $S_{ab} \neq 1$, the effective binding constant k_a^* will be dependent on C and therefore θ_a is not a simple Langmuir isotherm. Clearly, when we focus on site *a* and follow its average occupation as a function of λ (or C), the effective binding constant k_a^* is affected by what happens at site *b*.

When $S_{ab} = 1$, $k_a^* = k_a$. Both θ_a and θ_b become simple Langmuir isotherms. The BI of the entire system $\theta = (\theta_a + \theta_b)/2$ will, in general, not be a Langmuir isotherm, i.e.,

$$\theta = \frac{1}{2} \frac{(k_a + k_b)C + 2k_a k_b C^2}{1 + (k_a + k_b)C + k_a k_b C^2} = \frac{1}{2} \frac{k_a C}{1 + k_a C} + \frac{1}{2} \frac{k_b C}{1 + k_b C} \tag{4.4.8}$$

Thus, although each of θ_a and θ_b is a simple Langmuir isotherm, the average of the two has a different form. Figure 4.6 shows θ_a, θ_b, and θ for a *noncooperative* ($S_{ab} = 1$) system with $k_a = 1$ and $k_b = 10^2$, 10^3, and 10^4.

A glance at Fig. 4.6 reveals that the curves are the same as those in Fig. 3.4 (provided the corresponding binding constants are the same). The reason is the obvious fact that in each case the sites *a* and *b* are independent. Whether they are located on different molecules, as in Section 3.5, or on the same molecule, as in this section, they exhibit the same *binding* behavior. However, the two systems are different. For later generalization we write here the GPF of a system **a** of $2M$

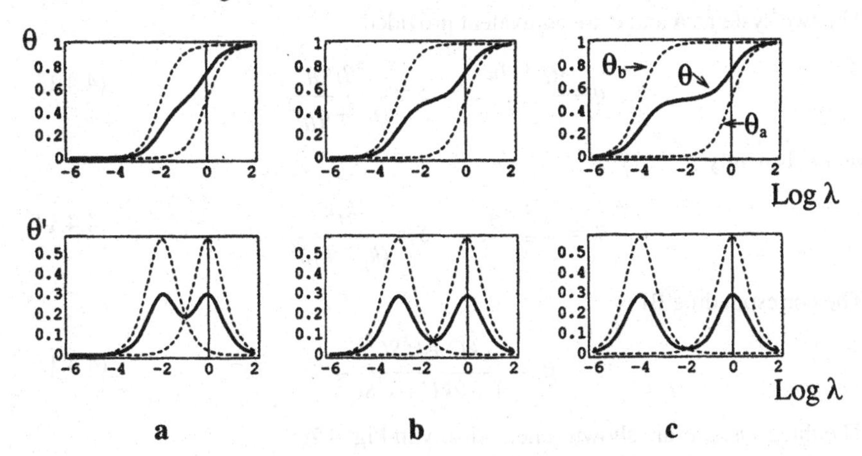

Figure 4.6. The BIs and their derivatives for a two-site system with different binding constants: (**a**) $k_a = 1$, $k_b = 100$; (**b**) $k_a = 1$, $k_b = 10^3$; (**c**) $k_a = 1$, $k_b = 10^4$. Here, *a* and *b* correspond to *L* and *H* in Fig. 3.4. The full line is θ [see Eq. 4.4.8)]. The dashed lines are θ_a and θ_b.

independent molecules each having a single site, $M_L = M$ of which are of type L and $M_H = M$ of type H,*

$$\Xi_{\mathbf{a}} = \frac{\xi_L^{M_L}\xi_H^{M_H}}{M_L!M_H!} = \frac{Q_L(0)^{M_L}(1+q_L\lambda)^{M_L}Q_H(0)^{M_H}(1+q_H\lambda)^{M_H}}{M_L!M_H!} \tag{4.4.9}$$

and of a system **b** of M *identical* and *independent* two-site molecules each having two *different* and *independent* sites,

$$\Xi_{\mathbf{b}} = \frac{Q(0,0)^M(1+(q_L+q_H)\lambda+q_Lq_H\lambda^2)^M}{M!} \tag{4.4.10}$$

Clearly the two systems are different, yet they have the same BI. The factors $Q_L(0)$, $Q_H(0)$, and $Q(0, 0)$ and the factorials $M_L!$, $M_H!$, and $M!$ do not affect the BI. It is evident that if $M_L = M$ and $M_H = M$, the two systems are *equivalent* in the sense of having the same BI. This follows immediately from the identity of the two polynomials

$$(1+q_L\lambda)(1+q_H\lambda) = 1+(q_L+q_H)\lambda+q_Lq_H\lambda^2 \tag{4.4.11}$$

The more remarkable and well-known fact is that the two systems **a** and **b** are also equivalent (in the above sense) to a system **c** of M *identical* and *independent* molecules, each having two *identical* but *dependent* sites, with one intrinsic constant k and pair correlation S. The GPF of this system **c** is

$$\Xi_{\mathbf{c}} = \frac{Q(0,0)^M(1+2q\lambda+q^2S\lambda^2)^M}{M!} \tag{4.4.12}$$

The two systems **a** and **c** are equivalent provided

$$q = \frac{q_L+q_H}{2}, \qquad S = \frac{4q_Lq_H}{(q_L+q_H)^2} \tag{4.4.13}$$

or, equivalently,

$$k = \frac{k_L+k_H}{2}, \qquad S = \frac{4k_Lk_H}{(k_L+k_H)^2} \tag{4.4.14}$$

The corresponding BI is

$$\theta_{\mathbf{c}} = \frac{kC+k^2SC^2}{1+2kC+k^2SC^2} \tag{4.4.15}$$

The three systems are shown schematically in Fig. 4.7.

*We use the notation L and H for a and b, respectively, to conform with the notation of Section 3.5 and Appendix F.

Note that we have switched from the notation a and b for the two sites to L and H. This has been done in order to emphasize the equivalence of the system **b** of this section with the system discussed in Section 3.5. There, L and H originated from "freezing-in" an equilibrated mixture $L \rightleftharpoons H$. Here, on the other hand, the PF in Eq. (4.4.9) pertains to a *mixture* of L and H, with $M_L = M_H = M$, not necessarily originating from "freezing-in" an equilibrated system.

From the experimental BI one cannot distinguish between the three systems **a**, **b**, and **c**. (Note, however, that other properties of the system, such as the pressure, energy, entropy, etc., are different.) This fact could, in some cases, lead to misinterpretation of the cooperative behavior of the system. If $k_L \neq k_H$, S defined in Eq. (4.4.14) must be smaller than unity. This follows from the identity

$$S = \frac{4k_L k_H}{(k_L + k_H)^2} = 1 - \frac{(k_L - k_H)^2}{(k_L + k_H)^2} \leq 1 \tag{4.4.16}$$

i.e., the BI of system **c** is negatively cooperative.

If we know that our system is either **a** or **b**, then observing the BI of the form of Fig. 4.6 is unlikely to mislead us into believing that our system is cooperative. The two peaks in the slope plot will be recognized as originating from the two *different* binding constants k_L and k_H. Therefore, the apparent cooperativity observed in either **a** or **b** must be *spurious*. It will be distinguished from the *genuine* cooperativity if the system is **c**. We shall see in Section 4.8 that in some cases such a clear-cut distinction is not easy, even when we know that we have a two-site system. Of course, if we observe a BI of the type of Fig. 4.6 without knowing the system, we cannot tell whether the system is cooperative or not.

The qualitative physical reason for the equivalence of system **a** (or **b**) and **c** is easily comprehended in the case where $k_H >> k_L$, in which case we observe an

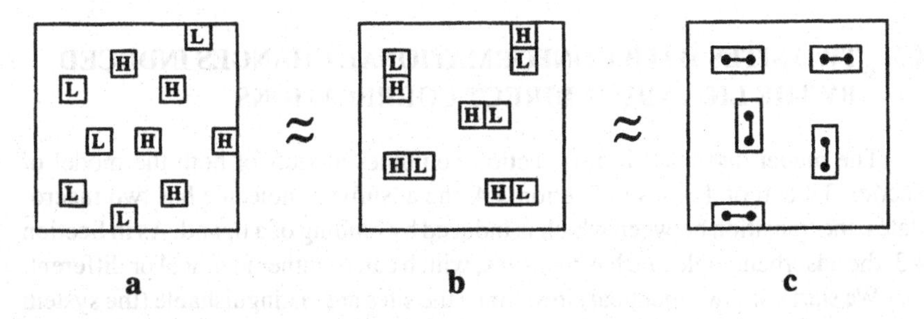

Figure 4.7. Schematic illustration of the three equivalent systems **a**, **b**, and **c**, corresponding to Eqs. (4.4.9), (4.4.10), and (4.4.12). System **a** consists of a mixture of single-site particles, in **b** we have double-site particle where the sites are *different* and *independent* and in **c** the two sites are *identical* but *dependent*.

apparent "saturation" at $\theta = 1/2$. The (negative) spurious cooperativity in this case is

$$S = \frac{4k_L k_H}{(k_L + k_H)^2} \approx 4\frac{q_L}{q_H} << 1 \tag{4.4.17}$$

A system **c**, with a very strong (genuine) negative cooperativity, will *first* be filled only at one of its sites on each molecule. The reason is that two ligands on the same molecule will strongly repel each other. Only at very high ligand activity will the ligands be forced to occupy the *second* site on the molecules. When the system is either **a** or **b**, then again, because $k_H >> k_L$, the sites of type H will first be filled—and sites of type L will remain empty—not because of *repulsion*, but simply because of the overwhelming preference to binding on H. Again, we shall observe an apparent "saturation" at $\theta = 1/2$. At very high ligand activities, when nearly all the preferred sites H have been occupied, the sites of type L will start to be occupied. Thus, we see that the patterns of occupation, and hence the BI, are the same [provided relations (4.4.14) are fulfilled], though the physical reasons for the apparent saturation at $\theta = 1/2$ are different.

We shall discuss in Section 4.6 and Appendix F a generalization of this behavior where higher-order spurious cooperativities can be observed in noncooperative systems.

The main moral of this section (as well as of Appendix F) is that cooperativity *cannot* be determined from the *form* of the BI. Unfortunately, this practice is still very popular among biochemists, who use the *form* of the Hill plot to determine cooperativity (sometimes referred to as *macroscopic* cooperativity). As we have already noted in Section 3.5, this could lead to the absurd reference to a single-site system as being cooperative, though cooperativity is not even defined in such systems.

4.5. TWO SITES WITH CONFORMATIONAL CHANGES INDUCED BY THE LIGANDS: INDIRECT CORRELATIONS

The model discussed in this section combines aspects of both the model of Section 3.4 and of 4.3. As in Section 3.4, the adsorbent molecule has two macrostates, the transition between which is induced by binding of a ligand. As in Section 4.3, the adsorbent molecule has two sites, which can be either identical or different.

We start with two identical sites. Since the sites are distinguishable (the system is localized), there are altogether eight states for the system, shown in Fig. 4.8. The GPF of a single system is

$$\xi = Q(0) + Q(1)\lambda + Q(2)\lambda^2 \tag{4.5.1}$$

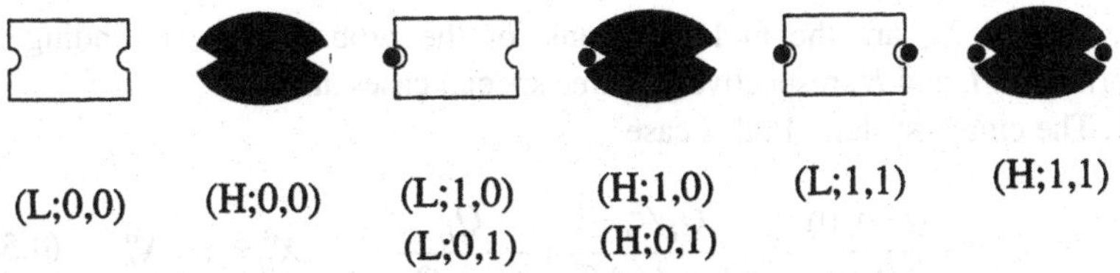

(L;0,0) (H;0,0) (L;1,0) (H;1,0) (L;1,1) (H;1,1)
(L;0,1) (H;0,1)

Figure 4.8. The eight possible configurations of a two-state system with two identical sites. The L form is represented by a rectangle (white) and the H form by an ellipse (black).

where, using obvious notation,

$$\left.\begin{aligned} Q(0) &= Q(L;\,0,\,0) + Q(H;\,0,\,0) \\ Q(1) &= Q(L;\,1,\,0) + Q(H;\,1,\,0) + Q(L;\,0,\,1) + Q(H;\,0,\,1) \\ Q(2) &= Q(L;\,1,\,1) + Q(H;\,1,\,1) \end{aligned}\right\} \tag{4.5.2}$$

By introducing the notation

$$\begin{aligned} q_L &= \exp(-\beta U_L), \quad q_H = \exp(-\beta U_H) \\ S_L &= \exp[-\beta U_L(1,\,1)], \quad S_H = \exp[-\beta U_H(1,\,1)] \end{aligned} \tag{4.5.3}$$

we can write the BI as

$$\begin{aligned} \theta = \frac{\bar{n}}{2} &= \frac{(Q_L q_L + Q_H q_H)\lambda + (Q_L q_L^2 S_L + Q_H q_H^2 S_H)\lambda^2}{(Q_L + Q_H) + 2(Q_L q_L + Q_H q_H)\lambda + (Q_L q_L^2 S_L + Q_H q_H^2 S_H)\lambda^2} \\ &= \frac{k_1 C + k_{11} C^2}{1 + 2k_1 C + k_{11} C^2} \end{aligned} \tag{4.5.4}$$

where the two intrinsic constants k_1 and k_{11} were defined in Section 2.2.

There are many probabilities that can be either read directly from, or constructed from, the corresponding terms in the GPF. For example,

$$P(L;\,0,\,0) = \frac{Q(L;\,0,\,0)}{\xi}, \qquad P(L;\,1,\,0) = \frac{Q(L;\,1,\,0)\lambda}{\xi}, \qquad P(L;\,1,\,1) = \frac{Q(L;\,1,\,1)\lambda^2}{\xi} \tag{4.5.5}$$

Some marginal probabilities are

$$\begin{aligned} X_L &= P(L) = (Q_L + 2Q_L q_L \lambda + Q_L q_L^2 S_L \lambda^2)/\xi \\ X_H &= P(H) = (Q_H + 2Q_H q_H \lambda + Q_H q_H^2 S_H \lambda^2)/\xi \end{aligned} \tag{4.5.6}$$

Here, X_L and X_H are the mole fractions, or the probabilities, of finding the conformation L and H, respectively. Three special cases are:

1. The empty system: In this case

$$X_L^0 = \frac{P(L;0,0)}{P(0,0)} = \frac{Q_L/\xi}{(Q_L+Q_H)/\xi} = \frac{Q_L}{Q_L+Q_H}, \qquad X_H^0 = 1 - X_L^0 \tag{4.5.7}$$

which is the conditional probability of finding L given that the system is empty. This is, of course, the same as the equilibrium mole fraction of the empty system as in Section 3.2.

2. The singly occupied system: The conditional probability of finding L given a specific site (say the lhs) occupied and the second site empty is

$$X_L^{(1)} = \frac{P(L;1,0)}{P(1,0)} = \frac{Q_L q_L \lambda/\xi}{(Q_L q_L + Q_H q_H)\lambda/\xi} = \frac{Q_L q_L}{Q_L q_L + Q_H q_H} \tag{4.5.8}$$

(When the sites are different, say a and b, we must distinguish between $X_L^{(a)}$ and $X_L^{(b)}$.)

3. The doubly occupied system: Finally, the conditional probability of finding L given that the system is doubly occupied is

$$X_L^{(2)} = \frac{P(L;1,1)}{P(1,1)} = \frac{Q_L q_L^2 S_L}{Q_L q_L^2 S_L + Q_H q_H^2 S_H} \tag{4.5.9}$$

Figure 4.9 shows $X_L(\theta)$, $X_H(\theta)$, and $K(\theta) = X_H(\theta)/X_L(\theta)$ for the case $Q_L = Q_H = 1$ and $q_L = 10$, $q_H = 1$, $S_L = S_H = S = 1$. Clearly, since $q_L/q_H > 1$, X_L increases with λ or with θ. The values of $X_L^{(0)}$, $X_L^{(1)}$, and $X_L^{(2)}$ are in this case 0.5, 0.909, and 0.99099, respectively (Fig. 4.9d). The value of $X_L = 0.99099$ is the limiting value of X_L at $\lambda \to \infty$, or $\theta \to 1$.

Two other useful probabilities are

$$\begin{aligned} P(1,1) &= (Q_L q_L^2 S_L + Q_H q_H^2 S_H)\lambda^2/\xi \\ P(1) = P(1,\_) &= P(\_,1) = (Q_L q_L + Q_H q_H)\lambda/\xi \end{aligned} \tag{4.5.10}$$

The first is the probability of finding a molecule doubly occupied—the conformational state of the molecule is unspecified. Clearly, this is the sum of two probabilities,

$$P(1,1) = P(L;1,1) + P(H;1,1) \tag{4.5.11}$$

The second is the probability of finding the molecule singly occupied—both the conformational state and the occupational state of the second site are unspecified.

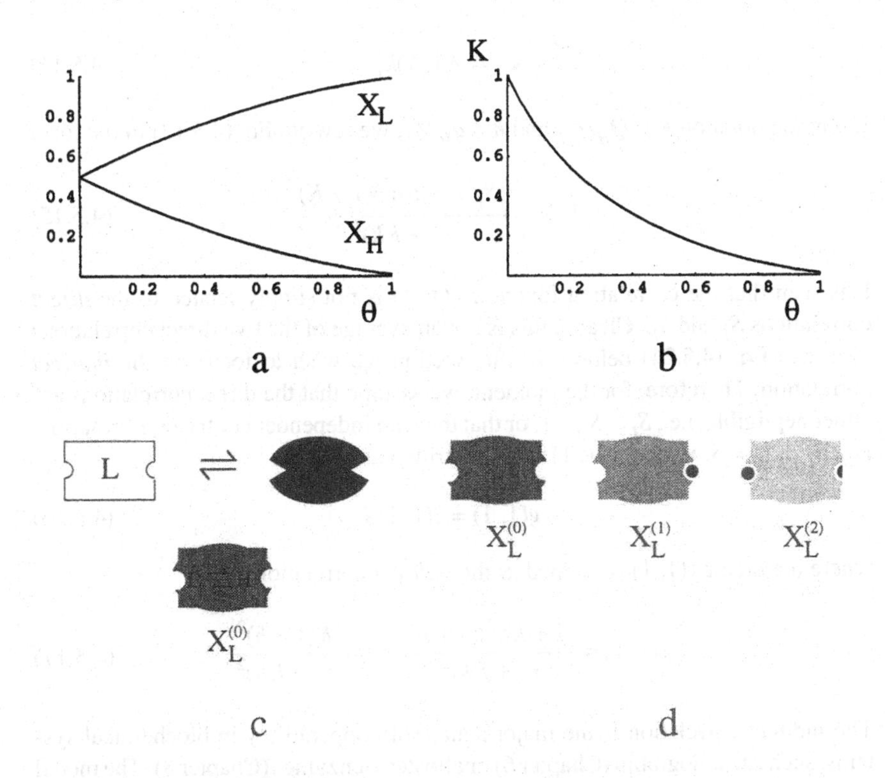

Figure 4.9. (a) $X_L(\theta)$ and $X_H(\theta)$ as given by Eq. (4.5.6). (b) $K(\theta) = X_H(\theta)/X_L(\theta)$. (c) The equilibrium mixture of L and H is represented by superposition of the rectangle and the ellipse (gray). (d) The three different occupancy states are represented by a mixture of L and H (varying shades of gray).

The pair correlation is defined by

$$g(1, 1) = \frac{P(1, 1)}{P(1, \_)P(\_, 1)} \tag{4.5.12}$$

As we have seen in Section 4.2, all we need is the $\lambda \to 0$ limit of this correlation function, which is

$$g^0(1, 1) = \lim_{\lambda \to 0} [g(1, 1)] = \frac{Q(1, 1)Q(0, 0)}{Q(1, 0)Q(0, 1)} \tag{4.5.13}$$

For the remainder of this book we shall always refer to $g^0(1, 1)$ as the pair correlation, and drop the superscript zero. We saw in Section 4.2. that this correlation measures the deviation of k_{11} from $(k_1)^2$, i.e.,

$$k_{11} = g(1, 1)k_1^2 \tag{4.5.14}$$

Using the notation $K = Q_H/Q_L$ and $h = q_H/q_L$, we rewrite Eq. (4.5.13) in the form

$$g(1, 1) = \frac{(S_L + S_H Kh^2)(1 + K)}{(1 + Kh)^2} \tag{4.5.15}$$

It is seen that the correlation function $g(1, 1)$ is not simply related to the *direct* correlations S_L and S_H. Clearly, this is not an average of the two direct correlations [see also Eq. (4.5.24) below]. In this section we wish to focus on the *indirect* correlation, Therefore, for the moment, we assume that the direct correlations are either negligible, i.e., $S_L \sim S_H \sim 1$, or that they are independent of the conformation, i.e., $S_L = S_H = S$. Hence, $g(1, 1)$ may be written as

$$g(1, 1) = y(1, 1)S \tag{4.5.16}$$

where the factor $y(1, 1)$ is defined as the *indirect* correlation

$$y(1, 1) = \frac{(1 + Kh^2)(1 + K)}{(1 + Kh)^2} = 1 + \frac{K(1 - h)^2}{(1 + Kh)^2} \tag{4.5.17}$$

The indirect correlation is the major source of cooperativity in biochemical systems, such as hemoglobin (Chapter 6) or allosteric enzymes (Chapter 8). The model treated in this section is the simplest binding model having indirect correlation.

We now examine some of the outstanding properties of the indirect correlation $y(1, 1)$.

First, we note that $y(1, 1) = 1$ (i.e., no indirect correlation) if, and only if, either $K = 0$ or $h = 1$. This follows directly from the equality on the rhs of Eq. (4.5.17). Incidentally, this necessary and sufficient condition for the occurrence of an indirect correlation is exactly the same as the necessary and sufficient condition for inducing conformational changes (i.e., $d_L \neq 0$; see Section 3.4). Therefore it follows that indirect correlation occurs if, and only if, the ligand induces conformational changes in the adsorbent molecules. Analytically, we may eliminate h from Eq. (3.4.5), and from Eq. (4.5.17), to obtain a relation between $y(1, 1)$ and d_L, namely,

$$y(1, 1) = 1 + d_L^2 \frac{(1 + K)^2}{K} \tag{4.5.18}$$

Thus, for each value of K (recall that we have restricted $0 \leq K \leq 1$ by the choice of $E_L < E_H$), $y(1, 1)$ is a parabolic function in d_L. Figure 4.10 shows this function for the four values $K = 0.2, 0.4, 0.6, 1$. The two functions $y(1, 1)$ and d_L were drawn as

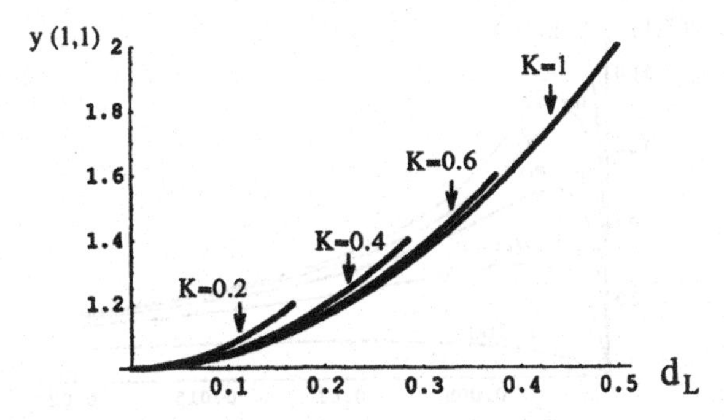

Figure 4.10. $y(1, 1)$ as a function of d_L [see Eq. (4.5.18)] for different values of K, indicated next to each curve. Each curve is a parametric plot of $y(K, h)$ and $d_L(K, h)$ for a fixed value of K and $0 \le h \le 1$.

parametric plots by fixing the value of K and letting $0 \le h \le 1$. Each of the curves starts with $y(1, 1) = 1$ at $d_L = 0$ for $h = 1$ and terminates at a limiting point for $h \to 0$. These limiting points are: for $K = 0.2$, (0.167, 1.2); for $K = 0.4$, (0.28, 1.4); for $K = 0.6$, (0.37, 1.6); and for $K = 1$, (0.5, 2). Note that $0 \le h \le 1$ means the ligand has preference for binding on L, hence d_L is always positive. Similar plots with negative values of d_L may be obtained for $1 \le h \le \infty$.

For any fixed value of h, $y(1, 1)$ has a maximum as a function of K at the point $K = h^{-1}$. Figure 4.11 presents $y(1, 1)$ as a function of K for several values of h. At the point $K = h^{-1}$, the maximal value of $y(1, 1)$ is

$$y_{max} = \frac{2 + h + h^{-1}}{4} = \frac{2 + K + K^{-1}}{4} \tag{4.5.19}$$

where we recall that $K = \exp[-\beta(E_H - E_L)]$ and $h = \exp[-\beta(U_H - U_L)]$. Thus, for very large h (or very small $K = h^{-1}$), y_{max} is determined either by the difference of the binding energies $U_H - U_L$, or by the difference of the energies $E_H - E_L$.

It is clear that always $y(1, 1) \ge 1$. This follows from Eq. (4.5.17). We also see from Eq. (4.5.19) that y_{max} can be very large, depending on the values of $E_H - E_L$ or $U_H - U_L$, but not on S or $U(1, 1)$. Thus, whenever there exists conformational change induced by the binding process we should find a *positive* contribution to the cooperativity. The physical reason for this behavior is quite simple. When the ligand has preference for one conformation, i.e., $h \ne 1$, the binding of the first ligand will shift the equilibrium $L \rightleftharpoons H$ toward the species that it favors. The second ligand approaching the second site will find a new equilibrium concentration of L and H, namely, $X_L^{(1)}$ and $X_H^{(1)}$ (this is shown schematically in Fig. 4.9d). Hence, the conditional probability of binding to the second site is larger than the (unconditional) probability of binding to the first site.

In terms of free energies we have $W(1, 1) = -k_BT \ln g(1, 1)$ and, assuming

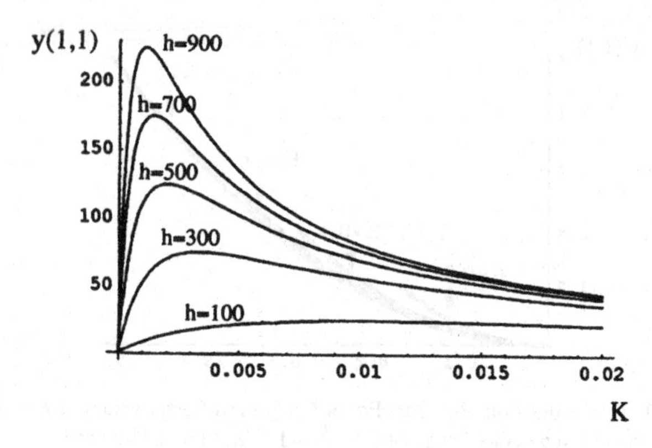

Figure 4.11. $y(1, 1)$ as a function of K for several values of h, as indicated.

$S_L = S_H = S$, we find

$$W(1, 1) = U(1, 1) + [A(1, 1) + A(0, 0) - A(0, 1) - A(1, 0)]$$

$$= U(1, 1) + \delta A \tag{4.5.20}$$

where $\delta A < 1$, δA being the indirect part of the work associated with the process

$$(0, 1) + (1, 0) \rightarrow (1, 1) + (0, 0) \tag{4.5.21}$$

The quantity δA can also be rewritten as

$$\delta A = [A(1, 1) - A(0, 1) - \mu] - [A(1, 0) - A(0, 0) - \mu]$$

$$= \Delta A^*(1/1) - \Delta A^*(1/0) < 0 \tag{4.5.22}$$

where $\Delta A^*(1/0)$ and $\Delta A^*(1/1)$ are the binding free energies on the first and second sites, respectively.

We now summarize the main difference between the direct and indirect correlations:

1. The sign of the direct correlation depends on the *direct* interaction between the ligands. The sign of the indirect correlation is *always* positive (for two identical sites; see below for two different sites), and is independent of $U(1, 1)$, but dependent on the difference of binding energies $U_H - U_L$, and the difference of energies of the two conformers $E_H - E_L$.

2. The direct interaction depends on the distance between the ligands and has the same range as the ligand–ligand pair potential. The indirect correlation, in this particular model, is *independent* of the ligand–ligand distance. It does depend on

the capacity of the ligand to induce conformational changes ($h \neq 1$), and on the responsiveness of the adsorbing molecule ($K \neq 0$).

3. $U(1, 1)$ is presumed to be temperature-independent,[†] while $W(1, 1)$ is in general temperature-dependent. In terms of correlations, the sign of the temperature dependence of S depends on the sign of $U(1, 1)$. If $U(1, 1) < 0$, then S decreases with T; when $U(1, 1) > 0$, S increases with T as shown in Fig. 4.12a. On the other hand, the dependence of $y(1, 1)$ on T (Fig. 4.12b) is determined by the enthalpy change associated with the process (4.5.21). In biochemical processes the *indirect* correlation can enormously increase the cooperativity, far beyond the normally weak, or even negligible, *direct* cooperativity that depends on the ligand–ligand interactions (say, two oxygen molecules in hemoglobin). The trick of the *indirect* route is to use the changes in energies of the macromolecule to produce a very large cooperativity in spite of the weak ligand–ligand interaction. This spectacular and sophisticated trick has been selected by the long and incessant evolutionary search for more efficient ways of regulating biochemical processes. We shall return to this aspect when discussing hemoglobin (Chapter 6) and regulatory enzymes (Chapter 8).

The temperature dependence of $y(1, 1)$ is

$$\frac{\partial \ln y}{\partial T} = \frac{-\Delta H^*}{k_B T^2} \sim \frac{-\Delta E^*}{k_B T^2} \tag{4.5.23}$$

where ΔH^* and ΔE^* are the enthalpy and the energy associated with the indirect part of the work $W(1, 1)$, i.e., with δA in Eq. (4.5.20). Figure 4.12 shows that $|S|$ is always a monotonically decreasing function of T, while y can increase or decrease with T for a given set of molecular parameters, h and K. This fact could serve as a diagnostic test for the existence of an indirect correlation. If the temperature dependence of the total correlation is of the type b in Fig. 4.12, then it is very likely that indirect correlations are operative.

We have started our discussion of the properties of $y(1, 1)$ with the assumption $S_L = S_H = S$. This allowed the factorization of $g(1, 1)$ as in Eq. (4.5.16). It is easy to see that all the properties of $y(1, 1)$ hold true also when $S_L \neq S_H$. In this case, Eq. (4.5.15) can be rewritten as

$$g(1, 1) = \frac{(1 + Kh^2)(1 + K)}{(1 + Kh)^2} \frac{(S_L + S_H Kh^2)}{(1 + Kh^2)} = y(1, 1) \langle S \rangle \tag{4.5.24}$$

where $y(1, 1)$ is exactly the same as in Eq. (4.5.17). The factor S in Eq. (4.5.16)

[†]Note that $U(1, 1)$ itself could be a free energy, hence temperature-dependent (see Appendix C). Here, as in most of the book, we have suppressed the free-energy character of $U(1, 1)$ to stress the emergence of new free energy, here δA, which brings an additional dependence on T.

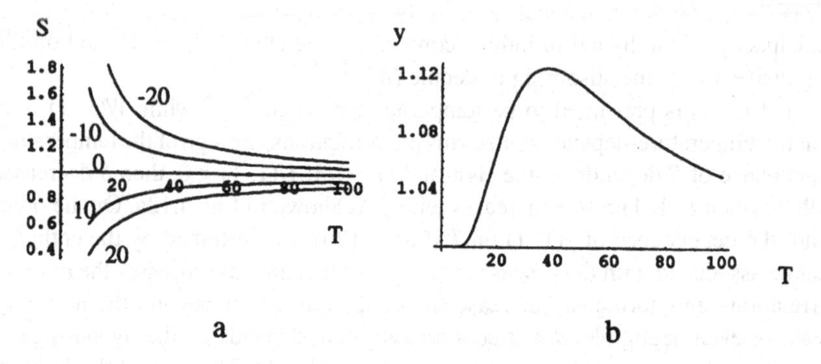

Figure 4.12. Temperature dependence of (a) the direct correlation S, for positive and negative values of $U(1, 1)$, and (b) of $y(1, 1)$ with $K = \exp(-100/T)$ and $h = \exp(-100/T)$ with T in dimensionless units.

is now replaced by an average of S_L and S_H, with weights $(1 + Kh^2)^{-1}$ and $Kh^2/(1 + Kh^2)$, respectively.*

We now briefly discuss the case of two *different* sites for which the PF is

$$\xi = (Q_L + Q_H) + (Q_L q_{La} + Q_H q_{Ha})\lambda + (Q_L q_{Lb} + Q_H q_{Hb})\lambda$$
$$+ (Q_L q_{La} q_{Lb} + Q_H q_{Ha} q_{Hb}) S\lambda^2 \tag{4.5.25}$$

where we have employed the notation

$$\begin{aligned} q_{La} &= \exp(-\beta U_{La}), \quad q_{Lb} = \exp(-\beta U_{Lb}) \\ q_{Ha} &= \exp(-\beta U_{Ha}), \quad q_{Hb} = \exp(-\beta U_{Hb}) \end{aligned} \tag{4.5.26}$$

The BI for this case is

$$\bar{n} = \frac{(k_a + k_b)C + 2k_a k_b g_{ab} C^2}{1 + (k_a + k_b)C + k_a k_b g_{ab} C^2} \tag{4.5.27}$$

where g_{ab} is defined by

$$g_{ab} = g(a, b) = \frac{Q(a, b)Q(0, 0)}{Q(a, 0)Q(0, b)} \tag{4.5.28}$$

*These may be interpreted as the mole fractions of L and H in a hypothetical doubly-occupied system, where $S_L = S_H = 1$. This is a hypothetical system, since in a real system the mole fractions of L and H are given by Eq. (4.5.9).

If we denote

$$K = \frac{Q_H}{Q_L}, \qquad h_a = \frac{q_{Ha}}{q_{La}}, \qquad h_b = \frac{q_{Hb}}{q_{Lb}} \tag{4.5.29}$$

and again assume that $S_L = S_H = S$, we can write $y(a, b)$ in the form

$$y(a, b) = \frac{(1 + h_a h_b K)(1 + K)}{(1 + h_a K)(1 + h_b K)} = 1 + \frac{(1 - h_a)(1 - h_b)K}{(1 + h_a K)(1 + h_b K)} \tag{4.5.30}$$

The important difference between this case and the previous one in Eq. (4.5.17) is that the indirect cooperativity can be either positive or negative, depending on whether the signs of $(1 - h_a)$ and $(1 - h_b)$ are the same or differ. Thus, if $h_a > 1$ and $h_b < 1$, binding on a will shift the equilibrium $L \rightleftharpoons H$ toward H (the favored conformation when the ligands bind to site a), but since $h_b < 1$, this means that the new conformational equilibrium will be less favorable for binding on site b. In this case

$$P(b/a) < P(b) \tag{4.5.31}$$

or, equivalently,

$$A(a, b) + A(0, 0) - A(a, 0) - A(0, b) > 0$$

or

$$\Delta A^*(b/a) - \Delta A^*(b) > 0 \tag{4.5.32}$$

4.6. SPURIOUS COOPERATIVITY IN TWO IDENTICAL-SITE SYSTEMS

In Section 3.5 we discussed the phenomenon of *spurious* cooperativity in single-site systems. Since *cooperativity*, as defined in this book, is undefinable for single-site systems, any apparent cooperative behavior must be due to the presence of *different* and *independent* sites. In Section 4.4 we encounter the same phenomenon in two-site systems with *different* sites. This was shown to be equivalent to the system in Section 3.5.

In this section we start with two-site systems, where *genuine* (positive or negative) cooperativity exists in each molecule. We explore the emergence of additional *spurious* cooperativity due to "freezing-in" of an equilibrium between two forms $L \rightleftharpoons H$.[†] As we shall see below, in this case it is not always possible to distinguish spurious from genuine cooperativity.

[†]This is also equivalent to a mixture of two different binding systems. However, here we stress the case of a mixture that is obtained from an equilibrated system. It is only in such a case that one might misinterpret spurious cooperativity as genuine; see Section 4.8 for an experimental example.

The model used here is essentially the same as that of Section 4.5. We define the GPF of state L and H by

$$\xi_L = Q_L + Q_L q_L \lambda + Q_L q_L^2 S_L \lambda^2 \quad \text{and} \quad \xi_H = Q_H + Q_H q_H \lambda + Q_H q_H^2 S_H \lambda^2 \tag{4.6.1}$$

The corresponding equilibrium concentrations of L and H are

$$X_L^{eq} = \xi_L/\xi \quad \text{and} \quad X_H^{eq} = \xi_H/\xi \tag{4.6.2}$$

and the corresponding BIs of pure L and H are

$$\theta_L = \frac{\lambda}{2}\frac{\partial \ln \xi_L}{\partial \lambda} \quad \text{and} \quad \theta_H = \frac{\lambda}{2}\frac{\partial \ln \xi_H}{\partial \lambda} \tag{4.6.3}$$

The equilibrated and "frozen-in" BIs are defined as in Section 3.5 by

$$\theta^{eq} = \frac{\lambda}{2}\frac{\partial \ln \xi}{\partial \lambda} = X_L^{eq}\theta_L + X_H^{eq}\theta_H \tag{4.6.4}$$

and

$$\theta^f = X_L^f \theta_L + X_H^f \theta_H \tag{4.6.5}$$

where $\xi = \xi_L + \xi_H$, while X_L^f and X_H^f are any arbitrarily chosen mole fractions of L and H, with $X_L^f + X_H^f = 1$. On the other hand, X_L^{eq} and X_H^{eq} are determined by the molecular parameters of the system, as well as the temperature, pressure, etc. [see Eq. (4.6.2)].

We shall discuss here a particular simple case where initially (i.e., at $\lambda = 0$) we have*

$$X_L^{eq}(\lambda = 0) = X_H^{eq}(\lambda = 0) = 1/2 \tag{4.6.6}$$

With this assumption we "freeze-in" the conversion $L \leftrightarrow H$ at $\lambda = 0$, hence $X_L^f = X_H^f = 1/2$, and we follow the BI of the equilibrated and the "frozen-in" system. We discuss the following three cases.

(i) No genuine cooperativity, i.e., $S_L = S_H = 1$: The following three systems are equivalent (in the sense of having the same BI); see Fig. 4.13.

(a) $2M$ independent double-site molecules, M of which are in state L and M in state H. There is no genuine cooperativity within each molecule. The

*See footnote p. 65 in Section 3.5. It is possible that the two forms L and H will be very similar, i.e., $Q_L = Q_H$, hence Eq. (4.6.6), but the binding properties of L and H differ widely. The analytical study of the case $X_L \neq X_H$ is possible but quite involved. In Appendix F we discuss this aspect for a single-site system.

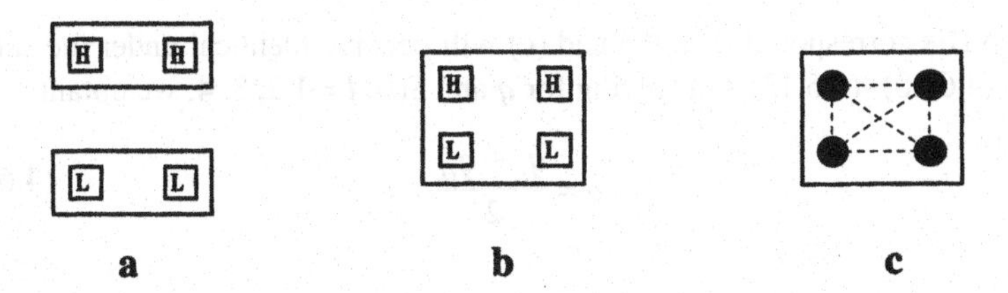

Figure 4.13. Schematic illustration of the three equivalent systems corresponding to Eqs. (4.6.7), (4.6.8), and (4.6.9): (**a**) double-site molecules of types H and L; (**b**) four-site molecules, two sites of type H and two of type L; (**c**) four-site molecules. The sites are identical but correlated, as indicated by the connecting dashed lines.

corresponding GPF is

$$\Xi_{\mathbf{a}} = \xi_L^M \xi_H^M = [(1 + 2q_L\lambda + q_L^2\lambda^2)(1 + 2q_H\lambda + q_H^2\lambda^2)]^M \tag{4.6.7}$$

where ξ_L and ξ_H are defined in Eq. (4.6.1), with $S_L = S_H = 1$, and we ignore factors such as Q_L, Q_H, and $(M!)^2$, which do not affect the BI.

(**b**) M independent quadruple-site molecules, each molecule of which has two sites of type H (with k_H) and two of type L (with k_L). The corresponding GPF is

$$\Xi_{\mathbf{b}} = [1 + (2q_L + 2q_H)\lambda + (4q_Lq_H + q_H^2 + q_L^2)\lambda^2 + (2q_L^2q_H + 2q_H^2q_L)\lambda^3 + q_L^2q_H^2\lambda^4]^M \tag{4.6.8}$$

(**c**) M independent quadruple-site molecules. All four sites are *identical* (with the same k) but *dependent*, with pair, triplet, and quadruplet cooperativities. The corresponding GPF is

$$\Xi_{\mathbf{c}} = 1 + 4q\lambda + 6q^2S(2)\lambda^2 + 4q^3S(3)\lambda^3 + q^4S(4)\lambda^4 \tag{4.6.9}$$

The equivalence of **a** and **b** is obtained simply by expanding the product of ξ_L and ξ_H in Eq. (4.6.7). The equivalence between (**b**) and (**c**) can be obtained by imposing the conditions

$$4q = 2q_L + 2q_H \tag{4.6.10}$$

$$6q^2S(2) = 4q_Lq_H + q_H^2 + q_L^2 \tag{4.6.11}$$

$$4q^3S(3) = 2q_L^2q_H + 2q_H^2q_L \tag{4.6.12}$$

$$q^4S(4) = q_L^2q_H^2 \tag{4.6.13}$$

The two BIs corresponding to (**b**) and (**c**) will become identical under the substitutions (4.6.10)–(4.6.13). On solving for q and $S(l)$, $l = 1\ 2, 3, 4$, we obtain

$$q = \frac{q_L + q_H}{2} \tag{4.6.14}$$

$$S(2) = 1 - \frac{(q_H - q_L)^2}{3(q_H + q_L)^2} \tag{4.6.15}$$

$$S(3) = 1 - \frac{(q_H - q_L)^2}{(q_H + q_L)^2} \tag{4.6.16}$$

$$S(4) = 1 - \frac{(q_H - q_L)^2(q_H^2 + 6q_Hq_L + q_L^2)}{(q_H + q_L)^4} \tag{4.6.17}$$

All the cooperativities are seen to be negative, i.e., $S(l) < 1$, $l = 2, 3, 4$.

Incidentally, since the newly defined coefficients in Eq. (4.6.9) are products of factors, there is more than one way to choose the new correlations. In the choice made in Eqs. (4.6.10)–(4.6.13), all sites are identical and all six pair correlations are identical. Another possible choice is obtained by replacing condition (4.6.11) by the requirement

$$q^2[4S_4(2) + 2] = 4q_Lq_H + q_H^2 + q_L^2 \tag{4.6.18}$$

in which case

$$S_4(2) = 1 - \frac{(q_H - q_L)^2}{2(q_H + q_L)^2} \tag{4.6.19}$$

Here, all sites have identical intrinsic constants but the pairs that originally belonged to the same molecules are left uncorrelated. All the pair correlations are here "assigned" to the newly formed four pairs of type LH.

In the above model our starting molecule had no genuine cooperativity, i.e., the coefficients of λ^2 had no factors of the type S_L and S_H. Figure 4.14 shows the BIs θ_L, θ_H, θ^f, and θ^{eq} of a system of double-site molecules having no genuine cooperativity. It is clear that the curve θ^f shows an apparent negative cooperativity. Since we have assumed no genuine cooperativity, this behavior must be due to spurious cooperativity. Note that the extent of the spurious cooperativity may be estimated by the distance between the two peaks of the slope curve, exactly as one would estimate genuine cooperativity from Fig. 4.5.

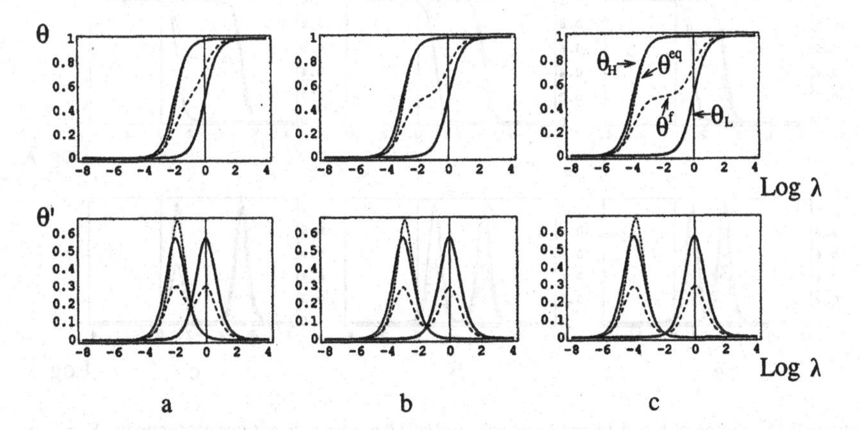

Figure 4.14. Binding isotherms and their derivatives for a system of double-site molecules with no genuine cooperativity. The full lines correspond to pure H (on the lhs) and pure L (on the rhs). The equilibrated BIs, θ^{eq}, are the dotted curves corresponding to the system described in Eq. (4.6.4). The "frozen-in" BI are the dashed curves (with $X_L^0 = X_H^0 = 1/2$): (**a**) $q_L = 1$, $q_H = 100$; (**b**) $q_L = 1$, $q_H = 1000$; (**c**) $q_L = 1$, $q_H = 10000$. The larger the ratio q_H/q_L, the larger the spurious cooperativity (as can be estimated from the distance between the two peaks of the slope curves on the lower panel).

(ii) A system with genuine positive cooperativity, i.e., $S_L > 1$, $S_H > 1$. The three systems considered here are similar to those described above, the only difference being that S_L and S_H in Eq. (4.6.1) are larger than unity.

As before, there are several ways of assigning new cooperativities in the equivalent system (**c**). One way is to assign the same pairwise cooperativity to all *six* pairs of sites, and the same triplet cooperativity to all four triplet sites. The analogue of Eq. (4.6.9) would then be

$$\Xi_c = 1 + 4q\lambda + 6q^2S(2) + 4q^3S(3) + q^4S(4)\lambda^4 \tag{4.6.20}$$

for which

$$S(2) = \frac{2}{3}\frac{4q_Hq_L + q_H^2S_H + q_L^2S_L}{(q_H + q_L)^2} \tag{4.6.21}$$

$$S(3) = \frac{4q_Hq_L}{(q_L + q_H)^2}\frac{q_LS_L + q_HS_H}{(q_L + q_H)} \tag{4.6.22}$$

$$S(4) = \frac{16q_H^2q_L^2}{(q_H + q_L)^4} \cdot S_LS_H \tag{4.6.23}$$

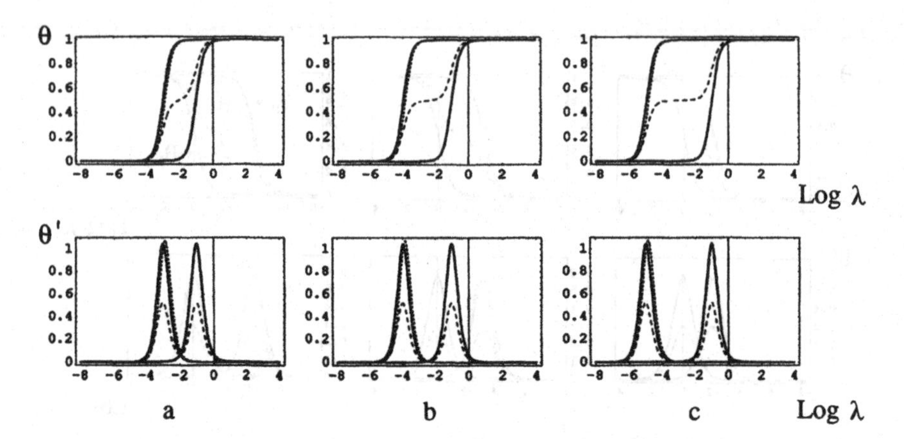

Figure 4.15. Same as Fig. 4.14, but here each molecule has a *genuine positive* cooperativity, $S_L = S_H = S = 100$. The notation and all other parameters are as in Fig. 4.14. Note that because of the genuine positive cooperativity, the slopes of the curves θ_L, θ_H, and θ^{eq} are sharper than in the corresponding curves in Fig. 4.14 (no genuine cooperativity). As in Fig. 4.14, the negative spurious cooperativity is larger, the larger q_H/q_L.

where $S(2)$ is now a combination of the genuine cooperatives, the sign of which depends on S_L and S_H; $S(3)$ has an average of the original cooperativities, times a factor that is smaller than unity; and $S(4)$ has the original correlations S_L and S_H, times a factor that is less than unity. Of course there are other possibilities of assigning correlations.

In Fig. 4.15 we show curves similar to those in Fig. 4.14, but for a system of molecules having *genuine positive* cooperativity. In this illustration $X_L^0 = X_H^0 = 1/2$, $S_L = S_H = 100$, and (a) $q_L = 1, q_H = 100$; (b) $q_L = 1, q_H = 1000$; (c) $q_L = 1, q_H = 10{,}000$. Note that the spurious *negative* cooperativity is roughly the same as in Fig. 4.14. The distance between the two peaks of the slope curves is similar to that in Fig. 4.14. As a result of the genuine positive cooperativity, the slopes of both θ_L and θ_H are larger than those of Fig. 4.14. Thus we see that even though the system has genuine *positive* cooperativity, and this could be very large, the system would appear to have *negative* cooperativity.*

*In the molecular approach to cooperativity, we have distinguished between direct and indirect sources of cooperativity. We shall also distinguish between pair, triplet, and higher-order cooperativities. All of these do not necessarily have the same sign. On the other hand, when relying on the *shape* or *form* of the BI to detect cooperativity, one can easily be misled to conclude that BIs of the form θ^f in Fig. 4.15 originate from *negative* cooperative systems, whereas in fact these systems, by construction, have positive cooperativity. This embarrassing conclusion has led some authors to define macroscopic and microscopic cooperativities. As we have pointed out in Section 3.5, it is quite awkward to refer to a system of *independent* molecules as being (macroscopically or microscopically) cooperative. It becomes both awkward and confusing to refer to systems which, by definition, are positively cooperative as being (macroscopically) negatively cooperative.

(iii) A system with genuine negative cooperativity, i.e., $S_L < 1, S_H < 1$: Perhaps the potentially most misleading case is that where each molecule has a genuine negative cooperativity, e.g., dicaboxylic acids; see Section 4.8. Here, the spurious cooperativity is superimposed on the genuine negative cooperativity and, in general, it would be difficult to distinguish between the two.

Figure 4.16 shows curves similar to those in Fig. 4.14 for a system of molecules having *genuine negative* cooperativity. For this demonstration we chose $S_L = S_H = 0.1$ and the three values $q_H = 10^2, 10^4$, and 10^6. Here, due to negative cooperativity, the derivatives of both θ_L and θ_H show two peaks, the distances between them corresponding to S_L and S_H, respectively. In addition, the θ^f curve now has *four* points of maximal slope. This is evident in the slope curves in Fig. 4.16. The separation between the two *pairs* of closed maxima is a measure of the spurious cooperativity. It depends on the ratio q_H/q_L and not on either S_L or S_H. Note also that from the form of the BI, θ^f, we could not have suspected that we have a quadruple-peaked derivative. The BI, θ^f, seems to be a typical BI of a system of two sites with varying degrees of negative cooperativities.

The results of Fig. 4.16 are closest to calculations that we shall present in Section 4.8. They demonstrate, in an idealized situation, our interpretation of the source of the outstandingly large negative cooperativities of some alkylated succinic acid. Whenever we have a mixture of two components, each having *genuine negative* cooperativity, the mixture as a whole has actually *four* binding constants. However, if the two pairs of binding constants, say k_{1L}, k_{2L}, and k_{1H}, k_{2H}, are well separated, as in Fig. 4.16d, then the BI (or equivalently the titration curve) of the mixture might seem as if it has only *two* binding constants, the separation between which is determined by the distance between the *pairs* of the maxima in the slope curve rather than by the distance between the two maxima within each pair.

The actual determination of a correlation function from experimental data depends on the method used to measure the binding constants. The most common method for dicarboxylic acids is from the limiting behaviors at $C \to 0$ (the high pH limit) and at $C \to \infty$ (the low pH limit). These two limiting behaviors of the BI are (see Section 2.2).

$$\theta(C) = k_1 C + O(C^2) \tag{4.6.24}$$

and

$$\theta(C) = 1 - \frac{1}{k_2}\frac{1}{C} + O\left(\frac{1}{C}\right)^2 \tag{4.6.25}$$

Here, k_1 and k_2 replace k_1 and $k_{1/1}$ in Section 2.2. These are the more common notations for the *first* and *second* intrinsic binding constants. The latter limiting behaviors enables one to determine the correlation function

$$g(1, 1) = k_2/k_1 \tag{4.6.26}$$

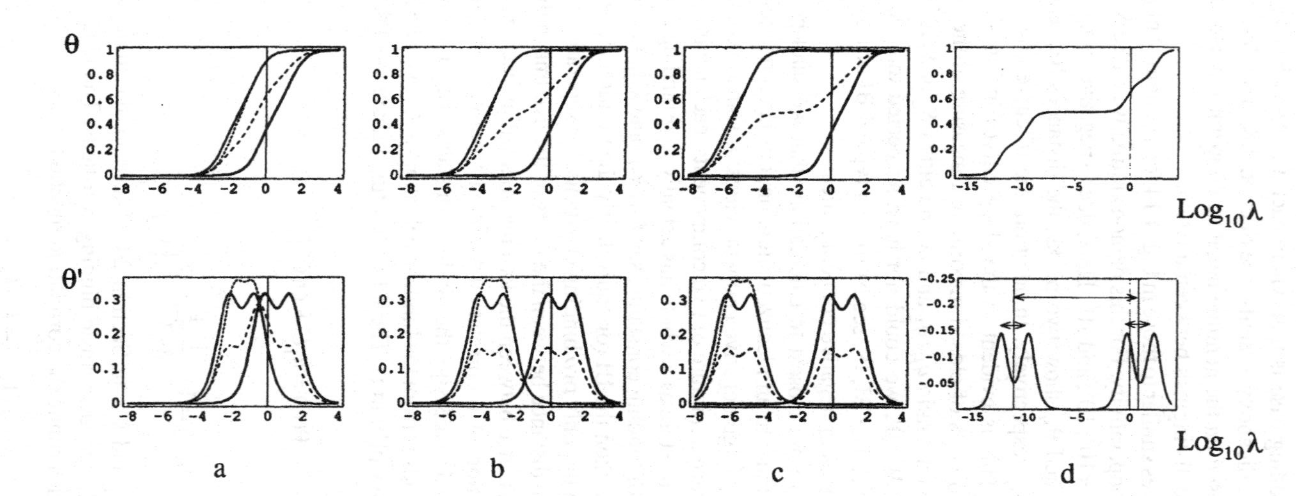

Figure 4.16. Same notation as in Figs. 4.14 and 4.15, but here each molecule has a *genuine negative* cooperativity with $S_L = S_H = 0.1$: (**a**) $q_L = 1$, $q_H = 100$; (**b**) $q_L = 1$, $q_H = 10^4$; (**c**) $q_L = 1$, $q_H = 10^6$. In (**d**), the short double-headed arrow corresponds to the genuine negative cooperativity ($S_L = S_H = 10^{-2}$) while the long double-headed arrow is a measure of the spurious cooperativity ($q_H/q_L = 10^{12}$).

However, by using this method for a *mixture* of two components, such as the curve θ^f in Fig. 4.16, we would actually obtain

$$\theta(C) \approx (X_L^0 k_{1L} + X_H^0 k_{1H})C + O(C^2) = k_{1eff}C + O(C^2) \tag{4.6.27}$$

and

$$\theta(C) \approx 1 - \left(\frac{X_L^0}{k_{2L}} + \frac{X_H^0}{k_{2H}}\right)\left(\frac{1}{C}\right) + O\left(\frac{1}{C}\right)^2 = 1 - \left(\frac{1}{k_{2eff}}\right)\frac{1}{C} + O\left(\frac{1}{C}\right)^2 \tag{4.6.28}$$

where the effective binding constants are *defined* in the last two equations.

The *effective* correlation function is determined from the *effective* binding constants by k_{2eff}/k_{1eff}, i.e.,

$$g_{eff} = (X_L^0 k_{1L} + X_H^0 k_{1H})^{-1}\left(\frac{X_L^0}{k_{2L}} + \frac{X_H^0}{k_{2H}}\right)^{-1} \tag{4.6.29}$$

This depends on the separation between the *pairs* of peaks in the slope curves rather than on the two peaks pertaining to θ_L and θ_H. We examine this dependence in a simple case where the genuine cooperativity within each conformer is negative but small, i.e.,

$$S_L \sim S_H \sim 1 - \varepsilon \quad (\varepsilon > 0) \tag{4.6.30}$$

If we set $h = q_H/q_L = k_{1H}/k_{1L}$, then the *effective* correlation is given by

$$g_{eff} \approx \frac{h}{(X_H^0 + hX_L^0)(hX_H^0 + X_L^0)} - \frac{h\varepsilon}{(X_H^0 + hX_L^0)(X_L^0 + hX_H^0)} + O(\varepsilon^3) \tag{4.6.31}$$

Thus, for any finite X_L^0 (and $X_H^0 = 1 - X_L^0$), g_{eff} is determined by the first term in Eq. (4.6.31), i.e.,

$$g_{eff} \sim \frac{h}{(X_H^0 + hX_L^0)(hX_H^0 + X_L^0)} = 1 - \frac{(h-1)^2 X_L^0 X_H^0}{(X_H^0 + hX_L^0)(hX_H^0 + X_L^0)} \tag{4.6.32}$$

It is clear that the effective cooperativity as measured by the limiting slopes [Eqs. (4.6.27) and (4.6.28)] will always be negative, and will be larger (i.e., $g_{eff} \to 0$) the larger the separation between k_{1H} and k_{1L} (i.e., either $h \to 0$ or $h \to \infty$). A particularly simple case is when $X_L^0 = X_H^0 = 1/2$, in which case

$$g_{eff} = 1 - \frac{(h-1)^2}{(h+1)^2} \tag{4.6.33}$$

In connection with the previous example, shown in Fig. 4.15, we commented that it is both awkward and confusing to refer to that system as (macroscopically) negatively cooperative. In the present examples, shown in Fig. 4.16, there is nothing awkward in referring to these systems as negatively cooperative since negative cooperativity has been built-in in the model. However, the extent of the negative cooperativity cannot be estimated from the distance between the two *pairs* of peaks in the slope curves. This distance could be very large, depending on $h = q_H/q_L$, but not on the genuine negative cooperativity S. The latter may be estimated from the distance between the two peaks belonging to the corresponding curves of each of the conformers.

4.7. TWO SITES ON TWO SUBUNITS: TRANSMISSION OF INFORMATION ACROSS THE BOUNDARY BETWEEN THE SUBUNITS

We extend the model of Section 4.5 by one aspect. The adsorbent molecules now consist of two identical subunits, each having one site. The subunit itself can be in one of two conformations, L or H. Hence, altogether there are *four* possible states for the entire empty adsorbent molecule: LL, LH, HL, and HH. Formally, this model extends the model of Section 4.5 in allowing the four states instead of two. In this respect all the results of the previous model apply also to this model, and in some special cases ($\eta \to 0$, see below) the two models actually become identical.

There is an important virtue in studying this model in detail beyond the fact that many real biological systems do consist of subunits, namely, we obtain an understanding of the mechanism by which information on the occupancy state of one site (i.e., on one subunit) is transmitted to the second site. We shall see in this and subsequent sections that the latter intriguing mechanism is prevalent in biochemical systems.

4.7.1. The Empty System

The system consists of two subunits, each of which can attain one of two states, denoted by L and H, having energies E_L and E_H, respectively. In addition, we have intersubunit interactions, which we denote by E_{LL}, $E_{LH} = E_{HL}$, and E_{HH}, depending on the state of the two subunits. Note that, in general, E_{LH} can differ from E_{HL} (Fig. 4.17). However, for simplicity, we assume that $E_{LH} = E_{HL}$. Denote

$$Q_\alpha = \exp(-\beta E_\alpha), \quad Q_{\alpha\beta} = \exp(-\beta E_{\alpha\beta}) \tag{4.7.1}$$

where subscripts α and β can be either L or H.

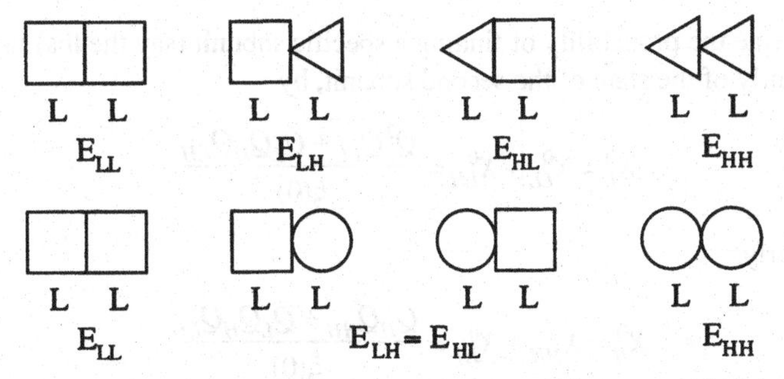

Figure 4.17. Two subunits with symmetrical (lower panel) and unsymmetrical (upper panel) interactions.

Let us first examine the equilibrium properties of this system in the absence of ligands. The GPF of a single system is

$$\xi(0) = Q_L^2 Q_{LL} + 2Q_L Q_H Q_{LH} + Q_H^2 Q_{HH} \tag{4.7.2}$$

Note that this is actually the canonical partition function for a single empty system. It is also the limit of the GPF of the system [Eq. (4.7.16)] obtained for $\lambda \to 0$.

The four states of the polymer are *LL*, *LH*, *HL*, and *HH* (see the first row of Fig. 4.18) with corresponding probabilities, or mole fractions,

$$X_{LL}^0 = \frac{Q_L^2 Q_{LL}}{\xi(0)}, \qquad X_{LH}^0 = X_{HL}^0 = \frac{Q_L Q_H Q_{LH}}{\xi(0)}, \qquad X_{HH}^0 = \frac{Q_H^2 Q_{HH}}{\xi(0)} \tag{4.7.3}$$

In Eq. (4.7.3) the superscript zero indicates the empty system, i.e., the absence of ligands. Note that X_{LH}^0 is the probability of finding, say, the rhs subunits in the *H* state and the lhs subunits in the *L* state. The mole fraction of the system such that any one of its subunits is in the *L* state and the other in the *H* state is the sum of X_{LH}^0 and X_{HL}^0, which in our case is simply $2X_{LH}^0$.

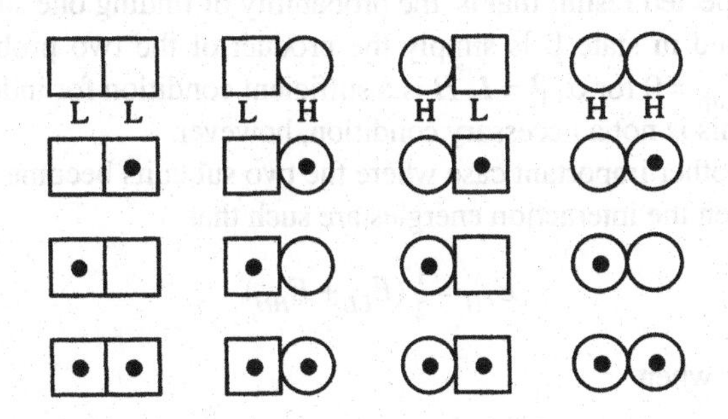

Figure 4.18. All sixteen configurations of a system with two subunits.

We define the probability of finding a specific subunit (say the lhs) in state L, independently of the state of the second subunit, by

$$X_L^0 = X_{LL}^0 + X_{LH}^0 = \frac{Q_L^2 Q_{LL} + Q_L Q_H Q_{LH}}{\xi(0)} \tag{4.7.4}$$

and, similarly,

$$X_H^0 = X_{HH}^0 + X_{HL}^0 = \frac{Q_H^2 Q_{HH} + Q_L Q_H Q_{LH}}{\xi(0)} \tag{4.7.5}$$

Clearly, if there are no interactions between the subunits, i.e., $E_{\alpha\beta} = 0$ for any α, β, then

$$X_L^0 \rightarrow \frac{Q_L^2 + Q_L Q_H}{(Q_L + Q_H)^2} = \frac{Q_L}{Q_L + Q_H} = X_L^{0,\infty} \tag{4.7.6}$$

$$X_H^0 \rightarrow \frac{Q_H^2 + Q_L Q_H}{(Q_L + Q_H)^2} = \frac{Q_H}{Q_L + Q_H} = X_H^{0,\infty} \tag{4.7.7}$$

The symbols $X_L^{0,\infty}$ and $X_H^{0,\infty}$ stand for the probability of finding a subunit in state L or H in the *absence of ligands* (0) and in the *absence* of *intersubunit interaction*. This may be obtained by separating the two subunits to infinite distance, hence the superscript ∞. Clearly, in this limit, the present model becomes identical with the model treated in Section 3.2.

It follows from the definition (4.7.3) that in the limit of infinite separation between the two subunits,

$$X_{\alpha\beta}^{0,\infty} \rightarrow \frac{Q_\alpha Q_\beta}{(Q_L + Q_H)^2} = X_\alpha^{0,\infty} X_\beta^{0,\infty} \tag{4.7.8}$$

which is the expected result; that is, the probability of finding one subunit in state α and the second in state β is simply the product of the two probabilities. We conclude that $E_{\alpha\beta} = 0$ for $\alpha, \beta = L, H$ is a sufficient condition for independence of the subunits. This is not a necessary condition, however.

There is another important case where the two subunits become independent. This occurs when the interaction energies are such that

$$E_{LH} = \frac{1}{2}(E_{LL} + E_{HH}) \tag{4.7.9}$$

or, equivalently, when

$$Q_{LH}^2 = Q_{LL} Q_{HH} \tag{4.7.10}$$

In this case the PF can be written as

$$\xi(0) = Q_L^2 Q_{LL} + 2Q_L Q_H \sqrt{Q_{LL} Q_{HH}} + Q_H^2 Q_{HH} = (Q_L \sqrt{Q_{LL}} + Q_H \sqrt{Q_{HH}})^2 \tag{4.7.11}$$

From Eqs. (4.7.3) and (4.7.4), we find for this case

$$X_L^0 X_L^0 = \frac{(Q_L^2 Q_{LL} + Q_L Q_H \sqrt{Q_{LL} Q_{HH}})^2}{(Q_L \sqrt{Q_{LL}} + Q_H \sqrt{Q_{HH}})^4}$$

$$= \frac{Q_L^2 Q_{LL}}{(Q_L \sqrt{Q_{LL}} + Q_H \sqrt{Q_{HH}})^2} = X_{LL}^0$$

and, similarly, for other combinations of indices we have

$$X_{LL}^0 = X_L^0 X_L^0, \qquad X_{LH}^0 = X_L^0 X_H^0, \qquad X_{HH}^0 = X_H^0 X_H^0 \tag{4.7.12}$$

Thus whenever the condition (4.7.9) or (4.7.10) is fulfilled, the probability of the states of the two subunits is the product of the probabilities of the states of each subunit. We see that the existence of interaction between the subunits does not imply dependence of the states of the subunits. In other words, the condition $E_{\alpha\beta} = 0$ is not a necessary condition for independence.

Consider now the reaction

$$(LL) + (HH) \rightarrow 2(LH) \tag{4.7.13}$$

For this reaction the equilibrium constant is

$$\eta = \frac{(X_{LH}^0)^2}{X_{LL}^0 X_{HH}^0} = \frac{(Q_L Q_H Q_{LH})^2}{Q_L^2 Q_{LL} Q_H^2 Q_{HH}} = \frac{Q_{LH}^2}{Q_{LL} Q_{HH}} = \exp[-\beta(2E_{LH} - E_{LL} - E_{HH})] \tag{4.7.14}$$

Thus, in the empty system the equilibrium constant η is determined only by the interaction energies $E_{\alpha\beta}$. Condition (4.7.9) is equivalent to the condition $\eta = 1$. We shall see in Section 4.7.3 that the equilibrium constant η is also responsible for transmitting information between the two ligands across the boundary between the two subunits.

It is easy to show that if Eqs. (4.7.12) hold, i.e. if $X_{\alpha\beta}^0$ is a product of X_α^0 and X_β^0 as defined in Eqs. (4.7.4) and (4.7.5), then $\eta = 1$. This follows directly from the definitions of $X_{\alpha\beta}^0$, and X_α^0 and X_β^0. Therefore $\eta = 1$ is a necessary and sufficient condition for independence of the *states* of the two subunits. We shall see that this statement holds true also for the *ligands* occupying the sites of the system.

4.7.2. The Binding Isotherm

We next introduce ligands into the system. The binding energies to states L and H are U_L and U_H, respectively, and we assume also direct interaction energy $U(1, 1)$ between the ligands (for simplicity, this is independent of the states of the subunits). We define

$$q_L = \exp(-\beta U_L), \quad q_H = \exp(-\beta U_H), \quad S = \exp[-\beta U(1, 1)] \tag{4.7.15}$$

The grand PF for the system is now

$$\begin{aligned}\xi &= Q(0; 0) + [Q(0; 1) + Q(1; 0)]\lambda + Q(1; 1)\lambda^2 \\ &= \sum_{\alpha\beta} Q_\alpha Q_\beta Q_{\alpha\beta} + \left(\sum_{\alpha\beta} Q_\alpha Q_\beta q_\alpha Q_{\alpha\beta} + \sum_{\alpha\beta} Q_\alpha Q_\beta q_\beta Q_{\alpha\beta}\right)\lambda + \sum_{\alpha\beta} Q_\alpha Q_\beta Q_{\alpha\beta} q_\alpha q_\beta S\lambda^2\end{aligned} \tag{4.7.16}$$

The sixteen terms in this equation correspond to the sixteen states of the system in Fig. 4.18. Each sum in Eq. (4.7.16) corresponds to one row in Fig. 4.18. We use the semicolon to separate the specifications of the left and right subunits. Note also that we assume, for simplicity, that $S_L = S_H = S$.

Another way of writing the PF in Eq. (4.7.16) is

$$\xi = \sum_{\alpha\beta} Q_\alpha Q_\beta Q_{\alpha\beta}[1 + (q_\alpha + q_\beta)\lambda + q_\alpha q_\beta S\lambda^2] \tag{4.7.17}$$

Here, each term with specific α and β, e.g., $\alpha = L$ and $\beta = L$, corresponds to one conformational state of the system, i.e., one *column* in Fig. 4.18. For instance,

$$\xi_{LL} = Q_L^2 Q_{LL}(1 + 2q_L\lambda + q_L^2 S\lambda^2) \tag{4.7.18}$$

is the PF for the system in conformational state LL; clearly, this is the same as the PF of a single-state system with two binding sites. The general form of the binding isotherm is

$$\theta = \frac{\bar{n}}{2} = \frac{\lambda}{2}\frac{\partial \ln \xi}{\partial \lambda} = \frac{\Sigma_{\alpha\beta} Q_\alpha Q_\beta Q_{\alpha\beta}(q_\alpha\lambda + q_\alpha q_\beta S\lambda^2)}{\xi} \tag{4.7.19}$$

Note that even when $S = 1$, the isotherm does not reduce to the Langmuir form. If, on the other hand, both $S = 1$ and $\eta = 1$, then we find that the PF is

$$\xi = \sum_{\alpha\beta} Q_\alpha Q_\beta \sqrt{Q_{\alpha\alpha} Q_{\beta\beta}}(1 + q_\alpha\lambda)(1 + q_\beta\lambda) = \left[\sum_\alpha Q_\alpha \sqrt{Q_{\alpha\alpha}}(1 + q_\alpha\lambda)\right]^2 \tag{4.7.20}$$

which is essentially the same PF as that of the model treated in Section 4.5 with

$S_L = S_H = 1$. We shall see in the following subsections that the important new features of this model arise when $\eta \neq 1$.

As in Section 4.5, we define the intrinsic binding constant by

$$k_1 = \frac{Q(0;\,1)}{Q(0;\,0)}\lambda_0 \tag{4.7.21}$$

and the correlation function

$$g(1,\,1) = \frac{Q(1;\,1)Q(0;\,0)}{Q(0;\,1)Q(1;\,0)} \tag{4.7.22}$$

and rewrite the BI as

$$\theta = \frac{k_1 C + k_1^2 g(1,\,1)C^2}{1 + 2k_1 C + k_1^2 g(1,\,1)C^2} \tag{4.7.23}$$

which is formally the same as in Section 4.3, but with k_1 replacing k and $g(1, 1)$ replacing S.

4.7.3. Correlation Function and Cooperativity

The correlation function was defined in Section 4.2 for any two events. In particular, if the events are "site one is occupied" and "site two is occupied," then the (λ-dependent) correlation function is

$$g(1,\,1) = \frac{P(1;\,1)}{P(1)^2} = \frac{Q(1;\,1)\lambda^2\xi}{[Q(0;\,1)\lambda + Q(1;\,1)\lambda^2]^2} \tag{4.7.24}$$

Note that here $P(1; 1)$ is the probability of finding the two sites occupied (independently of the states of the subunits) and $P(1)$ is the probability of finding a *specific* site (say, the lhs in Fig. 4.18) occupied (independently of the state of the subunits and of the occupational state of the second site); $P(1) = P(1; \_) = P(\_; 1)$.

As in Sections 4.2 and 4.5, we need only the $\lambda \to 0$ limit of this correlation function, which is the quantity defined in Eq. (4.7.22), to which we refer as the correlation function between the two events. For these particular *events* we also say that whenever there exists correlation [i.e., $g(1, 1) \neq 1$)], the two *ligands* cooperate; hence there exists cooperativity between the ligands, or simply, the system is cooperative.

We now focus on the *indirect* correlation. To do so, we can either factor S from $g(1, 1)$, or simply assume that $S = 1$ and examine the remaining correlation, denoted by $y(1, 1)$. Using the notations (4.7.14) and (4.7.15), and $K = Q_H/Q_L$ and $K' = Q_{HH}/Q_{LL}$, we note that K and $\sqrt{K'}$ always appear together. Hence, we set

$$K = K\sqrt{K'} = \frac{Q_H\sqrt{Q_{HH}}}{Q_L\sqrt{Q_{LL}}} \tag{4.7.25}$$

and write the indirect correlation in the simplified form

$$y(1, 1) = \frac{(1 + 2\bar{\eta}K + K^2)(1 + 2\bar{\eta}hK + h^2K^2)}{(1 + \bar{\eta}K + \bar{\eta}hK + hK^2)^2}$$

$$= 1 - \frac{(\bar{\eta}^2 - 1)(h - 1)^2K^2}{(1 + \bar{\eta}K + \bar{\eta}hK + hK^2)^2} \tag{4.7.26}$$

The second form on the rhs of this equation is very convenient for examining the condition under which indirect correlation exists. In Eq. (4.7.26) and in the following section, we put $\bar{\eta} = \sqrt{\eta}$.

Equation (4.7.26) should be compared with Eq. (4.5.17) in Section 4.5; we note that the former reduces to the latter when $\eta = 0$ and $K^2 = K$. In the model of Section 4.5 we found that "either $h = 1$ or $K = 0$" is a necessary and sufficient condition for $y(1, 1) = 1$. In the present model, we see from Eq. (4.7.26) that "either $h = 1$, or $K = 0$, or $\eta = 1$" is a necessary and sufficient condition for $y(1, 1) = 1$.

As in the model of Section 4.5, if $K = 0$ (or $K = \infty$), the system is not responsive to the binding process. Also, if $h = 1$, the ligand cannot induce conformational changes in the system. Here, in addition to the requirement $K \neq 0$ and $h \neq 1$, we need $\eta \neq 1$ in order to have indirect correlation. Moreover, the correlation $y(1, 1)$ in Section 4.5 is always positive [$y(1, 1) \geq 1$]. Here, the correlation may be either positive or negative, depending on whether $\eta < 1$ or $\eta > 1$. In the next subsection we shall see that this property is related to the way the conformational change induced in one subunit is transmitted to the second subunit. Clearly, if all $E_{\alpha\beta} = 0$, then there can be no communication between the two subunits, hence $y(1, 1) = 1$. However, this result is also obtained under much weaker conditions, namely, $2E_{LH} = E_{LL} + E_{HH}$, or, equivalently, $\eta = 1$. Thus, when E_{LH} equals the average of E_{LL} and E_{HH}, i.e., $\eta = 1$, then $y(1, 1) = 1$. The reverse is also true, as can be seen from Eq. (4.7.26).

To summarize, we see that in order to have indirect cooperativity, the following conditions must be fulfilled (see the schematic illustration in Fig. 4.19):

(a) The ligand must have a preference for one of the conformational states, i.e., it must be able to *induce* conformational change. The capacity to do so is measured by h. When $h = 1$, no conformational change can be induced.

(b) The subunit must be responsive. If $K = 0$ (or $K = \infty$, but we have chosen $0 \leq K \leq 1$), the conformational state will not respond to the binding process. The system remains in its most stable form (which was chosen to be the L form). To be responsive, the system must have a nonzero equilibrium constant, K.

(c) Whatever the change of conformation that has been induced in the subunit on which the ligand is bound, there must be another conformational change induced in the second subunit. The extent and direction of this change is measured by η. If $\eta < 1$, the change induced in the second subunit will be in the *same direction* as in

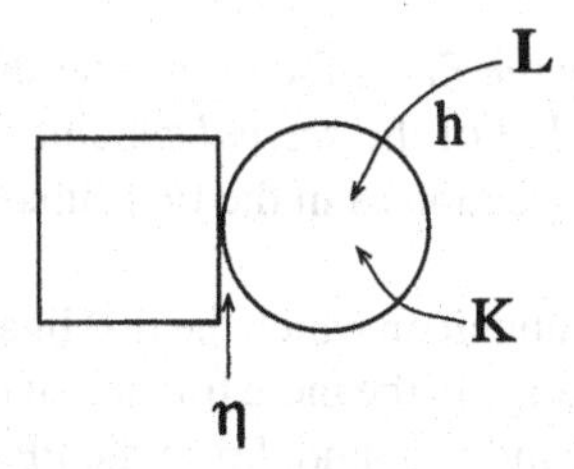

Figure 4.19. A schematic assignment of parameters h, K, and η, where h is a property of both the ligand and the subunit, K is a property of the conformations of the subunits, and η is determined by the subunit–subunit interactions.

the first (see the next subsection), hence the approaching second ligand will find a subunit, the state of which is the preferable state for binding, hence $y(1, 1) > 1$. The reverse is true when $\eta > 1$, leading to $y(1, 1) < 1$, i.e., negative cooperativity. When $\eta = 1$, there is no transmission of information across the boundaries, hence no indirect cooperativity. This is a very general scheme of the conditions that must be fulfilled to obtain indirect correlation. We shall repeatedly find these conditions in all the models discussed in the next chapters.

The generalization to the case of two different sites (but the subunits are still nearly the same, i.e., the same energies E_L, E_H and E_{LL}, E_{LH}, E_{HH}) is quite straightforward. The result is

$$g(a, b) = 1 - \frac{(\bar{\eta}^2 - 1)(h_a - 1)(h_b - 1)K^2}{[1 + \bar{\eta}(h_a + 1)K + h_a K^2][1 + \bar{\eta}(h_b + 1)K + h_b K^2]} \tag{4.7.27}$$

which is the generalization of Eq. (4.7.26). Here, the sign of the cooperativity depends on $(\bar{\eta}^2 - 1)$ as well as on the sign of the product $(h_a - 1)(h_b - 1)$.

4.7.4. Induced Conformational Changes in the Two Subunits

In Section 3.4 we analyzed the conformational changes induced in a *single* subunit (which was then the entire system) by a *single* ligand. In Section 4.5 we studied the induced conformational changes in a single subunit (the entire system) by *two* ligands. We found that the indirect correlation is intimately related to the capacity of the ligands to induce conformational changes. We also found that the ligand will always shift the equilibrium concentrations of L and H toward the state for which the binding energy is stronger (more negative). This produces positive cooperativity for two identical sites. The situation is more complex in the present model. Here, there are two subunits, in each of which there could be an induced conformational change. The conformational change induced on the subunit on which the ligand binds is in the same direction as in previous models. However, the induced shift experienced by the *first* subunit induces another shift in the equilibrium concentrations of L and H of the *second* subunit. The latter can be either in the same or in the opposite direction as in the first subunit. We shall now examine these effects quantitatively.

Let us first examine the subunit on which the first ligand binds, say the left-hand one. We denote by $d_{L;}^{(1)}$ the change in the mole fraction of the L form, in the left-hand subunit on which the *first* ligand is bound. From the PF of the system we find

$$d_{L;}^{(1)} = P(L; /1; 0) - P(L; /0; 0)$$

$$= \frac{Q_L q_L \Sigma_\alpha Q_\alpha Q_{L\alpha}}{\Sigma_{\alpha\beta} Q_\alpha Q_\beta Q_{\alpha\beta} q_\alpha} - \frac{Q_L \Sigma_\alpha Q_\alpha Q_{L\alpha}}{\Sigma_{\alpha\beta} Q_\alpha Q_\beta Q_{\alpha\beta}} \tag{4.7.28}$$

The quantity $P(L; /1; 0)$ denotes the conditional probability of finding the left subunit in state L, given that the left subunit is occupied and the right subunit is empty. Therefore, $d_{L;}^{(1)}$ is the difference in the probabilities (or mole fractions) of the left-hand subunit being in state L, before and after the binding on the left subunit [the sums in Eq. (4.7.27) are over $\alpha = L, H$ and $\beta = L, H$].

In terms of the notation of Section 4.7.3, this quantity can be written as

$$d_{L;}^{(1)} = \frac{(1-h)K(\bar{\eta} + K)(1 + \bar{\eta}K)}{(1 + 2\bar{\eta}K + K^2)[1 + \bar{\eta}(1+h)K + hK^2]} \tag{4.7.29}$$

We see that the sign of $d_{L;}^{(1)}$ is determined only by $(1 - h)$. It is positive when $h < 1$, i.e., the ligands prefer to bind on L. Note that when $\eta = 1$ and $K = K$, this reduces to the quantity d_L in Section 3.4.

The second quantity of interest, denoted by $d_{;L}^{(1)}$, is the change in the mole fraction of L in the *second* subunit (or the right-hand one) induced by binding on the *first* subunit (the left-hand one). This quantity is defined by

$$d_{;L}^{(1)} = P(; L/1; 0) - P(; L/0; 0)$$

$$= \frac{Q_L \Sigma_\alpha Q_\alpha Q_{\alpha L} q_\alpha}{\Sigma_{\alpha\beta} Q_\alpha Q_\beta Q_{\alpha\beta} q_\alpha} - \frac{Q_L \Sigma_\alpha Q_\alpha Q_{\alpha L}}{\Sigma_{\alpha\beta} Q_\alpha Q_\beta Q_{\alpha\beta}} \tag{4.7.30}$$

which may be simplified to

$$d_{;L}^{(1)} = \frac{(1-h)(1-\bar{\eta}^2)K^2}{(1 + 2\bar{\eta}K + K^2)[1 + \bar{\eta}(1+h)K + hK^2]} \tag{4.7.31}$$

Here, in addition to $(1 - h)$ the sign of $d_{;L}^{(1)}$ is determined also by $(1 - \bar{\eta}^2)$. If $\eta = 1$, then there is no transmission of information between the subunits, hence $d_{;L}^{(1)} = 0$.

Whatever the conformational changes induced in the left-hand subunit (on which the ligand is bound), the right-hand subunit does not respond.

When $\eta < 1$, then $d_{;L}^{(1)}$ has the *same* sign as that produced by $(1 - h)$, e.g., if the ligand increases the L form on the left subunit [$d_{L;}^{(1)} > 0$ in Eq. (4.7.29)] and $1 - \eta^2 > 0$, then also the right-hand side distribution (of L and H) is shifted in the same direction. This results in positive cooperativity, as found in Section 4.7.3. The opposite is true when $\eta > 1$, in which case a *positive* change on the left-hand subunit ($d_{L;}^{(1)} > 0$) will induce a *negative* change on the right-hand subunit ($d_{;L}^{(1)} < 0$).

For completeness, we write also the changes produced by the *second* ligand. These are

$$d_{;L}^{(2)} = P(;L/1;1) - P(;L/1;0)$$

$$= \frac{(1-h)K(\bar{\eta} + hK)(1 + \bar{\eta}hK)}{(1 + 2\bar{\eta}hK + h^2K^2)[1 + \bar{\eta}(1+h)K + hK^2]} \tag{4.7.32}$$

and

$$d_{L;}^{(2)} = P(L;/1;1) - P(L;/1;0)$$

$$= \frac{(1-h)(1-\bar{\eta}^2)hK^2}{(1 + 2\bar{\eta}hK + h^2K^2)[1 + \bar{\eta}(1+h)K + hK^2]} \tag{4.7.33}$$

The interpretation of $d_{;L}^{(2)}$ and $d_{L;}^{(2)}$ is similar to $d_{L;}^{(1)}$ and $d_{;L}^{(1)}$ but refers to the second ligand. A schematic illustration of the relevance of these quantities to the two subunits is shown in Fig. 4.20.

To summarize, when $\eta = 0$, $d_{L;}^{(1)}$ and $d_{;L}^{(1)}$ are identical (as are also $d_{L;}^{(2)}$ and $d_{;L}^{(2)}$). In this case the extent of conformational change induced in one subunit is the same, in both sign and magnitude, as in the second subunit. The two subunits act concertedly, and this model reduces to the model of Section 4.5.

When $0 < \eta < 1$, the change $d_{;L}^{(1)}$ has the same *sign* as $d_{;L}^{(1)}$ but $|d_{L;}^{(1)}| > |d_{;L}^{(1)}|$, i.e., the induced change diminishes from the first to the second subunit. The extreme case is when $\eta = 1$. Whatever the value of $d_{L;}^{(1)}$, the value of $d_{;L}^{(1)}$ becomes zero. There is no transfer of information across the boundary [we have already seen that in this case, $y(1, 1) = 1$]. When $1 < \eta < \infty$, we find a reversal of signs from $d_{L;}^{(1)}$ to $d_{;L}^{(1)}$, and in the limit $\eta \to \infty$ the two become identical in magnitude but have opposite signs.

We have already discussed the physical reason for the effect of the parameters h and K on the indirect correlation (Section 4.5) or, equivalently, on the induced conformational changes. But we still need an intuitive explanation of the effect of η. Why does $\eta = 1$ mean no "communication" between the sites? Why does $\eta < 1$ result in positive indirect cooperativity or, equivalently, induces conformational change in the second subunit in the same direction as in the first subunit? And why does $\eta > 1$ have the opposite effect?

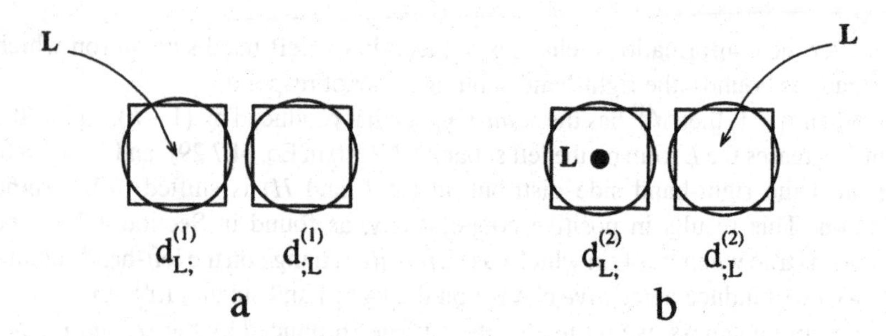

Figure 4.20. (a) Adding the *first* ligand **L** to the left subunit causes conformational changes $d^{(1)}_{L;}$ and $d^{(1)}_{;L}$. (b) Adding the *second* ligand to the right subunit causes conformational changes $d^{(2)}_{L;}$ and $d^{(2)}_{;L}$.

The answer to these questions may be obtained by viewing the *second subunit* as the adsorbent molecule and the *first subunit* as a "ligand" binding to the second subunit. This "ligand" is affected by the real ligand **L** through the parameter h. For concreteness, suppose that $h < 1$, i.e., the *real* ligand always favors the L conformer, hence when binding to the subunit, it shifts the equilibrium $L \rightleftharpoons H$ in favor of L.

Suppose the first subunit, the "ligand," approaches to "bind" to the second subunit (Fig. 4.21). Its "binding energies" to L and H are

$$\text{“}U_L^0\text{”} = X_L^{(0)} E_{LL} + X_H^{(0)} E_{HL}, \qquad \text{“}U_H^0\text{”} = X_L^{(0)} E_{HL} + X_H^{(0)} E_{HH} \tag{4.7.34}$$

and the corresponding h parameter, denoted "h^0," is

$$\text{“}h^0\text{”} = \exp[-\beta(\text{“}U_H^0\text{”} - \text{“}U_L^0\text{”})] \tag{4.7.35}$$

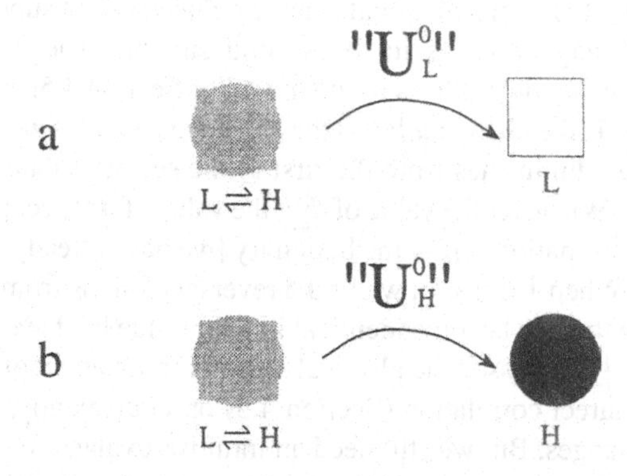

Figure 4.21. The "binding" of one subunit to a second. (a) An equilibrated subunit "binds" to a subunit L. The corresponding "binding energy" is "U_L^0." (b) An equilibrated subunit "binds" to a subunit H. The corresponding "binding energy" is "U_H^0." A superimposed square and circle represent an equilibrated mixture of L and H.

The superscript "0" indicates that the first subunit is empty. We can write the analogues of Eqs. (4.7.34) and (4.7.35) for the first subunit being occupied by a (true) ligand. When this is viewed as our new "ligand" (Fig. 4.22), we get the same equations as (4.7.34) and (4.7.35) except for replacing the superscript "0" by "1" (indicating that the first subunit is occupied). The ratio of "h^1" and "h^0," denoted "h," is the effective h-parameter of the (true) ligand **L** with respect to the *second* subunit. The ratio is

$$\text{“}h\text{”} = \frac{\text{“}h^1\text{”}}{\text{“}h^0\text{”}} = \exp[-\beta(X_L^{(1)} - X_L^{(0)})(2E_{LH} - E_{LL} - E_{HH})]$$

$$= \exp[(X_L^{(1)} - X_L^{(0)}) \ln \eta] \tag{4.7.36}$$

Let us first consider the case $\eta < 1$. If $h < 1$, i.e., the (true) ligand favors L, then $X_L^{(1)} - X_L^{(0)} > 0$, hence "$h$" < 1. Similarly, when $h > 1$, $X_L^{(1)} - X_L^{(0)} < 0$, hence "$h$" > 1. We see that in this case the ligand affects the second subunit in the *same* direction as the first subunit. The second case is when $\eta > 1$. If $h < 1$, then $X_L^{(1)} - X_L^{(0)} > 0$, hence "$h$" > 1. If $h > 1$, then $X_L^{(1)} - X_L^{(0)} < 0$, and "$h$" < 1. We see that in this case the (true) ligand affects the second subunit in the *opposite* direction to the effect on the first subunit.

As we have seen in the previous subsection, this reversal of the induced conformational changes, when $\eta > 1$, is also responsible for the change in the sign of the indirect cooperativity.

We shall see in the subsequent chapters that for systems with more than two subunits, the case $\eta > 1$ leads to an oscillatory dependence of the correlation as a function of the number of subunits.

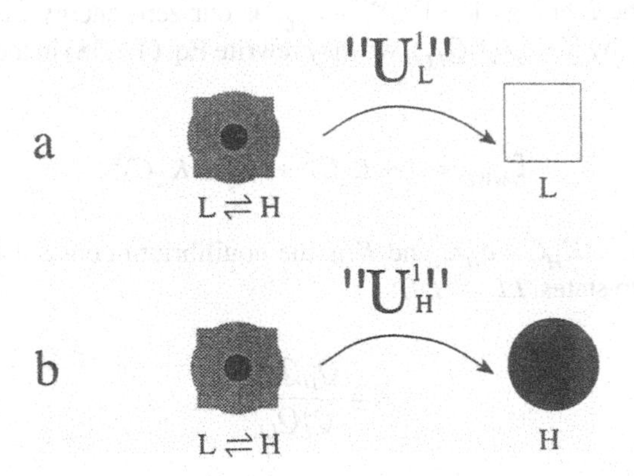

Figure 4.22. The same as in Fig. 4.21, but now the equilibrated subunit is occupied. The corresponding new "binding energies" are "U_L^1" and "U_H^1."

4.7.5. Two Limiting Cases

1. The Concerted Model

The first limiting case was suggested originally by Monod, Wyman, and Changeux (MWC) (Monod *et al.*, 1965). This model requires that the two subunits be either in the L or in the H state. The conformations of the two subunits change in a concerted way. This is equivalent to the consideration of the first and fourth columns in Fig. 4.18. Mathematically, we can obtain this limiting case by taking $\eta = 0$, which essentially means that the equilibrium concentrations of X_{LH} (and X_{HL}) are negligible.

The PF of the system may be obtained from Eq. (4.7.17) by substituting $Q^2_{\alpha\beta} = Q_{\alpha\alpha}Q_{\beta\beta}\delta_{\alpha\beta}$. The result is

$$\xi_{MWC} = \sum_{\alpha} Q^2_{\alpha}Q_{\alpha\alpha}(1 + 2q_{\alpha}\lambda + q^2_{\alpha}S\lambda^2) = \xi_{LL} + \xi_{HH} \tag{4.7.37}$$

This is essentially the same PF as that of the model treated in Section 4.5 with the replacement of Q_{α} everywhere by $Q^2_{\alpha}Q_{\alpha\alpha}$. In essence, the MWC model is equivalent to a two-state model with energies corresponding to $2E_L + E_{LL}$ and $2E_H + E_{HH}$. The fact that we have two subunits does not affect the formalism, except for the redefinitions of the energy levels.

Originally, the MWC was applied for systems with negligible direct interactions between the ligands (i.e., $S = 1$), in which case Eq. (4.7.37) reduces to

$$\xi_{MWC} = Q^2_L Q_{LL}(1 + q_L\lambda)^2 + Q^2_H Q_{HH}(1 + q_H\lambda)^2 \tag{4.7.38}$$

If we choose the L energy level, $2E_L + E_{LL}$, as our zero energy (or, equivalently, define a new PF by $\xi' = \xi/Q^2_L Q_{LL}$), we may rewrite Eq. (4.7.38) in the more familiar form

$$\xi'_{MWC} = (1 + K_L C)^2 + K(1 + K_H C)^2 \tag{4.7.39}$$

where $K_L C = q_L\lambda$, $K_H C = q_H\lambda$, and K is the equilibrium constant for conversion between the two states, $LL \rightleftharpoons HH$:

$$K = \frac{Q^2_H Q_{HH}}{Q^2_L Q_{LL}} \tag{4.7.40}$$

K is the same as K^2 in Eq. (4.7.25). From Eq. (4.7.26), we see that in the case $\eta = 0$ the indirect correlation function (which is the same as the total correlation function if $S = 1$) is

$$y(1, 1) = 1 + \frac{(1-h)^2 K}{(1+Kh)^2} \geq 1 \tag{4.7.41}$$

which is essentially the same as Eq. (4.5.17) with the replacement of K by K. In this case, as in the model of Section 4.5, we always have positive (indirect) cooperativity. The molecular reason for this result is quite clear in view of the analysis of the origin of the cooperativity as discussed in Section 4.5. Here, a conformational change on one subunit is *fully transmitted* to the second subunit. In other words, the two subunits respond in a concerted manner, as if they were a single subunit, i.e., as in the model of Section 4.5.

2. *The Sequential Model*

The second extreme case, suggested by Koshland, Nemethy, and Filmer (KNF) (Koshland *et al.*, 1966), is also known as the sequential model. The mathematical conditions required to obtain this limiting case are quite severe. First, it is assumed that, in the absence of a ligand, one of the conformations is dominant, say the *LL* form. In addition, it is assumed that a ligand binding to any subunit will change the conformation of that subunit into the H form. These assumptions lead to the consideration of only the four diagonal states of Fig. 4.18, for which the PF is

$$\xi_{KNF} = Q_L^2 Q_{LL} + 2Q_L Q_H Q_{LH} q_H \lambda + Q_H^2 Q_{HH} q_H^2 S \lambda^2 \tag{4.7.42}$$

The empty state is the *LL* state on the top left corner of Fig. 4.18. The binding of a ligand on any of the subunits will shift its conformation completely from L to H without affecting the conformation of the second subunit. Binding of the two ligands will shift the entire polymer to the state *HH*. Thus, in each binding process there is a total change of conformation of one subunit; hence the term sequential model.

The mathematical requirements necessary to obtain the KNF model from the general one can be stated as follows. Let K^0 be the equilibrium constant for the $H \leftrightarrow L$ conversion of each subunit when it is known to be empty. Likewise, let K^1 be the equilibrium constant when the subunit is known to be occupied. Both of these equilibrium constants are functions of λ, i.e.,

$$K^{(0)} = \frac{P(H, 0;)}{P(L, 0;)} = \frac{Q_H}{Q_L} F(\lambda) \tag{4.7.43}$$

and

$$K^{(1)} = \frac{P(H, 1;)}{P(L, 1;)} = \frac{Q_H q_H}{Q_L q_L} F(S\lambda) \tag{4.7.44}$$

where $P(H, 0;)$ is the probability of finding the left-hand subunits in state H and being empty. Similar meanings apply to the other probabilities in Eqs. (4.7.43) and (4.7.44). The functions $F(\lambda)$ and $F(S\lambda)$ may be obtained from the PF, through the relevant probabilities.

The mathematical requirement of the model is that $K = Q_H/Q_L \rightarrow 0$, i.e., L is infinitely more stable than H, but $Kh = Q_H q_H/Q_L q_L \rightarrow \infty$, i.e., the occupied subunit in *state* H is infinitely more stable than L. This means that we require a *total conversion* $L \rightarrow H$, upon binding in a system which has *infinite resistance* ($K \rightarrow 0$) to conformational changes. This can be achieved only when h is such a strong infinity that even after multiplication by $K \rightarrow 0$ the product Kh is still infinity.

4.8. BINDING OF PROTONS TO A TWO-SITE SYSTEM

4.8.1. Introduction, Notation, and Some Historical Perspectives

Perhaps the simplest two-site cooperative systems are small molecules having two binding sites for protons, such as dicarboxylic acids and diamines. Despite their molecular simplicity, most of these molecules do not conform with the modelistic assumptions made in this chapter. Therefore, their theoretical treatment is much more intricate. The main reasons for this are: (1) there is, in general, a continuous range of macrostates; (2) the direct and indirect correlations are both strong and intertwined, so that factorization of the correlation function is impossible. In addition, as with any real biochemical system, the solvent can have a major effect on the binding properties of these molecules.

The literature on dissociation (or ionization) constants of these and related compounds is immense. We present in this section only a few examples to illustrate various aspects of cooperativity in these systems.

As a rule, we shall use only *intrinsic binding* (or *association*) constants. The relevant experimental data are usually reported in terms of either the *thermodynamic* dissociation or association constant. The general relation between intrinsic and thermodynamic constants has been discussed in Section 2.3. It will be repeated below for the special cases of this section.

We denote by k_1 the *first* intrinsic binding constant, and by k_2 the *second* intrinsic binding constant. The latter is the same as $k_{1/1}$, i.e., the conditional binding on the second site, given that the first site is occupied.* Figure 4.23 shows two alternative, but equivalent, ways of describing the binding of protons to a dicarboxylic acid.

(a) The first view, referred to as the "macroscopic" view, recognizes *three* species: the empty, the singly occupied, and the doubly occupied molecules. The

*One should be careful to distinguish k_1 and k_2 from k_a and k_b for systems with *different* binding sites. Both k_a and k_b are for the *first* site. See Subsection 4.8.3.

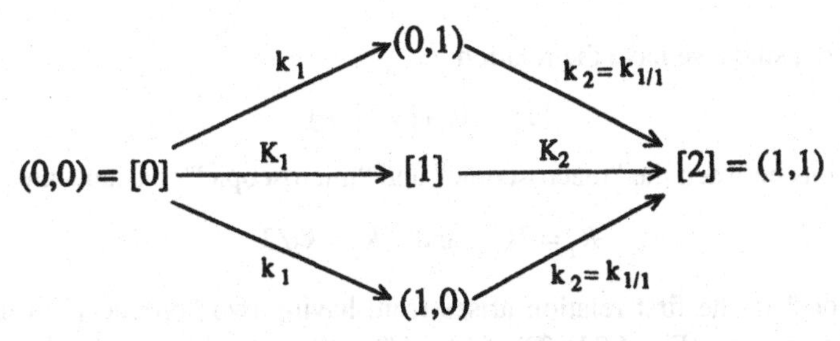

Figure 4.23. Two alternative ways of describing the binding of protons to a dicarboxylic acid. [0], [1], and [2] denote the di-ion, singly and doubly occupied acid, respectively. (0, 1) and (1, 0) indicate a molecule with a specific single site occupied. The *macroscopic* binding constants K_1 and K_2 are related to the reactions of adding successively the first and second proton to the di-ion, respectively. The *intrinsic* binding constants k_1 and k_2 refer to the same processes but for a *specific* site, as if the two (identical) binding sites were labelled, say the right and left one.

corresponding densities of the species are denoted by square brackets. The two ("macroscopic") equilibrium constants are

$$K_1 = \frac{[1]}{[0][H]}, \qquad K_2 = \frac{[2]}{[1][H]} \tag{4.8.1}$$

where $[H]$ is the proton density (in moles per unit of volume) in the solution.

(b) The second view, sometimes referred to as the "microscopic" view, recognizes *four* species, denoted by (0, 0), (0, 1), (1, 0), and (1, 1). Here, we "label" the two equivalent "sites," say the right and left sites, and distinguish between configurations (0, 1) and (1, 0). These configurations are in general indistinguishable (unless the molecules are localized, or if we consider two nonequivalent sites, say in salicylic acid or, of course, in amino acids).

The "microscopic" equilibrium constants are now defined by

$$k_1 = \frac{[0, 1]}{[0, 0][H]} = \frac{[1, 0]}{[0, 0][H]} \tag{4.8.2}$$

and

$$k_2 = \frac{[1, 1]}{[0, 1][H]} = \frac{[1, 1]}{[1, 0][H]} \tag{4.8.3}$$

Note that the terms "macroscopic" and "microscopic" constants do not imply that these quantities measure macroscopic or microscopic quantities, respectively. Here, in the "macroscopic" view we have simply grouped the two "microscopic" species (0, 1) and (1, 0) into one species denoted by (1). Both of these constants can be macroscopic or microscopic, depending on whether we study the binding per molecule or per one mole of molecules.

Clearly, since we have the relation

$$[1] = [0, 1] + [1, 0] \tag{4.8.4}$$

the relations between the "macroscopic" and "microscopic" constants are

$$K_1 = 2k_1 \quad \text{and} \quad K_2 = k_2/2 \tag{4.8.5}$$

The factor 2 in the first relation arises from having two "products" in the first association process (Fig. 4.23). The factor 1/2 in the second arises from having two "reactants" from which the doubly occupied acid is formed. (These factors are sometimes referred to as "statistical"; a better term would be combinatorial factors, the general origin of which is discussed in Sections 2.2 and 2.3.) The binding isotherm (or the equivalent titration curve) should not depend on the way we choose to view our species. If we denote by C the proton density $[H]$ in the solution (assuming that we are in the regime of very dilute solutions), then the binding isotherm is simply the ratio between the total bound protons and the total number of adsorbent molecules, i.e.,

$$n = \frac{[0, 1] + [1, 0] + 2[1, 1]}{[0, 0] + [0, 1] + [1, 0] + [1, 1]} = \frac{K_1 C + 2K_1 K_2 C^2}{1 + K_1 C + K_1 K_2 C^2}$$

$$= \frac{2k_1 C + 2k_1 k_2 C^2}{1 + 2k_1 C + k_1 k_2 C^2} \tag{4.8.6}$$

The distinction between K_i and k_i becomes clearer when the two sites are not equivalent (e.g., glycine), in which case $2k_1$ in Eq. (4.8.6) is the sum of two different intrinsic constants,* say the binding to the acidic group k_a and to the basic group, k_b, such as

$$2k_1 = k_a + k_b \tag{4.8.7}$$

From the experimental binding isotherm $\bar{n} = \bar{n}(C)$, one can determine the two "experimental" constants K_1 and K_2. If the two sites are identical, then one can convert from K_1 and K_2 into k_1 and k_2 via Eq. (4.8.5). This is not possible when the two sites are not identical in which case one cannot obtain the intrinsic constants, say k_a and k_b, solely from knowledge of K_1 and K_2 (see Subsection 4.8.3).

*Again, we note that when we refer to an *intrinsic* binding constant, we mean binding to a *specific* site. The corresponding binding free energy (or any pertinent thermodynamic quantity) *includes* the interaction of the proton with the *particular site* on which it binds as well as with the *entire* molecule. In some publications the term "intrinsic" refers to the interaction of the proton only with the particular site on which it is bound. This is, in general, not a well-defined quantity, unless other conditions are fulfilled. See also Appendix I.

The thermodynamic dissociation constants are related to the association constants by

$$K_1 = (K_{2,diss})^{-1} \text{ and } K_2 = (K_{1,diss})^{-1} \tag{4.8.8}$$

The pair correlation function for the two protons is given by

$$g_{11} = g(1, 1) = \frac{k_{1/1}}{k_1} = \frac{k_2}{k_1} = \frac{4K_2}{K_1} = \frac{4K_{2,diss}}{K_{1,diss}} \tag{4.8.9}$$

The corresponding work defined by

$$W_{11} = W(1, 1) = -k_B T \ln g_{11} \tag{4.8.10}$$

is the work associated with the process

$$(0, 1) + (1, 0) \rightarrow (1, 1) + (0, 0) \tag{4.8.11}$$

In this process we start from two singly occupied molecules, but at infinite separation from each other (and hence no correlation between the protons), and form one doubly occupied and one empty molecule. By taking *two* singly occupied *molecules* at infinite separation, we secure the independence of the protons at the same two sites as in the doubly occupied molecule, denoted (1, 1). The quantity W_{11} is computed from the pK values, usually reported in the literature, at $T = 298.15$ K (here K stands for degrees Kelvin),

$$W_{11} = -k_B T \ln \frac{4K_2}{K_1} = -k_B T\, 2.303 \log_{10} \frac{4K_2}{K_1}$$

$$= 1.364(pK_2 - pK_1 - \log_{10} 4) \tag{4.8.12}$$

where $pK = -\log K$ is the logarithm to base 10 of the constant K and $2.303\, k_B T = 1.364$ kcal/mol is the value for conversion to kcal/mol at $T = 298.15$ K.

In the next subsections we shall present some specific experimental data. In all cases known to us, W_{11} is positive, indicating negative cooperativity. Most molecular interpretations of this cooperativity have focused on the electrostatic interaction between the two protons. Formally, as shown below, one can always define a *microscopic* dielectric constant to account for the deviation in the experimental value of W_{11} from the value computed on the basis of purely Coulombic interaction. However, from the theoretical point of view, we can gain further insight into the molecular contributions to W_{11} by distinguishing between the following four sources of cooperativity: (1) *direct* proton–proton interaction; (2) *indirect* proton–proton correlation mediated by the solvent; (3) indirect correlation mediated by the adsorbent molecule at a fixed conformation; (4) indirect correlation, modified by the allowance of conformational changes in the adsorbent molecule.

Qualitatively, the four components of the correlation function can be visualized according to the following four steps: First, we bring the two protons from infinite separation to the final distance R_{HH} (where R_{HH} is assumed to be a *fixed* distance between the two protons on one of the conformational states of the adsorbent molecule). The interaction energy is simply the Coulombic interaction e^2/R_{HH}, where e is the charge of the proton. Next, we place the same two protons in a solvent. The interaction is now modified by the dielectric constant of the solvent. If the distance R_{HH} is large enough, one can use the macroscopic dielectric constant of pure water, which at 25 °C is 78.5. However, for smaller distances, the dielectric constant would depend on the distance R_{HH} (and, in general, also on the size and type of the charges on the two interacting species). Next, having selected a fixed conformation, we put the adsorbent molecule between the two protons. This will modify the correlation between the two protons. Finally, a further modification will be obtained by relaxing the requirement of a fixed conformation.

Perhaps the first attempt to interpret the proton–proton correlation, based on electrostatic interactions, was made by Bjerrum (1923). If e is the proton charge and R_{HH} the proton–proton distance in the diacid, then the interaction free energy was assumed to be given by

$$W_{11} = \frac{e^2}{DR_{HH}} \tag{4.8.13}$$

where D is the macroscopic dielectric constant (78.5 for water). One can fit the experimental values of W_{11} to an equation of the form (4.8.13) only for long α,ω-dicarboxylic acids where the protons are far apart, as illustrated in the next subsection. For the shorter diacids, W_{11} is much larger than the value predicted with the fixed dielectric constant D.

Clearly, if one takes a smaller value of D, one gets a higher value of W_{11}, for a given distance R_{HH}. Kirkwood and Westheimer (1938), Westheimer and Kirkwood (1938), and Westheimer and Shookhoff (1939) indeed argued that one should take a much smaller dielectric constant, since the intervening medium between the two protons more closely resembles a hydrocarbon liquid rather than water. In fact, for any dicarboxylic acid one can *define* an effective dielectric constant D_E to fit the experimental value of W_{11} by an equation of the form (4.8.13), with D_E being dependent on the proton–proton distance, the type and size of the acid and the solvent.

In 1959, Eberson (1959, 1992) found that a family of derivatives of succinic acid shows a remarkably large negative cooperativity, i.e., $g_{11} << 1$, which is difficult to explain on the basis of electrostatic theories only. We shall discuss these compounds in Subsection 4.8.6. At present, there is no satisfactory molecular interpretation of these findings. One of the more popular ideas, originally suggested by Jones and Soper (1936) and further elaborated upon by McDaniel and Brown (1953), is that an intramolecular hydrogen bond would facilitate the first dissociation of the proton, i.e., $K_{1,diss}$ becomes smaller (or K_2 becomes larger). Also, the second proton will dissociate with more difficulty. The net effect would be a decrease in the correlation function g_{11}. McCoy (1967) has shown that the values of g_{11} correlate well with the O···O distances, for a series of acids with nearly fixed O···O distance. The stronger the hydrogen bond, the smaller the correlation function, g_{11}, or the larger the value of W_{11}. There are other factors that can possibly contribute to the correlation function, one being specific solvent effects, such as formation of a hydrogen-bond bridge by solvent molecules. We shall mention this possibility in Chapter 9.

4.8.2. Two Identical Sites: Dicarboxylic Acids and Diamines

Table 4.1 shows the values of k_1, k_2 ($= k_{(1/1)}$), the correlation function $g(1, 1)$, and the corresponding free energy $W(1, 1)$, for a series of α,ω-dicarboxylic acids. We first note that all the correlations $g(1, 1) < 1$, i.e., negative cooperativity. As expected, the cooperativity is strongest for oxalic acid for which the proton–proton distance is about 3 Å; the value of $g(1, 1)$ gradually increases toward 1 for the larger molecules. For the largest dicarboxylic acids reported here—the azelaic acid, where seven methylene groups separate the two carboxylic groups—we still find substantial correlation between the two protons. The largest distance is about 10.6 Å. Note that, in the limit of very large *n*, the *second* binding constant should approach the value of *k* for the monocarboxylic acid, $k \approx 7.2 \times 10^4$ (see Section 2.6); the value of k_2 for azelaic acid is 7.12×10^4.

In Fig. 4.24, we plot the experimental free energies $W(1, 1)$ as a function of the proton–proton distance R_{HH}. We also plot the theoretical curve for the Coulombic interaction between the two protons, as modified by the *macroscopic* dielectric constant of water $D = 78.54$. It is clear that the values of $W(1, 1)$ for the larger molecules follow closely the theoretical curve with a fixed value of *D*. Large

Table 4.1
First and Second Intrinsic Binding Constants (in liter/mol), Pair Correlations, and Corresponding Work $W(1, 1)$ (in kcal/mol) for α,ω-Dicarboxylic Acid $COOH(CH_2)_nCOOH$ at 25 °C[a]

Acid	k_1	k_2	$g(1, 1)$	$W(1, 1)$
Oxalic acid ($n = 0$)	7.8×10^4	3.39×10^1	4.34×10^{-3}	3.22
Malonic acid ($n = 1$)	2.46×10^5	1.34×10^4	5.45×10^{-3}	3.09
Succinic acid ($n = 2$)	1.50×10^5	3.12×10^4	2.08×10^{-1}	0.93
Glutaric acid ($n = 3$)	1.31×10^5	4.41×10^4	3.35×10^{-1}	0.65
Adipic acid ($n = 4$)	1.29×10^5	5.23×10^4	4.05×10^{-1}	0.53
Pimelic acid ($n = 5$)	1.33×10^5	6.10×10^4	4.60×10^{-1}	0.46
Suberic acid ($n = 6$)	1.26×10^5	6.51×10^4	5.15×10^{-1}	0.39
Azelaic acid ($n = 7$)	1.30×10^5	7.12×10^4	5.48×10^{-1}	0.36

[a]Based on data from Gane and Ingold (1931).

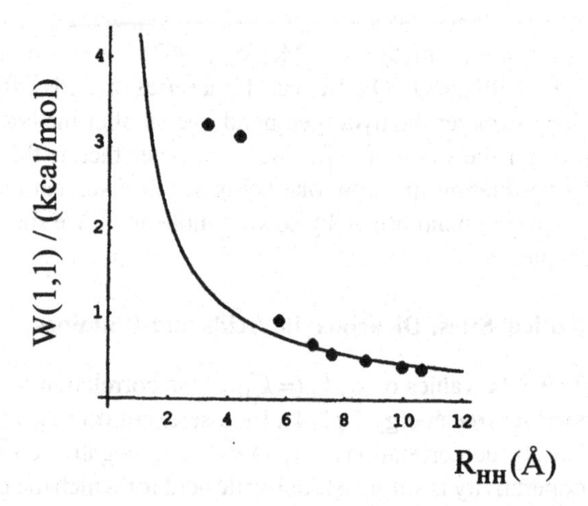

Figure 4.24. Experimental values of $W(1, 1)$, based on data from Gane and Ingold (1931) on α,ω-dicarboxylic acids. The proton–proton distances were calculated for the fully extended linear acids. The full curve is $W(1, 1) = e^2/78.5R_{HH}$ in kcal/mol.

deviations occur for the smaller molecule—which may be interpreted as due to a much smaller effective dielectric constant.

Table 4.2 presents some values of k_1, k_2 $(= k_{1/1})$, $g(1, 1)$, and $W(1, 1)$ for some αω-alkane diamines. Again, we see that as the proton–proton distance (in the di-ionized molecule) increases, the value of $g(1, 1)$ increases toward unity, but even for the 1,8-octane diamine the correlation is still quite large (0.504). In contrast to the case of dicarboxylic acid, the α,ω-diamines cannot be fitted on a Coulombic curve of the form (8.13) with the macroscopic dielectric constant of water. The best one can do is to fit these data to a Coulombic curve with an effective dielectric constant of $D_E \approx 28$. As in the case of dicarboxylic acids, we find here that for the

Table 4.2
First and Second Intrinsic Binding Constants (in liter/mol), Pair Correlations, and Corresponding Work $W(1, 1)$ (in kcal/mol) for α,ω-Alkane Diamines at 20 °C[a]

Diamine	k_1	k_2	$g(1, 1)$	$W(1, 1)$
1,2-Ethane diamine ($n = 2$)	6.15×10^9	2.00×10^7	3.25×10^{-3}	3.34
1,3-Propane daimine ($n = 3$)	2.08×10^{10}	8.73×10^8	4.19×10^{-2}	1.85
1,4-Butane diamine ($n = 4$)	3.15×10^{10}	4.48×10^9	1.42×10^{-1}	1.14
1.5-Pentane diamine ($n = 5$)	5.61×10^{10}	1.10×10^{10}	1.96×10^{-1}	0.95
1.8-Octane diamine ($n = 8$)	5.00×10^{10}	2.52×10^{10}	5.04×10^{-1}	0.40

[a]Based on data from p. 526 of Robinson and Stokes (1959).

long-chain diamine the first binding constant is about 5×10^{10}, very close to the limit of the k values of the normal mono-amines, which is about 4.4×10^{10} (Section 2.6).

4.8.3. Two Different Sites: Amino Acids

The BI of a system with two different sites, a and b, is

$$\theta = \frac{1}{2}\frac{K_1C + 2K_1K_2C^2}{1 + K_1C + K_1K_2C^2}$$

$$= \frac{1}{2}\frac{K_{1,diss}C + 2C^2}{K_{1,diss}K_{2,diss} + K_{1,diss}C + C^2} \tag{4.8.14}$$

where, in the first form, we have used the thermodynamic (or macroscopic) *association* constants (K_1, K_2) and, in the second form, we have used the thermodynamic *dissociation* constants ($K_{1,diss} = K_2^{-1}$, $K_{2,diss} = K_1^{-1}$). It is the latter quantities that are normally reported in the literature. The corresponding *intrinsic binding* (or association) constants k_a, k_b, k_c, and k_d (see diagram in Fig. 4.25) are related to the macroscopic constants by

$$K_1 = K_{2,diss}^{-1} = k_a + k_b, \qquad K_2^{-1} = K_{1,diss} = k_c^{-1} + k_d^{-1} \tag{4.8.15}$$

Note that k_a and k_b are the intrinsic *binding* constants on a and b, respectively; k_c and k_d are the conditional intrinsic binding constants, i.e.,

$$k_c = k_{b/a}, \qquad k_d = k_{a/b} \tag{4.8.16}$$

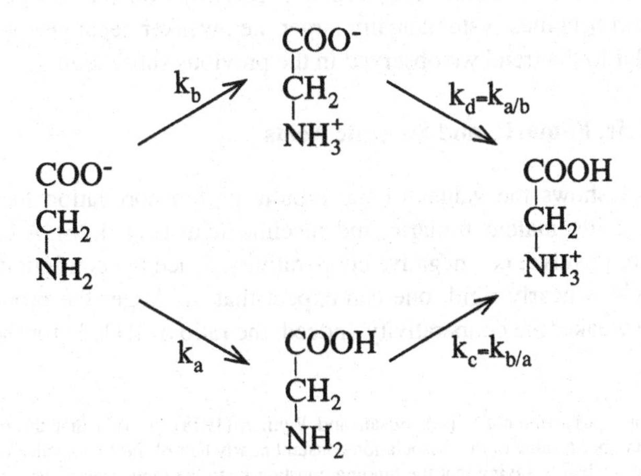

Figure 4.25. The intrinsic binding constants for an amino acid.

The correlation function g_{ab} is given by

$$g_{ab} = \frac{k_{b/a}}{k_b} = \frac{k_c}{k_b} = \frac{k_{a/b}}{k_a} = \frac{k_d}{k_a} \tag{4.8.17}$$

Clearly, from the experimental values of $K_{1,diss}$ and $K_{2,diss}$ one cannot obtain the three independent intrinsic constants k_a, k_b, and k_c ($k_d = k_a k_c / k_b$).

The traditional approximation made to solve for the instrinsic constants is to assume that the effect of the (unionized) carboxylic group on the dissociation of the NH_3^+ group is not much changed upon esterification of the carboxyl group.* If we denote by $k_{E,a}$ the binding (or association) constant to the NH_2 group of the esterified amino acid, then the approximation

$$k_c \cong k_{E,a} = k_{E,diss}^{-1} \tag{4.8.18}$$

is made, where $k_{E,diss}$ is the dissociation constant of the ester, normally reported in the literature. Using this assumption, one can calculate k_a, k_b, and the correlation function from the experimental quantities $K_{1,diss}$, $K_{2,diss}$, and $k_{E,diss}$:

$$k_a = \frac{K_{E,diss}}{K_{1,diss} K_{2,diss}}, \qquad k_b = \frac{K_{1,diss} - k_{E,diss}}{K_{1,diss} K_{2,diss}}, \qquad k_c = k_{E,diss}^{-1}$$

$$g(a, b) = \frac{K_{1,diss} - k_{E,diss}}{K_{1,diss}^{-} K_{2,diss}^{-}} \tag{4.8.19}$$

Table 4.3 shows the values of k_a, k_b, k_c, and $g(a, b)$ for some amino acids. Note that the correlation is always negative. The variation in $g(a, b)$ within the α-amino acids is quite small. When the amino group is displaced away from the carboxylic group, the correlation becomes systematically larger, i.e., weaker negative cooperativity. This is similar to the trend we observed in the previous subsection.

4.8.4. Maleic, Fumaric, and Succinic Acids

Table 4.4 shows the values of the proton–proton correlation for the three dicarboxylic acids: maleic, fumaric, and succinic acids (Fig. 4.26). All the values of $W(1, 1)$ are positive, i.e., negative cooperativity. Since the configuration of the first two acids is nearly rigid, one can expect that the larger the proton–proton distance, the weaker the cooperativity. Indeed, the ratio of $W(1, 1)$ for the first two

*It has been found experimentally [see Edsall and Wyman (1958), p. 485] that the monoester of dicarboxylic acid has a value of the dissociation constant nearly half of the $K_{1,diss}$ value of the dibasic acid. This is equivalent to saying that the binding constant k_E to the monoester is nearly the same as the *second* intrinsic binding constant for the dicarboxylic acid, i.e., $k_E \approx k_{1/1}$.

acids is

$$\frac{W(\text{maleic})}{W(\text{fumaric})} = \begin{cases} 5.04 \text{ (for water at 25 °C)} \\ 4.93 \text{ (for 50\% aqueous ethanol at 20 °C)} \end{cases} \tag{4.8.20}$$

Note that the ratio is almost the same for the two solvents. These ratios are, however, far from the inverse ratio of the proton–proton distances, which is about

$$\frac{R_{HH}(\text{fumaric})}{R_{HH}(\text{maleic})} \approx 1.94 \tag{4.8.21}$$

Thus, one cannot assume a relation of the form

$$W(1, 1) = \frac{e^2}{DR_{HH}} \tag{4.8.22}$$

for the two isomers, with the same dielectric constant. As we shall see below, the ratio of the $W(1, 1)$ values may be easily accounted for by assuming two different dielectric constants.

The case of succinic acid cannot be discussed in terms of Coulombic interactions alone. Here, conformational changes induced by the binding process can contribute significantly to the correlation. Note also that $g(1, 1)$ [or $W(1, 1)$] of succinic acid is not an "average" of the correlations in maleic and fumaric acids. This could be partially due to the configurational changes in the succinic acid, induced by the binding process. We shall discuss below a simple two-state model for succinic acid, and a continuous model in the next subsection.

Consider a simplified two-state model for succinic acid as depicted in Fig. 4.27. In this model we choose the C_α–C_α distance to be 1.54 Å for a single bond and 1.34 Å for a double bond. All the C–C–C bond angles are chosen to be of a tetrahedral angle $\theta_T = 109.47°$. We replace the carboxylate group by a point negative charge placed on the line extending the C_α–C bond at a distance of 1.25 sin(π/3) = 0.625 Å from the carboxylate carbon atom (1.25 Å is the C–O distance in the carboxylate group while the O–C–O angle is 120°). The protons, having a unit positive charge, are placed at a distance of $r_0 = 0.7$ Å from the negative point charge. Only two configurations are allowed for the dihedral angle ϕ, $\phi = 0$ for the cis and $\phi = \pi$ for the trans.

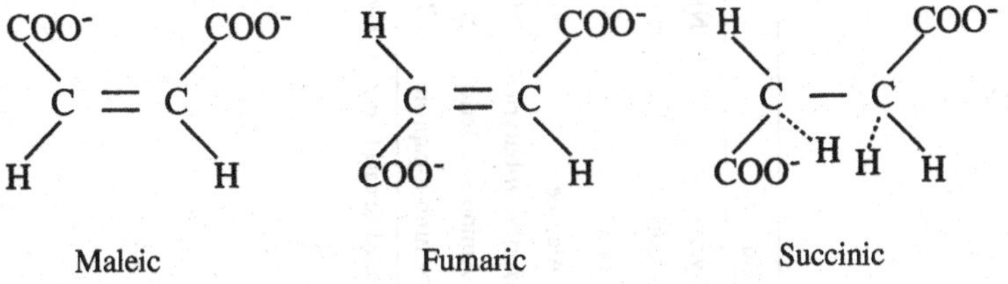

Maleic Fumaric Succinic

Figure 4.26. The structural formulas for maleic, fumaric, and succinic acids.

Table 4.3

Values of k_a, k_b, and k_c, $g(a, b)$ and $w(a, b)$ for Some Linear Amino Acids[a]

Acid	Formula	k_a	k_b	k_c	$g(a, b)$	$w(a, b)$
Glycine	$NH_2–CH_2COOH$	1.95×10^4	5.25×10^9	5.37×10^7	1.02×10^{-2}	2.71
α-Alanine	$CH_3\ CH(NH_2)\ COOH$	2.04×10^4	5.25×10^9	6.31×10^7	1.20×10^{-2}	2.62
Leucine	$(CH_3)_2\ CHCH_2\ CH(NH_2)COOH$	2.24×10^4	4.37×10^9	4.27×10^7	9.77×10^{-3}	2.74
β-Alanine	$NH_2–(CH_2)_2–COOH$	4.57×10^4	1.55×10^{10}	1.35×10^9	8.71×10^{-2}	1.45
γ-Amino-*n*-butyric	$NH_2–(CH_2)_3–COOH$	8.91×10^4	2.69×10^{10}	5.13×10^9	1.91×10^{-1}	0.98
δ-Amino-*n*-valeric	$NH_2–(CH_2)_4–COOH$	7.76×10^4	5.89×10^{10}	1.41×10^{10}	2.40×10^{-1}	0.85
ε-Amino-*n*-caproic	$NH_2–(CH_2)_5–COOH$	6.46×10^4	5.62×10^{10}	2.34×10^{10}	4.17×10^{-1}	0.52

[a]Based on data from p. 452 of Edsall and Wyman (1958).

Table 4.4

Experimental Values of $g(1, 1)$ and $w(1, 1)$ (in kcal/mol) for Maleic, Fumaric, and Succinic Acid. Below Are Some Computed Values for the Model Compounds Described in Section 4.8.4

	In water at 25 °C[a]		In 50% aqueous ethanol at 20 °C[b]	
	$g(1, 1)$	$w(1, 1)$	$g(1, 1)$	$w(1, 1)$
Maleic	1.52×10^{-4}	5.21	1.0×10^{-6}	8.18
Fumaric	1.75×10^{-1}	1.034	6.05×10^{-2}	1.66
Succinic	2.05×10^{-1}	0.938	4.92×10^{-2}	1.78
		Computed values		
"Maleic"	$D_C = 19.6$	$w = 5.225$ ($w^* = 0$)	$D_c = 12.5$	$w = 8.18$ ($w^* = 0$)
"Fumaric"	$D_T = 5.10$	$w = 1.035$ ($w^* = 0$)	$D_T = 31.8$	$w = 1.66$ ($w^* = 0$)
"Succinic"	$D_C = 19.6$	$w = 1.191$ ($w^* = -3.6$)	$D_C = 12.5$	$w = 1.704$ ($w^* = -6.05$)
$R(C_\alpha–C_\alpha) = 1.34$ Å	$D_T = 51.0$		$D_T = 31.8$	
"Succinic"	$D_C = 19.6$	$w = 1.208$ ($w^* = -3.3$)	$D_C = 12.5$	$w = 1.721$ ($w^* = -5.56$)
$R(C_\alpha–C_\alpha) = 1.54$ Å	$D_T = 51.0$		$D_T = 31.8$	

[a]From Eberson (1992).

[b]From Eberson (1959).

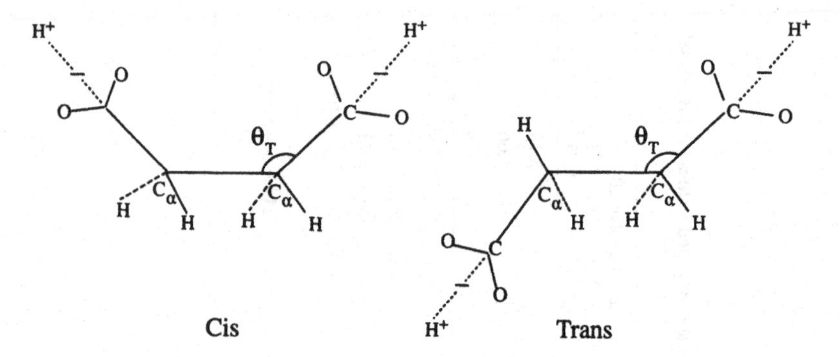

Figure 4.27. A two-state model for succinic acid: Cis with $\phi = 0$ and Trans with $\phi = \pi$.

For the analogues of maleic and fumaric acid, we choose the C_α–C_α distance to be 1.34 Å and fix the angle $\phi = 0$ and $\phi = \pi$, respectively. If it is assumed that for these molecules the proton–proton correlation is purely electrostatic, then we can write

$$W(\text{"maleic"}) = \frac{e^2}{D_C R_C} \tag{4.8.23}$$

and

$$W(\text{"fumaric"}) = \frac{e^2}{D_T R_T} \tag{4.8.24}$$

where $e^2 = 332.833$ kcal mol^{-1} Å; D_C, R_C, and D_T, R_T are the dielectric constant and proton–proton distance for the "maleic" (cis) and "fumaric" (trans) models, respectively.

One can easily adjust the values of the dielectric constants D_C and D_T to obtain the experimental values of W, as in Table 4.4. With a choice of $D_C = 19.6$ and $D_T = 51.0$ for water, and $D_C = 12.5$ and $D_T = 31.8$ for 50% water–ethanol, we obtain the experimental values of W. We now compute the total correlation function for the two-state model for succinic acid. Here the correlation cannot be computed as an average correlation of the two configurations (see Section 4.5). The total correlation of the *equilibrated* two-state model is

$$\begin{aligned} g(1,1) &= \frac{Q(1,1)Q(0,0)}{Q(1,0)Q(0,1)} \\ &= \frac{(S_C + S_T K h^2)(1+K)}{(1+Kh)^2} \\ &= (S_C X_C^* + S_T X_T^*)\,\frac{(1+Kh^2)(1+K)}{(1+Kh)^2} = \langle S \rangle y(1,1) \end{aligned} \tag{4.8.25}$$

where

$$K = \frac{Q_T}{Q_C}, \qquad h = \frac{q_T}{q_c} \qquad \text{and} \qquad X_C^* = \frac{Q_C q_C^2}{Q_C q_C^2 + Q_T q_T^2}, \qquad X_T^* = 1 - X_C^*$$

The quantity $\langle S \rangle$ is an average of the direct correlations S_C and S_T in a hypothetical system, where the mole fractions of the two configurations are X_C^* and X_T^*, respectively (see Section 4.5). As such, $\langle S \rangle$ is bound by S_C and S_T, but $y(1, 1)$ must be larger than unity and is not bound by S_C and S_T. In Table 4.4 we see that both the experimental and calculated values of $W(1, 1)$ are closer to the fumaric rather than the maleic values. One could argue that since the ionized succinic acid would be most of the time in the trans configuration, we should expect that the value of W for the succinic acid be closer to the fumaric acid value. The fact that the W (succinic) is indeed intermediate between W (maleic) and W (fumaric) is quite accidental.† The value of W (succinic) is determined by both the *negative* correlation $\langle S \rangle$ and the positive correlation $y(1, 1)$.

The calculation of W (succinic) was carried out twice, for $R(C_\alpha = C_\alpha) = 1.3$ Å and $R(C_\alpha - C_\alpha) = 1.54$ Å. If the correlation were determined by the electrostatic interaction alone, we should have weakened the interaction upon increasing the proton–proton distance. In Table 4.4, we see that by increasing the $C_\alpha - C_\alpha$ distance (keeping all other parameters fixed) we actually *increase* the value of W. This is clearly due to the relatively large effect of the indirect positive correlation in the equilibrated system. Unfortunately, the relative contribution of the direct and indirect correlations cannot be determined from the experimental data. The computed values of $W^* = -k_B T \ln y(1, 1)$ are shown in brackets in Table 4.4. It is seen that a large negative contribution to $W(1, 1)$ is due to the indirect cooperativity.

4.8.5. A Fully Rotating Electrostatic Model

We extend the model of Section 4.8.4 in two aspects. First, we allow a full rotational degree of freedom about the $C_\alpha - C_\alpha$ bond. Second, we remove the symmetry in the molecule. The $C_\alpha - C_\alpha$ distance is still 1.54 Å and all bond angles are taken as tetrahedral. As in the model of Fig. 4.27, the carboxylate group is replaced by a negative point charge at a distance of 2.16 Å from the C_α atom.‡ For the amine group, we place a point with no charge at a distance of 1.3 Å from the C_α atom. Figure 4.28 shows the new model schematically. The analogues of the following three molecules are considered: (1) the "succinic acid" with $e_a = e_b = -e$, with e the proton charge; (2) the "β-alanine" with $e_a = -e$ and $e_b = 0$; and (3) the "ethane diamine" with $e_a = e_b = 0$.

†This is true for the 50% water aqueous solutions. For water, the value of W (succinic) is even smaller than that of W (fumaric).

‡Similar models were developed by Hill (1943, 1944).

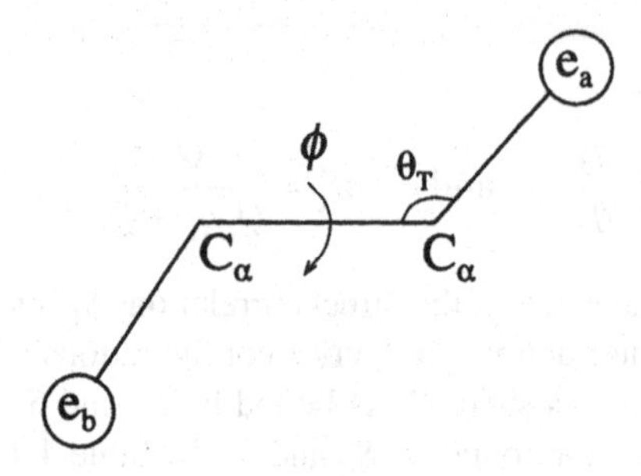

Figure 4.28. The skeleton model for succinic acid, β-alanine, and ethane diamine. The model is essentially the same as that described in Fig. 4.27. Instead of a two-state model, we allow a continuous range of variation, $0 \leq \phi \leq 2\pi$. Also, e_a and e_b can be either negative or zero for a carboxylate or an amine group, respectively.

Clearly, these models are quite far from any of the real molecules. We have chosen to develop these simplified models for two reasons: first, to show how the direct and indirect parts of the correlation are interwoven in such a way that there is no way of factorizing these two parts; second, to prepare the basic building block for the more elaborate model discussed in the next subsection.

The PFs of these molecules are constructed from the following elements:

The internal energy of the molecule is taken as the sum of three terms, all depending on the angle ϕ,

$$E_0(\phi) = U_{LJ}(\phi) + V_{rot}(\phi) + \frac{e_a e_b}{D(\phi) R_{ab}(\phi)} \tag{4.8.26}$$

where $U_{LJ}(\phi)$ is a Lennard-Jones potential function between the two functional groups: the carboxylate with a radius of 1.54 Å and the amine with a radius of 1.36 Å. The internal rotational potential $V_{rot}(\phi)$ is of the form

$$V_{rot}(\phi) = -1 + \cos(3\phi) \tag{4.8.27}$$

giving preference to configurations with $\phi = \pi/3$ and $-\pi/3$, as in butane. This function will be used only in the next subsection. Here, we simply put $V_{rot}(\phi) = 0$. The last term is the charge–charge interaction, depending on the charge–charge distance $R_{ab}(\phi)$ and the varying dielectric constant of the form

$$D(\phi) = \frac{1}{2}\{D_C[1 + \cos(\phi)] + D_T[1 + \cos(\phi)]\} \tag{4.8.28}$$

where D_C and D_T are the two extreme values of D for the cis ($\phi = 0$) and trans ($\phi = \pi$) configurations, respectively.

The binding energies of the proton with charge e_H (= +e) to the two sites are

$$U_a(\phi) = C + \frac{e_H e_a}{r_a} + \frac{e_H e_b}{R_{Hb}(\phi)} \tag{4.8.29}$$

and

$$U_b(\phi) = C + \frac{e_H e_b}{r_b} + \frac{e_H e_a}{R_{Ha}(\phi)} \tag{4.8.30}$$

where C is a constant, which is the nonelectrostatic part of the proton–site interaction; its value does not affect the correlation function. The second term on the rhs of Eqs.(4.8.29) and (4.8.30) is the electrostatic interaction between the proton and the charge on the sites to which it is bound, at distance $r_a = 0.7$ Å and $r_b = 0.1$ Å. The last term is the electrostatic interaction between the proton and the charge on the second site at a distance of $R_{Hb}(\phi)$ or $R_{Ha}(\phi)$. Finally, the proton–proton interaction is taken to be purely electrostatic,

$$U(1,1) = \frac{e_H^2}{D(\phi) R_{HH}(\phi)} \tag{4.8.31}$$

with $D(\phi)$ the varying dielectric constant given in Eq. (4.8.28) and $R_{HH}(\phi)$ the proton–proton distance.

The proton–proton correlation can be calculated from

$$g(a,b) = \frac{Q(a,b)Q(0,0)}{Q(a,0)Q(0,b)}$$

$$= \frac{\int_0^{2\pi} \exp[-\beta E_0(\phi)]d\phi \int_0^{2\pi} \exp[-\beta E_0(\phi) - \beta U(1,1) - \beta U_a(\phi) - \beta U_b(\phi)]d\phi}{\int_0^{2\pi} \exp[-\beta E_0(\phi) - \beta U_a(\phi)]d\phi \int_0^{2\pi} \exp[-\beta E_0(\phi) - \beta U_b(\phi)]d\phi} \tag{4.8.32}$$

where all the integrals are over the entire range of the rotational angle, $0 \leq \phi \leq 2\pi$.

With this description of the model we have adjusted the two dielectric constants $D_C = 55$ and $D_T = 78$ to compute the correlation functions of the three model compounds.

Table 4.5 shows the experimental values of $g(a, b)$ for the three (real) molecules. The major trend reflects both the change in the proton–proton distances (hence the *direct* part of the correlation) and the indirect part of the correlation, depending on the configurational changes induced in the molecule upon binding

Table 4.5
Experimental and Computed Values of Proton–Proton Correlation in Succinic Acid, β-Alanine, and Ethane Diamine and Their Model Analogues

	Experimental	Model calculation	
Succinic acid	0.205	"–,–"	0.201
β-Alanine	0.071	"0,–"	0.111
Ethane diamine	0.0032	"0,0"	0.082

of protons. Table 4.5 also presents the computed values for the three models. The values of the "–,–" model for succinic acid was obtained by fitting the two parameters D_C and D_T. The value of the total correlation decreases from the "–,–" to the "0,–" to the "0,0" models, roughly similar to the trends in succinic acid, β-alanine, and ethane diamine.

As we pointed out above, there are two main reasons for the differences in the correlations in these molecules. To analyze the structural changes induced in the molecule, we define the "cis" mole fraction as the mole fraction of molecules in the range $-\pi/2 \le \phi \le \pi/2$ and the "trans" isomer in the range $\pi/2 \le \phi \le 3/2\pi$. With these definitions we find for the "–,–" model the following mole fractions of the different stages of occupation of the molecule,

$$X_C^{(0)} = 0.25, \quad X_C^{(a)} = X_C^{(b)} = 0.46, \quad X_C^{(2)} = 0.47 \tag{4.8.33}$$

Thus, the empty molecule is mostly in the trans configuration. Binding one proton to either a or b sites (here identical) introduces a considerable change in the mole fraction of the cis form. Binding the second proton causes only a minor change in X_C.

For the "0,–" model, we have

$$X_C^{(0)} = 0.46, \quad X_C^{(a)} = 0.46, \quad X_C^{(b)} = 0.67, \quad X_C^{(2)} = 0.46 \tag{4.8.34}$$

Here, binding to the "acidic" group does not change the distribution. However, binding to the basic group (b) shifts the equilibrium in favor of the cis configuration.

Finally, for the "0,0" model we have

$$X_C^{(0)} = 0.25, \quad X_C^{(a)} = X_C^{(b)} = 0.25, \quad X_C^{(2)} = 0.14 \tag{4.8.35}$$

Here, only when the two protons bind to the molecule do we observe a large shift in favor of the trans configuration.

These configurational changes will affect the total pair correlation in the model compounds. In addition, in the real molecules we have also a large effect due to changes in proton–proton distances (as well as differences in solvent effects and intramolecular hydrogen bonding).

4.8.6. Spurious Cooperativity in Some Alkylated Succinic Acids

Table 4.6 shows some experimental data on α–α' dialkyl succinic acids. The most remarkable finding is that some of these molecules have a very large negative cooperativity, far beyond what could be explained by electrostatic theories. These molecules exist in two isomers—the meso and racemic forms. The latter exists in two optically active enantiomers that are mirror images of each other—only one of these is shown in Fig. 4.29.

Table 4.6 shows that, while the correlation functions of the meso form change within less than one order of magnitude when the alkyl group **R** increases from methyl to tert-butyl, the corresponding values in the racemic series decrease by four to six orders of magnitude for the same variations in the alkyl group.

As we have noted in previous subsections, at present it is not possible to reproduce the experimental values of k_1, k_2, and $g(1, 1)$ for these compounds. The main difficulty is to account for the solvent effects, which cannot be ignored in these molecules. In spite of this limitation we shall see below that the phenomenon of spurious cooperativity can explain the two major observations: first, the decrease in $g(1, 1)$ by five to six orders of magnitude upon increasing the alkyl substituent in the racemic series, and second, that these changes occur only in the racemic and not in the meso form.

The model used for the α–α' dialkyl succinic acid is essentially an extension of the model used in the previous subsection. Here, instead of the two hydrogens on the C_α-carbones we have one alkyl group **R** on each of the C_α carbones, as shown in Fig. 4.30.

Table 4.6
Thermodynamic Dissociation Constants for Alkylated Succinic Acids in 50% Aqueous Ethanol at 20 C[a]

Compound	$pK_{1,diss}$	$pK_{2,diss}$	$g(1, 1)$	$W(1, 1)$
Succinic	5.44	7.35	4.92×10^{-2}	1.78
Racemic α,α'-dimethyl	5.04	8.17	2.97×10^{-3}	3.45
Meso α,α'-dimethyl	4.97	7.58	9.82×10^{-3}	2.74
Racemic α,α'-diethyl	4.76	9.22	1.39×10^{-4}	5.26
Meso α,α'-diethyl	5.37	7.43	3.48×10^{-2}	1.99
Racemic α,α'-diisopropyl	3.66	11.44	6.64×10^{-8}	9.79
Meso α,α'-diisopropyl	5.98	8.10	3.03×10^{-2}	2.07
Racemic α,α'-di-tert-butyl	3.58	13.12	1.15×10^{-9}	12.2
Meso α,α'-di-tert-butyl	6.43	8.29	5.52×10^{-2}	1.72

[a]From Eberson (1959). Values of the racemic isomer are underlined.

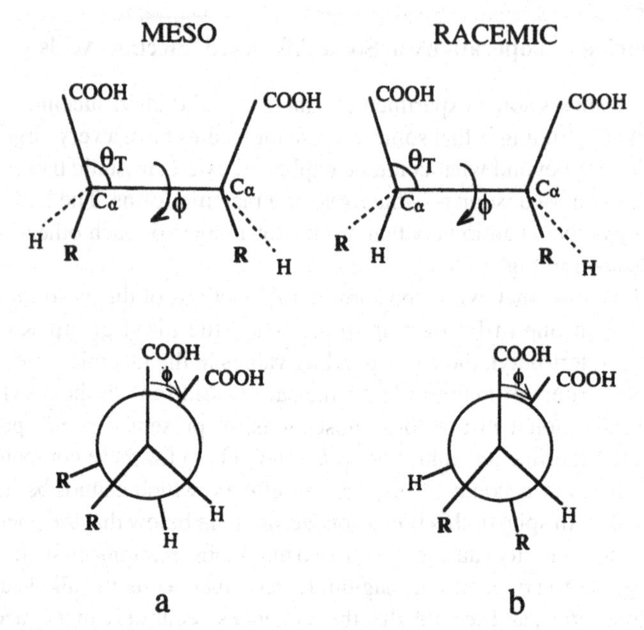

Figure 4.29. The meso (**a**) and racemic (**b**) forms of α–α′ dialkyl (**R**) succinic acid. The racemic form exists in two optically active enantiomers; one is shown on the rhs, the other is the mirror image of this form.

The essential simplification of the model is to replace the alkyl substituents, from methyl to tert-butyl, by Lennard-Jones spheres of increasing diameters. In this model we cannot calculate the exact values of k_1 and k_2 (in fact, these cannot be calculated even if we had the true rotational potential; the main missing information is the solvation Gibbs energies of the molecules involved). Nevertheless, we can demonstrate with this simplified model the two major experimental findings regarding the proton–proton correlation in these series of molecules, as shown in Table 4.6.

The canonical PFs of the fully rotational model are

$$\left.\begin{aligned} Q(0) &= A_w \int_0^{2\pi} \exp[-\beta E_0(\phi) - \beta\Delta\mu_0^*(\phi)]d\phi \\ Q(1) &= 2A_w \int_0^{2\pi} \exp[-\beta E_0(\phi) - \beta U(\phi) - \beta\Delta\mu_1^*(\phi)]d\phi \\ Q(2) &= A_w \int_0^{2\pi} \exp[-\beta E_0(\phi) - \beta 2U(\phi) - \beta U_{HH}(\phi) - \beta\Delta\mu_2^*(\phi)]d\phi \end{aligned}\right\} \quad (4.8.36)$$

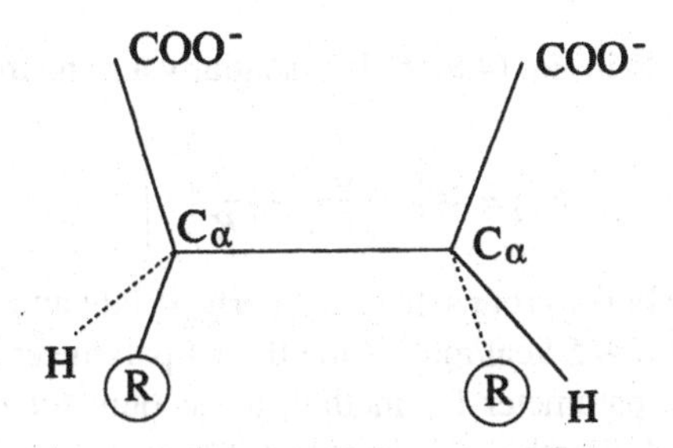

Figure 4.30. Molecular model for dialkyl succinic acid. The C_α–C_α distance is 1.54 Å; the C–C–C angles are tetrahedral, $\theta_T = 109.47$ °. The center of the negative charge was placed at a distance of 1.25/2 = 0.625 Å from the carboxyl carbon atom, and the positive charge at a distance of 0.7 Å from the negative charge. The effective radii for the various alkyl groups were computed from the van der Waals volumes given by Bondi (1968). These are: methyl, 1.75 Å; ethyl, 1.78 Å; isopropyl, 2.0 Å; tert-butyl, 2.2 Å. These radii were used to construct the Lennard-Jones potentials between the various groups [see Eq. (4.8.46)].

where A_w is a factor that depends only on the properties of the pure solvent (internal PF and $N!$ for the pure solvent) and will be cancelled out when computing the binding constants or the correlation function. The quantity $E_0(\phi)$ is essentially the rotational potential energy of the empty molecule, i.e., the doubly ionized acid, as given in Eqs. (4.8.26); $V_{rot}(\phi)$ in Eq. (4.8.26) is the rotational potential energy of ethane (Eliel and Wilen, 1994) and is given by

$$V_{rot}(\phi) = \frac{1}{2} V[1 - \cos(3\phi)] \tag{4.8.37}$$

with $V = -2$ kcal/mol giving preference to the gauche configuration for ethane. The charge–charge Coulombic interaction between the two charges on the carboxylate ions is

$$E_{EL}(\phi) = \frac{e^2}{D(\phi)R(\phi)} \tag{4.8.38}$$

where $R(\phi)$ is the charge–charge distance for each angle ϕ and $D(\phi)$ is an angular-dependent dielectric constant of the form

$$D(\phi) = 0.5\{D_C[1 + \cos(\phi)] + D_T[1 - \cos(\phi)]\} \tag{4.8.39}$$

when $\phi = 0$ (the "cis" configuration for the two carboxylic groups), $D(0) = D_C$; and when $\phi = \pi$ (the "trans" configuration for the two carboxylic groups), $D(\pi) = D_T$. The two extreme values D_C and D_T were fitted to obtain the values of k_1 and k_2 for succinic acid; see below.

The function $U_{LJ}(\phi)$ in Eq. (4.8.26) is essentially a Lennard-Jones potential

$$U_{LJ} = 4\varepsilon\left[-\left(\frac{\sigma}{R}\right)^{6} + \left(\frac{\sigma}{R}\right)^{12}\right] \tag{4.8.40}$$

operating between the bulky groups, here the carboxylate and the alkyl group **R**; ε was taken equal to 0.2945 kcal/mol, while the LJ parameter for methane σ was set initially as the LJ parameter for methyl, to account for the methyl–methyl interaction in the α–α' dimethyl succinic acid. Later, it is gradually increased to simulate the change in the rotational potential (essentially the repulsive part) when proceeding from methyl, to ethyl, isopropyl, and tert-butyl substituents.

The interaction energy $U(\phi)$ between the proton and the empty (di-ionized) acid comprises three parts:

$$U(\phi) = U_0 + U_{HB}(\phi) + U_{EL}(\phi) \tag{4.8.41}$$

where U_0 is the interaction of the proton with the site on which it binds, excluding the electrostatic interaction of the proton with the negative charge on the second site. The latter, $U_{EL}(\phi)$, is angle-dependent and is defined in Eq. (4.8.43) below. In addition, we add an effective hydrogen-bonding contribution $U_{HB}(\phi)$, which takes the value −6 kcal/mol whenever $\phi \approx \pm 60°$ and zero elsewhere. This gives additional stabilization to the gauche configurations. One can assign this additional stabilization to either an intramolecular hydrogen-bonding of the monoprotonated acid, or the specific hydrogen-bonded bridge by a water molecule (Ben-Naim, 1992). Hence we have

$$U_{HB}(\phi) = \begin{cases} -6 & \text{for } 50 \le \phi \le 70 \\ -6 & \text{for } -50 \ge \phi \ge -70 \\ 0 & \text{elsewhere} \end{cases} \tag{4.8.42}$$

and

$$U_{EL}(\phi) = \frac{-e^2}{D(\phi)R_{\pm}(\phi)} \tag{4.8.43}$$

where $R_{\pm}(\phi)$ is the distance between the proton (+) and the negative charge (−) on the second site. The quantity U_0 is independent of the angle ϕ and could be absorbed in the quantity λ_0, and has no effect on the correlation function.

The quantity $\Delta\mu_i^*(\phi)$ is the solvation Gibbs (or Helmholtz) energy for the molecule with occupation number i ($i = 0, 1, 2$) and specific angle ϕ. Although we believe some specific solvent effects might contribute significantly to the correlation function (see also Chapter 9), we did not include solvent effects in the present calculations (apart from the dielectric constants D_C and D_T, as indicated above).

Recall that λ_0 includes the solvation Gibbs energy of the proton, which is not known. Actually, λ_0 is not needed to calculate the correlation function g_{11}. However, to compute k_1 and k_2 one needs a value for λ_0. Therefore, λ_0 together with D_C and D_T were fitted to obtain the values of k_1 and k_2, hence of g_{11} of succinic acid. Once λ_0, D_C, and D_T are fixed, we proceed to compute the relevant quantities for the dialkylated succinic acid.

Finally, we note that in $Q(2)$ we have the proton–proton interaction U_{HH}, which depends on ϕ,

$$U_{HH}(\phi) = \frac{e^2}{D(\phi)R_{HH}(\phi)} \tag{4.8.44}$$

where $R_{HH}(\phi)$ is the proton–proton distance and $D(\phi)$ is the varying dielectric constant, as defined in Eq. (4.8.39).

With these specifications of the parameters we can compute k_1 and k_2 with the aid of Eqs. (4.8.36) and the correlation function g_{11} from

$$g_{11} = g(1, 1) = \frac{k_2}{k_1} = \frac{4K_2}{K_1} = 4\frac{Q(2)Q(0)}{Q(1)]^2} \tag{4.8.45}$$

Using the model described above with dielectric constant $D_T = 34.2$ and $D_C = 12$, and with $\lambda_0 = 3.62 \times 10^5$, we can reproduce the experimental values of k_1 and k_2 for succinic acid. These are shown in Table 4.7. Note that the dielectric constant for the trans form is very close to the macroscopic dielectric constant of 50% mixture of water and ethanol (Harned and Owen, 1958) (for which $D = 49$ at 25 °C; the experimental values of pK_1 and pK_2 reported in Table 4.6 are at 20 °C). As expected, the fitted dielectric constant for the cis configuration is far smaller than the macroscopic dielectric constant. Once we determined D_C, D_T, and λ_0, they were kept fixed for the rest of the calculations. The only varying parameter is now the diameter of the Lennard-Jones sphere that replaces the alkyl groups of varying size. The van der Waals contribution to the rotational potential is computed for each ϕ as the sum

$$U_{vdW} = \sum_{i=1}^{3}\sum_{j=4}^{6} U_{LJ}(i, j) \tag{4.8.46}$$

where U_{LJ} is a Lennard-Jones (LJ) potential (4.8.40) computed for all nine pairs of groups ($i = 1, 2, 3$ on the first C_α-carbon atom and $j = 4, 5, 6$ on the second C_α-carbon atom).

Figure 4.31 shows the total rotational potential for the meso (a) and racemic (b) forms of α–α' "di-tert-butyl" succinic acid, where the tert-butyl is replaced by

Table 4.7
Values of $g_{11} = g(1, 1)$ and $W_{11} = W(1, 1)$ for Succinic Acid and Its Model Derivatives as Described in Section 4.8.6[a]

	Succinic acid		$\sigma = 3.8$ Å		$\sigma = 4.0$ Å		$\sigma = 4.2$ Å	
	Exp	Calc.	Meso	Racemic	Meso	Racemic	Meso	Racemic
k_1	1.119×10^7	1.114×10^7	1.00×10^7	2.76×10^7	7.56×10^6	1.44×10^8	6.73×10^6	9.23×10^8
k_2	5.508×10^5	5.511×10^5	5.56×10^5	6.38×10^5	5.15×10^5	6.75×10^5	4.94×10^5	6.83×10^5
g_{11}	4.92×10^{-2}	4.95×10^{-5}	5.56×10^{-2}	2.31×10^{-2}	6.81×10^{-2}	4.67×10^{-3}	7.35×10^{-2}	7.39×10^{-4}
W_{11}	1.754	1.751	1.683	2.195	1.565	3.125	1.521	4.199

[a]The experimental values in the first column are from Eberson (1959).

a Lennard-Jones sphere of diameter 4.2 Å. The essential difference between the meso and racemic forms is the following. As we increase σ, the van der Waals repulsive part of the interactions between the bulky groups becomes so large as to be insurmountable. In the meso form there exists only *one* such insurmountable barrier, at $\phi = 0$. This is due both to the electrostatic repulsion between the two carboxylate groups and to the van der Waals repulsion between the two alkyl groups which are eclipsed at $\phi = 0$. Thus, although this barrier becomes insurmountable for large σ (or large alkyl groups in the real molecule), most of the rotational angles, say between 20°, to 340°, are still accessible and will contribute to the partition functions $Q(0)$, $Q(1)$, and $Q(2)$. The situation differs dramatically for the racemic form. Here, as we increase σ, *two*, rather than one, insurmountable barriers develop: one, at 120°, due to the van der Waals repulsion between the two alkyl groups, and a second, at 240°, due to the pair of repulsions between the alkyl and carboxylate groups. Clearly, once these two barriers become insurmountable, the molecule will split into two conformational isomers, one including the range −120° to 120° and the second including the range 120° to 240°. (The actual range will, of course, be smaller, but in defining the range of the two isomers we can take the entire range between the maxima. Those angles for which the potential is very high will have very low probability and hence will not contribute to the integrals defining the various binding constants.)

The values shown in Table 4.7 are those computed with the *entire* rotational partition functions. Thus, no matter how high the barriers, as long as we integrate over the *entire* range for ϕ between 0°–360°, we actually allow all possible configurations (with appropriate Boltzmann weights according to the values of the potential function). The main results of these computations are the following. First, we note that the values of g_{11} are almost unchanged when we increase the diameter σ in the meso series. In fact, there is a slight *increase* in g_{11} as we proceed from $\sigma =$ 3.8 Å to $\sigma = 4.2$ Å. This is similar to the general trend observed in the experimental values reported in Table 4.6. Regarding the racemic series, we observe a gradual *decrease* in g_{11} from 2.3×10^{-2} to about 7.4×10^{-4} for $\sigma = 4.2$ Å. With further increments in σ one finds that g_{11} reaches an almost limiting value of about 2.0×10^{-5}. Thus, while the correlation function remains nearly constant as we vary σ for the meso series, the corresponding values for the racemic series decrease by about *three* orders of magnitude for the same change in σ. This is still quite different from the experimental data, where we observe a decrease of almost *six* orders of magnitude, starting with succinic acid and ending with the α–α' di-tert-butyl succinic acid. This discrepancy can be settled by recognizing the occurrence of additional negative spurious cooperativity. The qualitative argument is the following. In all the calculations shown in Table 4.7, the *full* partition functions of the *equilibrated* systems has been used. In the meso series this procedure is justified, since the entire rotational range of 360° is indeed accessible. This is true despite the insurmountable barrier at $\phi = 0°$. The fully equilibrated PF will automatically

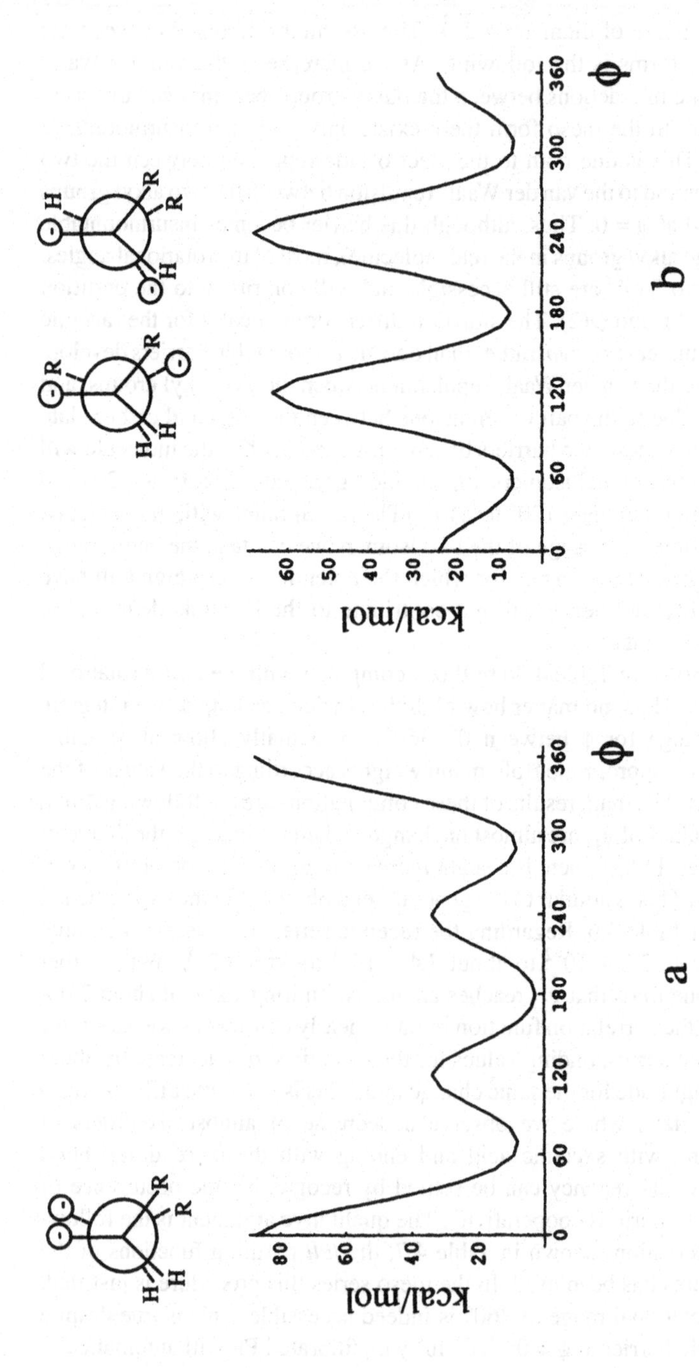

Figure 4.31. The total rotational potential, in kcal/mol, as a function of the dihedral angle φ for (a) the meso form and (b) the racemic form, of α–α' "di-tert-butyl" succinic acid; the tert-butyl is replaced by a Lennard-Jones sphere of diameter 4.2 Å.

give negligible probability to the region around $\phi \approx 0°$. This procedure is, however, invalid for the racemic series. Again, if we use the fully equilibrated PF we allow the entire rotational range. This means that as we perform a titration experiment, we allow the whole range of rotations to *respond* to changes in pH (or λ in our formulation of the theory). For instance, when $\lambda \to 0$, the diacid is fully ionized. The most probable configuration is the trans, $\phi \approx 180°$. On the other hand, when λ is very large, the acid becomes fully protonated, lending a larger probability to the cis–gauche configuration, say $\pm 60°$. Thus, whatever the heights of the rotational barriers, the computation of a binding isotherm or, equivalently, a titration curve (see Appendix G) from the *equilibrated* PF allows the entire distribution of rotational conformers to respond to λ (or to the pH). This is not what happens in practice. As the *two* barriers in the rotational potential become very high, there is a moment when the molecules will "freeze-in" in one of the two isomers, say the trans (ϕ around 180°) and the cis–gauche ($-60° \lesssim \phi \lesssim 60°$). Once this happens, the conformational distribution responds differently to changing λ. It now behaves as a mixture of two isomers, and the correct PF from which we must calculate the titration curve is one of the "frozen-in" PFs, as discussed in Section 4.6. The corresponding binding isotherm now has the form

$$\theta^f = X_L^0 \theta_L + X_H^0 \theta_H \tag{4.8.47}$$

where L and H represent the trans and cis–gauche configurations. Note that when we perform a titration experiment, each of the PFs of L and H allow the transition between all states *within* the corresponding rotational ranges of L and H, but not transition *between* the regions L and H.

The calculation of θ_L and θ_H uses the same integrals as in Eqs. (4.8.36), but only the range of integrations were limited to the regions corresponding to θ_L and θ_H. For instance,

$$\left.\begin{aligned} Q_T(0) = Q_L(0) &= \int_{2\pi/3}^{4\pi/3} [\text{same integrand as in Eq. (4.8.36)}] \\ Q_C(0) = Q_H(0) &= \int_{-2\pi/3}^{2\pi/3} [\text{same integrand as in Eq. (4.8.36)}] \end{aligned}\right\} \tag{4.8.48}$$

and similar integrals for $Q_L(1)$, $Q_H(1)$, $Q_L(2)$, and $Q_H(2)$. (The subscripts T and C correspond to trans and cis isomers.)

Table 4.8 shows the calculated values of k_{1C} and k_{2C} for the cis–gauche and k_{1T} and k_{2T} for the trans isomers, both belonging to the racemic form with $\sigma = 3.8$ and $\sigma = 4.4$ Å. It is clearly seen that both isomers have *genuine* negative coopera-

Table 4.8

Values of $g_{11} = g(1, 1)$ and $W_{11} = W(1, 1)$ for the cis–gauche and trans Isomers for the Model Compound Described in Section 4.8.6, with Two Diameters of the Substituted Alkyl Groups: $\sigma = 3.8$ Å and $\sigma = 4.4$ Å

Racemic $\sigma = 3.8$ Å				Racemic $\sigma = 4.4$ Å			
cis–gauche		trans		cis–gauche		trans	
k_{1C}	3.43×10^{13}	k_{1T}	6.68×10^{6}	k_{1C}	3.70×10^{13}	k_{1T}	6.68×10^{6}
k_{2C}	2.03×10^{10}	k_{2T}	4.93×10^{5}	k_{2C}	2.05×10^{10}	k_{2T}	4.93×10^{5}
g_{11C}	5.53×10^{-4}	g_{11T}	7.38×10^{-2}	g_{11C}	5.53×10^{-4}	g_{11T}	7.38×10^{-2}
W_{11C}	4.328	W_{11T}	1.518	W_{11C}	4.37	W_{11T}	1.518

tivity on the order of 6×10^{-4} for the cis–gauche and 7.38×10^{-2} for the trans isomers. Both values of g_{11} are, however, far larger than the experimental correlations, on the order of 10^{-7}, as cited in Table 4.6.

It was noted at the end of Section 4.6 that when the separation between the two sets of binding constants becomes large, we expect to observe spurious negative cooperativity. This does not need to be related to either k_{2C}/k_{1C} or k_{2T}/k_{1T}, but will be determined roughly by the separation between the average of k_{1C} and k_{2C} on the one hand, and the average of k_{1T} and k_{2T} on the other (see Section 4.6). The computed separation between the pKs is about 3 units on the pH scale for the cis–gauche and about one unit for the trans forms. The separation between the two *pairs* of binding constants is about 6–7 units on the pH scale or, equivalently, about 7.5–9 kcal/mol for W_{11}. This brings us very close to the experimental values reported in Table 4.6 for the bulkier alkyl groups.

The actual values of g_{11} determined experimentally depend, of course, on the method used to extract the effective k_1 and k_2 from the titration curve. Figure 4.32 shows the BI (transformed into $-2\theta + 2$ to conform with the titration curve, see Appendix G) and its derivative, along with the corresponding titration curves for the values of k_{1C}, k_{2C}, k_{1T}, and k_{2C} with $X_C^0 = X_T^0 = 1/2$. It is clear that the BI has four points at which the slope is maximal. Note that the experimental value of X_C^0 is unknown. It depends on the PFs of the cis and trans forms, as well as on the temperature and the pH of the solution in which the compounds are synthesized. Therefore, it is impossible to reproduce the experimental titration curve of the mixture of the two isomers without arbitrary assumptions. Nevertheless, from the plots shown in Fig. 4.32, it is clear that the separation between the average of the two peaks pertaining to the cis–gauche isomer and the average pertaining to the trans isomer is much larger than the separation within each pair. This gives an additional spurious negative cooperativity to the system, which agrees qualitatively with the experimental values reported in Table 4.6.

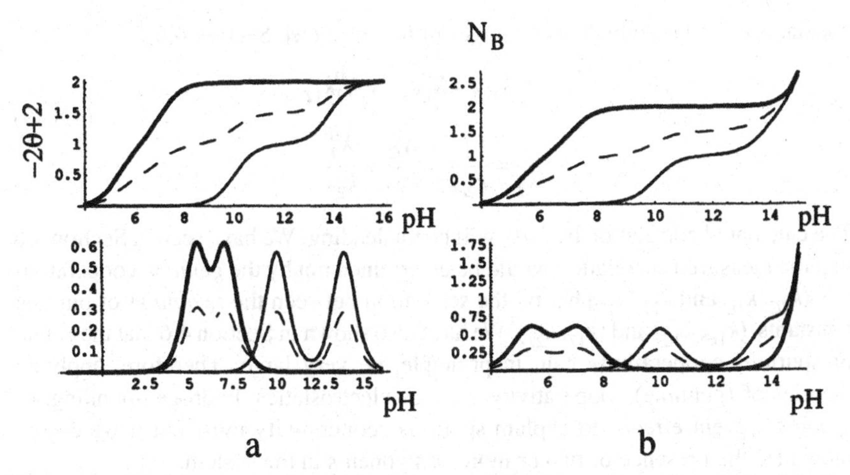

Figure 4.32. Binding isotherms (a) and titration curves (b), and the corresponding slopes of θ_L, θ_H, and θ^f, for values of the binding constants calculated for the racemic form with k values from Table 4.8 (here, L is the trans form and H the cis–gauche form). Note the four peaks in the slope curves, corresponding to the four binding constants k_{1T}, k_{2T}, k_{1C}, and k_{2C}. The binding isotherms are plotted in the form of $-2\theta + 2$ as a function of pH (see Appendix G).

4.8.7. Conclusion

If one looks at a binding isotherm or the equivalent titration curve which shows a well-resolved *pair* of binding constants, one cannot tell whether the system consists of a mixture of single-site molecules or a cooperative double-site molecule. However, if we know in advance that the system consists of single-site molecules, then observing a binding isotherm such as that in Fig. 4.6 will immediately tell us that the system is a mixture, and k_a and k_b belong to *different* single-site molecules. Thus, the occurrence of spurious cooperativity in single-site systems is unlikely to deceive us into thinking that the system is genuinely cooperative.

This is not the case when we know that the molecules in our system have two sites (e.g., succinic acid) and that there exists genuine negative cooperativity (mainly due to electrostatic interactions). Here, observing a BI with *two* widely-resolved binding constants might be deceptive. The computed correlation k_2/k_1 might be interpreted as a genuine cooperativity, although in fact it might be the result of a combination of genuine and spurious cooperativity.

We have shown in Section 4.8.6 that the dialkyl succinic acid in the racemic form might exist as a mixture of two isomers, referred to as cis–gauche and trans. If such a mixture actually exists, then the genuine cooperativities depend on k_{2C}/k_{1C} and k_{2T}/k_{1T} for the two isomers, respectively. However, using standard methods of extracting binding constants, say at the low and high percentage of neutralization regions (Eberson, 1959), one actually measures *effective* binding constants k_1 and k_2, which are averages of the form (see Section 4.6):

$$k_1 = k_{1eff} = X_C^0 k_{1C} + X_T^0 k_{1T}$$
$$\frac{1}{k_2} = \frac{1}{k_{2eff}} = \frac{X_C^0}{k_{2C}} + \frac{X_T^0}{k_{2T}} \tag{4.8.49}$$

The calculated correlation by k_2/k_1 will be misleading. We have seen in Section 4.6 that the measured correlation would be determined not by the *genuine* cooperativities (k_{2C}/k_{1C} and k_{2T}/k_{1T}) but by the separation between the *two pairs* of binding constants, (k_{1C}, k_{2C}) and (k_{1T}, k_{2T}). We have also shown in Section 4.6 that the extent of spurious cooperativity can, in principle, be very large. Therefore, applying theories of (genuine) cooperativity, such as electrostatics, hydrogen-bonding, or specific solvent effects, to explain spurious cooperativity must fail if we do not recognize the presence of two or more components in the system.

In the case of dialkylated succinic acid, we have seen that, due to the occurrence of two barriers in the rotational potential of the racemic form (and not of the meso form) with the bulkier alkyl groups (and not the smaller ones), it is likely that the system will "freeze-in" into a mixture of two components. This is exactly where we observed very large negative cooperativity in the experimental data shown in Table 4.6. One cannot avoid the conclusion that at least a substantial part of the observed cooperativity is spurious.

Finally, we comment on the possibility of the occurrence of spurious cooperativity in biochemical systems. It is well known that strong cooperative macromolecules play a decisive role in regulating many vital processes in biochemical systems. It is also known that a large part of this cooperativity originates from the capacity of the macromolecules (e.g., hemoglobin, Chapter 6, and allosteric enzymes, Chapter 8) to change their conformation in response to ligand binding (e.g., oxygen or a substrate). It therefore seems that the potential for the occurrence of spurious cooperativity already exists in such systems. We can speculate that under certain conditions, such as changes in temperature or pH, the system may switch between *equilibrated* and "frozen-in" mixtures. Such switches can turn on and off the cooperativity of the macromolecule. The net effect would be a regulation of the regulatory mechanism itself.

5

Three-Site Systems: Nonadditivity and Long-Range Correlations

5.1. INTRODUCTION

In this chapter we discuss three-site systems. We extend the three models treated in Chapter 4: direct correlation, indirect correlation mediated through the adsorbent molecule, and indirect correlation mediated by a chain of communicating subunits. Here, we discuss separately two possible structures of the system, a linear and a triangle arrangement of the sites (Fig. 5.1). Two fundamentally new features are discussed in considerable detail: the nonadditivity of the triplet correlation and the possibility of long-range correlations.

5.2. GENERAL FORMULATION OF THE PARTITION FUNCTION

We start by writing the GPF of a single system in various forms:

$$\begin{aligned}\xi &= Q(0) + Q(1)\lambda + Q(2)\lambda^2 + Q(3)\lambda^3 \\ &= Q(0, 0, 0) + [Q(a, 0, 0) + Q(0, b, 0) + Q(0, 0, c)]\lambda \\ &\quad + [Q(a, b, 0) + Q(0, b, c) + Q(a, 0, c)]\lambda^2 + Q(a, b, c)\lambda^3 \\ &= Q(0, 0, 0) + 3Q_s(1) + [Q(1, 1, 0) + Q(0, 1, 1) + Q(1, 0, 1)]\lambda^2 + Q(1, 1, 1)\lambda^3 \\ &= Q(0, 0, 0) + 3Q_s(1) + 3Q_s(2)\lambda^2 + Q(3)\lambda^3 \qquad (5.2.1)\end{aligned}$$

In the first equality the GPF is written in the most general form (for the case $m = 3$). Here, $Q(l)$ is the (canonical) PF for a system with l ligands. The second form is used whenever the sites are all *different*, here denoted by a, b, and c. Clearly, in this case

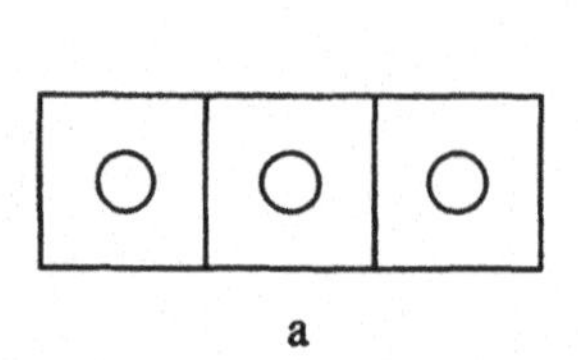

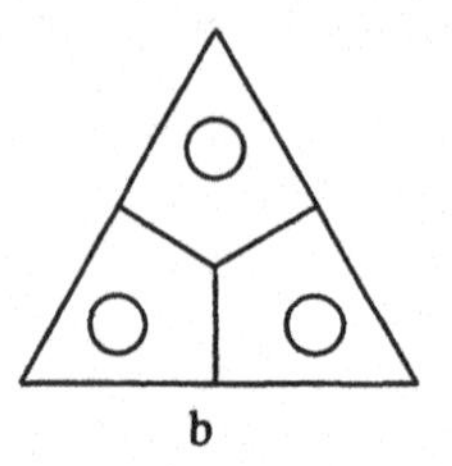

Figure 5.1. (a) Three linearly arranged sites. When the sites are chemically identical and there is no conformational change induced by ligands, the three sites are identical in a weak sense, i.e., there is only one intrinsic binding constant but two pair correlations. (b) Three identical sites arranged in an equilateral triangle. Due to symmetry of the system, there is only one binding constant and only one pair correlation. The three sites are identical in a strict sense.

there are three different PFs for singly occupied systems, denoted by $(a, 0, 0)$, $(0, b, 0)$, and $(0, 0, c)$ and three different PFs for doubly occupied systems. The third form is used when the three sites are *identical* in a *weak sense*. By identical in a weak sense we mean that the canonical PF of the singly occupied system is independent of the specific site on which the ligand is bound. Thus, instead of writing the *specific* PFs, $Q(1, 0, 0)$, $Q(0, 1, 0)$, and $Q(0, 0, 1)$, which have the same value, denoted by $Q_s(1)$, we simply take three times this quantity. The PFs for doubly occupied systems are assumed to be different for the different specific arrangements. A simple example of a system with three identical sites in the weak sense is a linear arrangement of three chemically identical sites with direct correlation only; see Section 5.3. The only requirement here is that the intrinsic binding constant on the first site be the same for any specific sites. Usually, this requires that the binding energy be the same for each of the sites.*

We stress that *chemically* identical sites do not guarantee the identity of the first binding constants. For example, benzene-1,2,3- or 1,2,4-tricarboxylic acid (Section 5.9) will have, in general, two or three different intrinsic binding constants, although the interaction of the proton with the carboxylate group itself is almost the same. The difference in k_1 will arise because of the electrostatic interaction between the proton and the charges on the other carboxylate groups.†

*Note that we always assume that the ligand is very small relative to the adsorbent molecule. Otherwise, even when the binding energy is the same for each site, the PFs for singly occupied molecules could differ, e.g., the rotational PF of a molecule having a ligand bound to the center or to the edge will be different. See also Section 5.10 and Appendix B.

†This is sometimes referred to as a perturbation effect. In this book we shall not use this term. In general, we shall include in the binding free energy the interactions of the ligand at site j with the *entire* molecule. See also Section 2.2 and Appendix I.

The last form on the rhs of Eq. (5.2.1) is used whenever the three sites are identical in a *strict* sense.* This is defined whenever the PF for *any specific* arrangement of l ligands has the same value, denoted by $Q_s(l)$. Hence, we simply write $3Q_s(1)$ and $3Q_s(2)$ without specifying the specific arrangement of the ligands.

An example that conforms to this definition would be benzene-1,3,5-tricarboxylic acid (Section 5.9). Clearly, owing to the symmetry of the molecule there is only one intrinsic binding constant k_1, and only one intrinsic binding constant for pairs k_{11} or, equivalently, only one pair correlation function g_{11}.†

The binding constants and correlation functions are defined as in Chapter 4. For instance,

$$k_a = \frac{Q(a, 0, 0)}{Q(0, 0, 0)} \lambda_0 \tag{5.2.2}$$

$$k_{ab} = \frac{Q(a, b, 0)}{Q(0, 0, 0)} \lambda_0^2 \tag{5.2.3}$$

$$g_{ab} = \frac{k_{ab}}{k_a k_b} = \frac{Q(a, b, 0)Q(0, 0, 0)}{Q(a, 0, 0)Q(0, b, 0)} \tag{5.2.4}$$

$$g_{abc} = \frac{k_{abc}}{k_a k_b k_c} = \frac{Q(a, b, c)Q(0, 0, 0)^2}{Q(a, 0, 0)Q(0, b, 0)Q(0, 0, c)} \tag{5.2.5}$$

Specific examples will be discussed in subsequent sections.

5.3. DIRECT INTERACTION ONLY

This is an extension of the model discussed in Section 4.3. The assumption is made that the binding of a ligand does not affect the state of the adsorbent molecule, hence all correlations are due to *direct* ligand–ligand interaction. For ligand–ligand interaction we usually assume pairwise additivity, i.e.,

$$U(a, b, c) = U(a, b, 0) + U(0, b, c) + U(a, 0, c) \tag{5.3.1}$$

*This is sometimes referred to as the "identical-symmetrical" case. Symmetry in itself is not enough to distinguish between a weak and strict sense. (Both linear and triangle, and similarly square and tetrahedral, models are symmetric.) Perhaps the requirement of the arrangement with highest symmetry better characterizes the identity in the strict sense.

†Note, however, that k_1 depends on the interaction of the proton with the other two charges on the two carboxylate groups. Hence, it is different from k_1 of benzoic acid; see Section 5.9. Similarly, k_{11} depends on the interaction with the third carboxylate group—hence it differs from k_{11} for 1,3-dibenzoic acid.

where $U(a, b, c)$ is the change in energy associated with the process of bringing the three ligands to the same configuration as on sites a, b, and c, but in vacuum. A similar meaning applies to $U(a, b, 0)$ for two ligands at a and b. When the sites are identical in the strict sense, we have

$$U(1, 1, 1) = 3U(1, 1) \tag{5.3.2}$$

where $U(1, 1, 1)$ and $U(1, 1)$ are the triplet and pair interactions, respectively. For the linear system of three sites, one usually assumes that, due to the short range of the ligand–ligand interaction, the interaction between ligands occupying nonneighboring sites, say a and c, is negligible. Therefore we write

$$U(a, b, c) \cong U(a, b) + U(b, c) \tag{5.3.3}$$

The BI for the general case with three different sites is

$$\begin{aligned}\theta &= \frac{1}{3}\lambda\frac{\partial \ln \xi}{\partial \lambda} \\ &= \frac{1}{3}\frac{(k_a + k_b + k_c)C + 2(k_a k_b S_{ab} + k_a k_c S_{ac} + k_b k_c S_{bc})C^2 + 3k_a k_b k_c S_{abc}C^3}{1 + (k_a + k_b + k_c)C + (k_a k_b S_{ab} + k_a k_c S_{ac} + k_b k_c S_{bc})C^2 + k_a k_b k_c S_{abc}C^3} \\ &= \frac{kC + 2k^2SC^2 + k^3S^3C^3}{1 + 3kC + 3k^2SC^2 + k^3S^3C^3}\end{aligned} \tag{5.3.4}$$

where, in the second form on the rhs of Eq. (5.3.4), we put $k_a = k_b = k_c = k$ and $S_{ab} = S_{ac} = S_{bc} = S$, $S_{abc} = S^3$. This is the most common BI for three identical sites with pairwise additive ligand–ligand interaction. Clearly, when $S = 1$ the BI reduces to a simple Langmuir isotherm.

In the general case there are three different correlations in this system. Presuming additivity [see Eq. (5.3.1)], we write

$$S_{abc} = S_{ab}S_{ac}S_{bc} \tag{5.3.5}$$

When the sites are identical in the strict sense,

$$S_{ab} = S_{ac} = S_{bc} = S, \; S_{abc} = S^3 \tag{5.3.6}$$

When the sites are identical and arranged linearly, so that long-range correlations may be neglected,

$$S_{ab} = S_{bc} = S, \; S_{ac} \cong 1, \; S_{abc} \cong S^2 \tag{5.3.7}$$

In all these cases the triplet correlation is expressible in terms of the pair correlations and the temperature dependence of the correlation is predictable, knowing the ligand–ligand interactions. This is, in general, not the case for systems with indirect correlations, discussed in the following sections.

5.4. THREE STRICTLY IDENTICAL SITES: NONADDITIVITY OF THE TRIPLET CORRELATION

We extend here the model of Section 4.5. The adsorbent molecule can be in one of two conformations, L and H (Fig. 5.2). For simplicity, and in order to highlight the new features of this model, we assume that the *direct* interactions are additive [see Eq. (5.3.1)] and are also independent of the conformation of **P** (i.e., the ligand–ligand distances are invariant under conformational changes). Hence

$$U(1, 1, 1) = 3U(1, 1) \tag{5.4.1}$$

The coefficients $Q(l)$ of the GPF in Eq. (5.2.1) are, in this case,

$$\left.\begin{aligned} Q(0) &= Q_L + Q_H \\ Q_s(1) &= Q_L q_L + Q_H q_H \\ Q_s(2) &= (Q_L q_L^2 + Q_H q_H^2)S \\ Q(3) &= (Q_L q_L^3 + Q_H q_H^3)S^3 \end{aligned}\right\} \tag{5.4.2}$$

Note that $Q_s(l)$ refers to l *specific* sites being occupied. The BI is

$$\theta = \frac{1}{3}\lambda\frac{\partial \ln \xi}{\partial \lambda}$$

$$= \frac{(Q_L q_L + Q_H q_H)\lambda + 2(Q_L q_L^2 + Q_H q_H^2)S\lambda^2 + (Q_L q_L^3 + Q_H q_H^3)S^3\lambda^3}{(Q_L + Q_H) + 3(Q_L q_L + Q_H q_H)\lambda + 3(Q_L q_L^2 + Q_H q_H^2)S\lambda^2 + (Q_L q_L^3 + Q_H q_H^3)S^3\lambda^3}$$

$$= \frac{k_1 C + 2k_{11}C^2 + k_{111}C^3}{1 + 3k_1 C + 3k_{11}C^2 + k_{111}C^3} \tag{5.4.3}$$

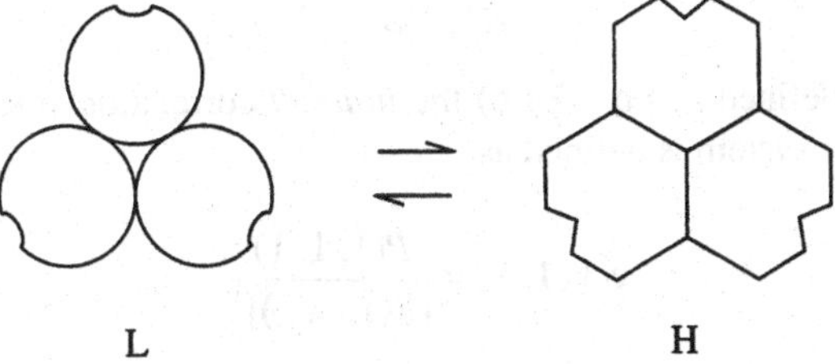

Figure 5.2. Two-state model for the adsorbent molecule. The three sites are identical in a strict sense, and the ligand–ligand interaction is assumed to be pairwise additive and independent of the conformational state. Note that the sites differ for the two states L and H.

where

$$\left.\begin{aligned} k_1 &= \frac{Q_L q_L + Q_H q_H}{Q_L + Q_H} \lambda_0 = (X_L^0 q_L + X_H^0 q_H)\lambda_0 \\ k_{11} &= \frac{Q_L q_L^2 + Q_H q_H^2}{Q_L + Q_H} \lambda_0^2 = (X_L^0 q_L^2 + X_H^0 q_H^2)\lambda_0^2 \\ k_{111} &= \frac{Q_L q_L^3 + Q_H q_H^3}{Q_L + Q_H} \lambda_0^3 = (X_L^0 q_L^3 + X_H^0 q_H^3)\lambda_0^3 \end{aligned}\right\} \tag{5.4.4}$$

Note that these are the intrinsic binding constants, i.e., they pertain to *specific* one, two, and three sites.

The pair correlation is defined, in general, as a ratio of the probabilities,*

$$g(1, 1) = \frac{P(1, 1, \_)}{P(1, \_, \_)^2} \tag{5.4.5}$$

where $P(1, 1, \_)$ denotes the probability of finding two specific sites occupied while the state of occupation of the third site is unspecified, i.e, it could be either empty or occupied. Likewise, $P(1, \_, \_)$ refers to the probability of finding one specific site occupied and the other two unspecified. This correlation is, in general, dependent on the ligand activity (see also Section 4.2). We shall only need the $\lambda \to 0$ limit of this correlation, which is given by

$$\begin{aligned} g(1, 1) &= \frac{Q(1, 1, 0)Q(0, 0, 0)}{[Q(1, 0, 0)]^2} \\ &= \frac{k_{11}}{k_1^2} = \frac{(X_L^0 q_L^2 + X_H^0 q_H^2)}{(X_L^0 q_L + X_H^0 q_H)^2} S \\ &= y(1, 1)S \end{aligned} \tag{5.4.6}$$

where we have defined in Eq. (5.4.6) the *indirect* correlation $y(1, 1)$. The triplet correlation in our system is defined as

$$g(1, 1, 1) = \frac{P(1, 1, 1)}{[P(1, \_, \_)]^3} \tag{5.4.7}$$

*There are other possibilities for defining pair correlations, such as $g(1, 1/0)$ or $g(1, 1/1)$, where the third site is *required* to be empty or occupied, respectively. That defined in Eq. (5.4.5) leaves the occupation state of the third site unspecified.

and we need only the $\lambda \to 0$ limit of this correlation, which we also denote by $g(1, 1, 1)$,

$$g(1,1,1) = \frac{Q(1,1,1)Q(0,0,0)^2}{[Q(1,0,0)]^3} = \frac{k_{111}}{k_1^3}$$

$$= \frac{X_L^0 q_L^3 + X_H^0 q_H^3}{(X_L^0 q_L + X_H^0 q_H)^3} S^3 = y(1,1,1)S^3 \tag{5.4.8}$$

where we have introduced the *indirect* part of the triplet correlation $y(1, 1, 1)$.

In terms of the correlation functions we write the BI as

$$\theta = \frac{k_1 C + 2k_1^2 g(1,1)C^2 + k_1^3 g(1,1,1)C^3}{1 + 3k_1 C + 3k_1^2 g(1,1)C^2 + k_1^3 g(1,1,1)C^3} \tag{5.4.9}$$

There are several important properties of the indirect correlation that are distinctly different from the direct correlation. First, the sign of the direct correlation depends only on the pair interaction $U(1, 1)$. Thus, $S \gtrless 1$ whenever $U(1, 1) \lessgtr 0$. On the other hand, the indirect correlations, both $y(1, 1)$ and $y(1, 1, 1)$, are always positive (i.e., larger than unity). This can be immediately seen by expressing the indirect correlations in terms of the parameters

$$K = Q_H/Q_L, \qquad h = q_H/q_L \tag{5.4.10}$$

$$y(1,1) = \frac{(1+Kh^2)(1+K)}{(1+Kh)^2} = 1 + \frac{K(h-1)^2}{(1+Kh)^2} \tag{5.4.11}$$

$$y(1,1,1) = \frac{(1+Kh^3)(1+K)^2}{(1+Kh)^3} = 1 + \frac{K(h-1)^2(2Kh+K+h+2)}{(1+Kh)^3} \tag{5.4.12}$$

Note that "either $K = 0$ or $h = 1$" is a necessary and sufficient condition for $y(1, 1) = 1$, as well as for $y(1, 1, 1) = 1$. It also follows that whenever $y(1, 1) > 1$, $g(1, 1, 1) > 1$ as well, and vice versa. As with the pair correlation function (see Section 4.5), the indirect triplet correlation goes through a maximum as a function of K (Fig. 5.3). The values of $y(1, 1, 1)$ are unbounded, e.g., taking $K = h^{-1}$ and letting $h \to 0$.

We have seen in Section 4.5 that the conditions for having indirect pair and triplet correlations are the same as those for conformational changes induced by the binding process. As in Sections 3.4 and 4.5, the change in the mole fraction of the L form is the same also in this model, i.e.,

$$d_L^{(1)} = \frac{K(1-h)}{(1+Kh)(1+K)} \tag{5.4.13}$$

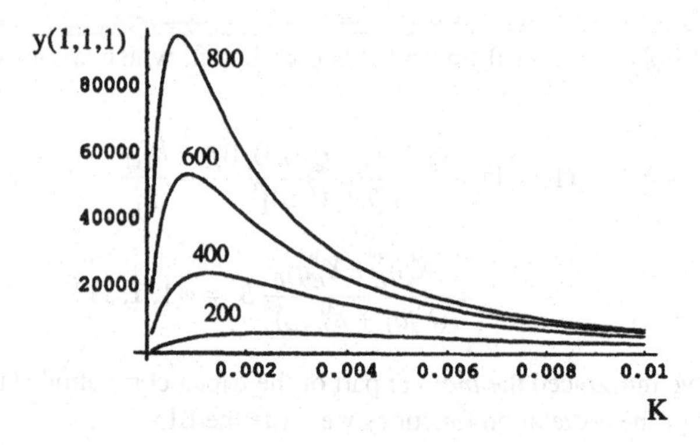

Figure 5.3. The triplet indirect correlation $y(1, 1, 1)$ as a function of K for various values of h, indicated next to each curve.

The changes induced by the second and third ligands, denoted, by $d_L^{(2)}$ and $d_L^{(3)}$, are obtained by replacing K in Eq. (5.4.13) by Kh and Kh^2, respectively. The reason is that K, Kh, and Kh^2 are the equilibrium constants for the reaction $L \rightleftharpoons H$ on the arrival of the first, second, and third ligand, respectively.

Second, perhaps the most important property of $y(1, 1, 1)$ is that it is nonadditive in the sense

$$y(1, 1, 1) \neq y(1, 1)^3 \tag{5.4.14}$$

In fact, there is a relationship between $y(1, 1, 1)$ and $y(1, 1)$, but it is not even approximately similar to the additivity of the kind in Eq. (5.3.6). For any given K, we eliminate h from Eqs. (5.4.11) and (5.4.12) to obtain the relationship

$$y(1, 1, 1) = 3y(1, 1) - 2 \pm \frac{[y(1, 1) - 1]^{3/2}(K - 1)}{\sqrt{K}} \tag{5.4.15}$$

Figure 5.4 shows $y(1, 1, 1)$ as a function of $y(1, 1)$ for the exact relation in Eq. (5.4.15) (with the minus sign and $K = 10^{-3}$) and for the approximate case of additivity $y(1, 1)^3$. A particularly simple form is obtained for $K = 1$, i.e., $Q_H = Q_L$ (which is the case of maximum responsiveness of the system). In this case we have

$$y(1, 1, 1) = 3y(1, 1) - 2 \tag{5.4.16}$$

Clearly, both Eqs. (5.4.15) and (5.4.16) do not indicate any additivity of the kind in Eq. (5.3.6).

Finally, the temperature dependence of S^3 is determined by U. It increases monotonically with T for $U > 0$, and decreases monotonically with T for $U < 0$. On

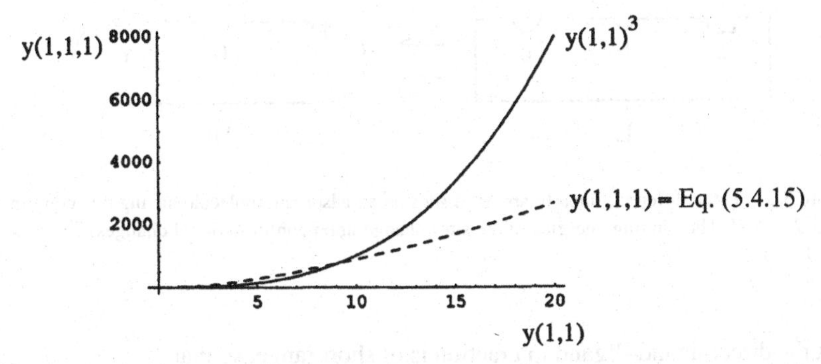

Figure 5.4. The triplet correlation as a function of the pair correlation for the approximate additive case $y(1, 1, 1) = y(1, 1)^3$ and for the exact case given by Eq. (5.4.15) (with the minus sign and $K = 10^{-3}$).

the other hand, $y(1, 1, 1)$ may change sign as a function of T. Both S^3 and $y(1, 1, 1)$ must eventually tend to unity for $T \to \infty$. Figure 5.5 demonstrates the typical temperature dependence of the triplet, direct and indirect correlations.

5.5. THREE DIFFERENT, LINEARLY ARRANGED SITES: LONG-RANGE CORRELATIONS

The model treated in this section is essentially the same as in the previous section, except that the sites are all different and are arranged linearly (Fig. 5.6). In this particular model we distinguish between nearest-neighbor (*nn*) sites, such as a and b, or b and c, and second-nearest-neighbor (*sn*) sites, here the sites a and c.

As in Section 5.3, we assume for simplicity that the distance between any two *nn* sites is the same, independent of the conformational state of **P**. Also, we assume

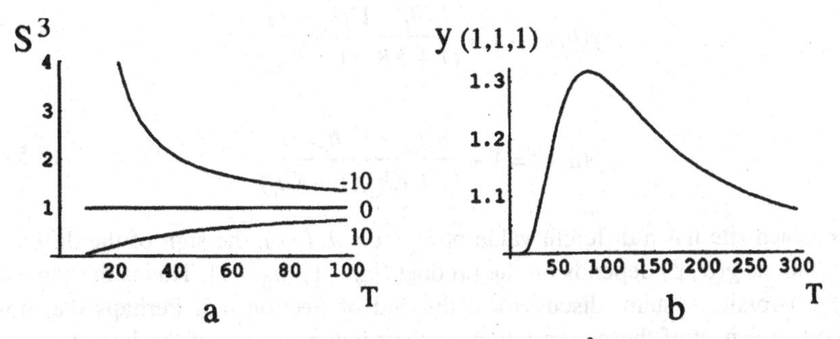

Figure 5.5. Temperature dependence of (a) the direct correlation S^3 with three different values of $U(1, 1)$, as indicated, and (b) the indirect correlation, with $E_H - E_L = -100$ and $U_H - U_L = -100$ and $k_B = 1$.

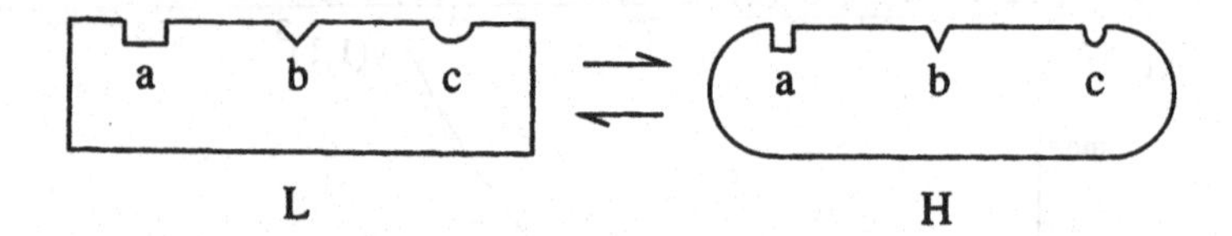

Figure 5.6. Three different, linearly arranged sites on an adsorbent molecule having two conformations, L and H. The binding energies to the sites change upon conformational changes, i.e., $h_a \neq 1$, $h_b \neq 1$, $h_c \neq 1$.

that the direct ligand–ligand interaction is of short range, so that

$$S_{ab} \sim S_{bc} = S, \quad S_{ac} \approx 1, \quad S_{abc} \approx S_{ab} S_{bc} \tag{5.5.1}$$

There are three pair correlations in this system, namely $g(a, b)$, $g(b, c)$, and $g(a, c)$, defined by

$$g(a, b) = \frac{Q(a, b, 0)Q(0, 0, 0)}{Q(a, 0, 0)Q(0, b, 0)} = y(a, b)S \tag{5.5.2}$$

and similar definitions for $g(a, c)$ and $g(b, c)$. We define the following constants,

$$K = \frac{Q_H}{Q_L}, \quad h_a = \frac{q_{Ha}}{q_{La}}, \quad h_b = \frac{q_{Hb}}{q_{Lb}}, \quad h_c = \frac{q_{Hc}}{q_{Lc}} \tag{5.5.3}$$

where h_α is defined as h in Section 5.3, but for each different site. We obtain the following expressions for the *indirect* pair correlations,

$$y(a, b) = \frac{(1 + Kh_a h_b)(1 + K)}{(1 + Kh_a)(1 + Kh_b)} = 1 + \frac{K(h_a - 1)(h_b - 1)}{(1 + Kh_a)(1 + Kh_b)} \tag{5.5.4}$$

$$y(b, c) = 1 + \frac{K(h_b - 1)(h_c - 1)}{(1 + Kh_b)(1 + Kh_c)} \tag{5.5.5}$$

$$y(a, c) = 1 + \frac{K(h_a - 1)(h_c - 1)}{(1 + Kh_a)(1 + Kh_c)} \tag{5.5.6}$$

Since each site has a different value of h_α ($\alpha = a, b, c$), the sign of the different correlations $g(\alpha, \beta)$ depends on the product $(h_\alpha - 1)$ $(h_\beta - 1)$. This is the same as in the two-site system, discussed at the end of Section 4.5. Perhaps the most important aspect of these correlations is their independence of the ligand–ligand distance. This is true for $y(a, b)$ and $y(b, c)$, as well as for $y(a, c)$. In fact, when the sites are identical in the weak sense, then $h_a = h_b = h_c = h$. In this case all the indirect

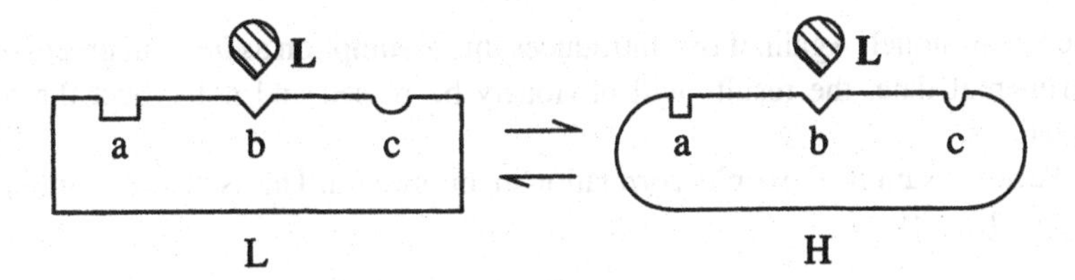

Figure 5.7. Same as Fig. 5.6, but site b is unchanged upon conformational changes, i.e., $h_a \neq 1$, $h_c \neq 1$ but $h_b = 1$.

correlations become identical. The direct correlations [see Eq. (5.5.1)] are not necessarily equal.

Since the extent of the indirect correlations depends on the parameters K and h_α, and not on the ligand–ligand distance the "long-range" correlation $y(a, c)$ could be even larger than the "short-range" correlations $y(a, b)$ and $y(b, c)$. As an extreme example, consider the case when $K = 10^{-3}$, $h_a = h_c = 10^3$, but $h_b = 1$. A schematic illustration of this case in shown in Fig. 5.7. Clearly, since $h_b = 1$, the binding of ligands on site b cannot induce conformational changes in **P**, hence there is no indirect correlation between site b and any other site. On the other hand, binding on either site a or c induces very large conformational changes. Therefore, in this case

$$y(a, b) = y(b, c) = 1, \quad y(a, c) \sim 250 \tag{5.5.7}$$

We conclude that the "long-range" correlation could be large even in the absence of "short-range" correlations. Such a situation is usually impossible for the direct correlation. This example, although extreme, should serve as a warning signal when processing experimental data. Based on the normal behavior of the *direct* correlation Eq. (5.5.1), one tends to expect similar behavior from the total correlation, and hence assumes that $g(a, b) \neq g(b, c) \neq 1$, but $g(a, c) \approx 1$. If these assumptions are *used* in the determination of the correlations from the experimental data, one could miss the "long-range" correlation, simply because $g(a, c) = 1$ is introduced as an *input* in processing the experimental data. A numerical example is shown in Appendix H.

A second warning signal, which applies to any three (or more) site systems, is the nonadditivity of the triplet correlations. Again, based on the normal behavior of the direct correlations in Eq. (5.4.1), one expects similar behavior from the total correlation, and writes

$$g(a, b, c) \sim g(a, b)g(a, c)g(b, c) \tag{5.5.8}$$

which reduces to

$$g(a, b, c) \sim g(a, b)g(b, c) \tag{5.5.9}$$

in the linear model. Again, if one introduces this assumption as *input* in processing experimental data, the results will obviously be biased and not reflect the true correlations.

We next examine the triplet correlation in this system. This is defined, similarly to Eq. (5.4.7), by

$$g(a, b, c) = \frac{P(a, b, c)}{P(a, \_, \_)P(\_, b, \_)P(\_, \_, c)} \tag{5.5.10}$$

and in terms of parameters (5.5.3) we have for the ($\lambda \to 0$ limit) indirect triplet correlation

$$y(a, b, c) = \frac{(1 + Kh_a h_b h_c)(1 + K)^2}{(1 + Kh_a)(1 + Kh_b)(1 + Kh_c)} \tag{5.5.11}$$

which is the generalization of Eq. (5.4.12) for the case of three different sites.

As shown in Section 5.4, $y(a, b, c)$ is not factorized into a product of the form (5.5.8) or (5.5.9). In fact, generalization of Eq. (5.4.15) yields

$$y(a, b, c) = y(a, b) + y(b, c) + y(a, c) - 2 \pm \frac{(K - 1)[y(a, b) - 1]^{1/2}[y(a, c) - 1]^{1/2}[y(b, c) - 1]^{1/2}}{\sqrt{K}} \tag{5.5.12}$$

The minus sign is correct for $1 \le K \le 1$. Thus, for any given K there exists a relationship between the triplet correlation and all the pair correlations, and it is not simple like Eq. (5.5.8) or (5.5.9). For $K = 1$, this reduces to the simple relationship

$$y(a, b, c) = y(a, b) + y(b, c) + y(a, c) - 2 \tag{5.5.13}$$

Note that when the sites become identical, i.e., $a = b = c$, hence $h_a = h_b = h_b = h$, then, if there is no direct correlation, the sites are identical in the strict sense, i.e., there is only one binding constant and one pair correlation. If, on the other hand, there are also direct correlations, then the three sites become identical only in a weak sense, i.e., there is only one binding constant, but two pair correlations g_{nn} and g_{sn}. It is also easy to generalize the result from Section 3.4 regarding the extent of induced structural changes by the first, second, and third ligands, i.e.,

$$\left.\begin{aligned} d_L^{(1)}(h, K) &= \frac{K(1 - h)}{(1 + Kh)(1 + K)} \\ d_L^{(2)}(h, K) &= d_L^{(1)}(h, Kh) \\ d_L^{(3)}(h, K) &= d_L^{(2)}(h, Kh) \end{aligned}\right\} \tag{5.5.14}$$

where, in the expression for $d_L^{(l)}(h,K)$, we simply substitute hK for K in $d_L^{(l-1)}$. The reason is the same as discussed in Sections 3.4 and 5.4, namely, that whenever the lth ligand approaches, it sees the system with an equilibrium constant of Kh^{l-1}, as determined by the $l-1$ bound ligands.

Finally, we note that the fact that we have four different correlations in this system, some possibly of different signs, renders meaningless the characterization of the "cooperativity" of the system by a single number (as is frequently done using the Hill coefficient, see Section 4.3). We shall introduce in Section 5.8 a measure of the average cooperativity in a system, a quantity that may vary widely, even in its sign, as the binding process proceeds.

5.6. THREE LINEARLY ARRANGED SUBUNITS: CORRELATION TRANSMITTED ACROSS THE BOUNDARIES BETWEEN THE SUBUNITS

This is an extension of the model of Section 4.7. Instead of two subunits we now have three linearly arranged subunits. Each subunit can attain two configurations, L and H. Altogether we have eight configurations for the empty adsorbent molecule **P** (Fig. 5.8). If the sites are different, we have altogether 64 different configurations of the system with ligands.

We use here the same notation for Q_α, $q_{\alpha a}$, S_{ab}, etc. The only new notation is $Q_{\alpha\gamma}^{ab}$ for the quantity

$$Q_{\alpha\gamma}^{ab} = \exp(-\beta E_{\alpha\gamma}^{ab}) \tag{5.6.1}$$

where $E_{\alpha\gamma}^{ab}$ is the interaction energy between two subunits a and b in conformational states α and γ. The definitions of the binding constants and the various correlations in terms of the canonical PF are the same as in Section 5.2, with appropriate reinterpretations of the various terms:

$$Q(0,0,0) = \sum_{\alpha\beta\gamma} Q_\alpha Q_\beta Q_\gamma Q_{\alpha\beta}^{ab} Q_{\beta\gamma}^{bc} \tag{5.6.2}$$

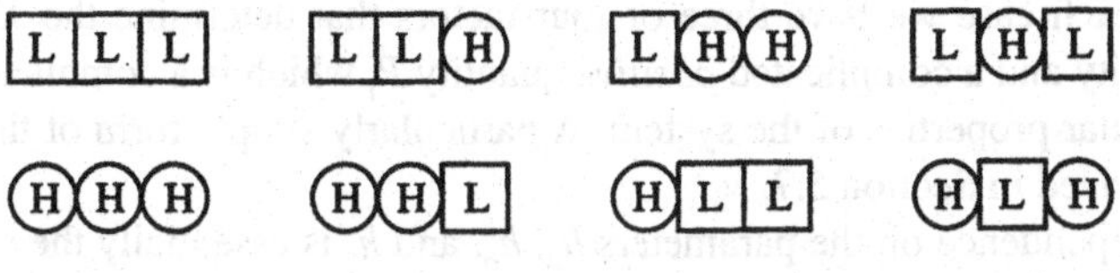

L L L   L L H   L H H   L H L

H H H   H H L   H L L   H L H

Figure 5.8. Three linearly arranged subunits, each subunit can attain either one of the conformations, L or H. Altogether there are eight configurations for the empty adsorbent molecule.

$$Q(a,0,0)=\sum_{\alpha\beta\gamma} Q_\alpha Q_\beta Q_\gamma Q^{ab}_{\alpha\beta} Q^{bc}_{\beta\gamma} q_{\alpha a} \tag{5.6.3}$$

and similarly for $Q(0, b, 0)$ and $Q(0, 0, c)$:

$$Q(a,b,0)=\sum_{\alpha\beta\gamma} Q_\alpha Q_\beta Q_\gamma Q^{ab}_{\alpha\beta} Q^{bc}_{\beta\gamma} q_{\alpha a} q_{\beta b} S_{ab} \tag{5.6.4}$$

and similarly for $Q(a, 0, c)$ and $Q(0, b, c)$:

$$Q(a,b,c)=\sum_{\alpha\beta\gamma} Q_\alpha Q_\beta Q_\gamma Q^{ab}_{\alpha\beta} Q^{bc}_{\beta\gamma} q_{\alpha a} q_{\beta b} q_{\gamma c} S_{abc} \tag{5.6.5}$$

We also introduce two new quantities, defined by

$$\eta_{ab}=\frac{(Q^{ab}_{LH})^2}{Q^{ab}_{LL}Q^{ab}_{HH}} \quad \text{and} \quad \eta_{bc}=\frac{(Q^{bc}_{LH})^2}{Q^{bc}_{LL}Q^{bc}_{HH}} \tag{5.6.6}$$

We shall always assume that $Q_{LH} = Q_{HL}$ for any pair of subunits and that the direct correlation can be factored from the total correlation. Also, to keep track of the pathway of transmission of information, we shall use the notation η_{ab} and η_{bc} even when the sites or subunits are identical. In such a case η_{ab} and η_{bc} will have equal magnitude, but will still be referred to as the transmission coefficient between the *first* and *second* or between the *second* and *third* subunits.

The general expressions for the indirect correlations are fairly complicated. Nevertheless, we can find conditions under which there exist pair and triplet correlations by factoring out the factors that determined the sign of the indirect correlations. In terms of the parameters h_a, h_b, h_c, and η_{ab}, η_{bc}, the pair correlations may be written as

$$y_{ab}=1-(\eta_{ab}-1)(h_a-1)(h_b-1)P_1 \tag{5.6.7}$$

$$y_{bc}=1-(\eta_{bc}-1)(h_b-1)(h_c-1)P_2 \tag{5.6.8}$$

$$y_{ac}=1+(\eta_{ab}-1)(\eta_{bc}-1)(h_a-1)(h_c-1)P_3 \tag{5.6.9}$$

where in each case we have three or four factors that determine the *sign* of the cooperativity and a complicated *positive* quantity P_i which is a combination of all the molecular properties of the system. A particularly simple form of these equations is derived in Section 5.7.

The dependence on the parameters h_a, h_b, and h_c is essentially the same as in the previous models. For instance, if either $h_a = 1$ or $h_b = 1$, there will be no indirect correlation between a and b, i.e., $y_{ab} = 1$. The new feature of this model is the effect of the parameters η_{ab} and η_{bc} that appear in Eqs. (5.6.7)–(5.6.9). We have seen in Section 4.7 that $\eta_{\alpha\beta}$ is a parameter which measures the extent of transmission of information between the two sites α and β. In contrast to the model of Section 5.5, where $K \neq 0$, $h_a \neq 1$ and $h_b \neq 1$ were sufficient conditions for having $y_{ab} \neq 1$ [see Eq. (5.5.4)]. Here, in addition to the condition $K \neq 0$, $h_a \neq 1$, and $h_b \neq 1$, we also need $\eta_{ab} \neq 1$ to obtain $y_{ab} \neq 1$. This means that even when the ligand *does* induce conformational changes when binding to site a ($h_a \neq 1$) and to site b ($h_b \neq 1$), the emergence of indirect correlation between a and b depends on the transmission of information across the boundary between the two subunits a and b. When $\eta_{ab} = 1$, there is no transmission of information across this boundary. (As we have seen in Section 4.7, $\eta_{ab} = 1$ is equivalent to the independence of subunits a and b.) When $\eta_{ab} = 0$ and $\eta_{bc} = 0$, this model essentially reduces to the model of Section 5.5, i.e., there is *total* transmission of information across the boundaries. (As we have seen in Section 4.7, when $\eta_{ab} \to \infty$, there is also total transmission of information, however of opposite sign. We shall not examine this case here.) In this section we consider only the case $0 \leq \eta_{ab} \leq 1$. In the next section, we present a simple model where the positive constants denoted by P_1, P_2, and P_3 in Eqs. (5.6.7)–(5.6.9) can be written explicitly. Similar considerations apply to y_{bc}. The behavior of y_{ac} (the "long-range" correlation) is quite different. In order to have indirect correlation between a and c one needs, in addition to $K \neq 0$, $h_a \neq 1$, and $h_c \neq 1$, also the fulfillment of the *two* conditions $\eta_{ab} \neq 1$ and $\eta_{bc} \neq 1$. This means that in order to have correlation between a and c, the information (on the occupancy state of the sites) must be transmitted across *both* the boundaries of ab and of bc. If any one of η_{ab} or η_{bc} is unity, the communication between a and c becomes "short" and hence there is no indirect correlation.

It is interesting to note that while in the model of Section 5.5 we found $y_{ab} = y_{bc} = y_{ac}$ whenever $h_a = h_b = h_c = h$, i.e., the long-range correlation was the same as the short-range correlation, here, even when $h_a = h_b = h_c = h$, the long-range correlation y_{ac} differs from the short-range correlations y_{ab} and y_{bc}. It is clear from Eq. (5.6.9) that when $\eta_{ab} < 1$ and $\eta_{bc} < 1$, there exists a "distance" dependence of y_{ac} but, in contrast to the direct correlation S_{ab} that depends on the actual *distance* between the ligands, here the indirect correlation depends on the "distance" only in the sense of the number of boundaries between the subunits a and b. (This is true also for any linear system of subunits; see also Sections 5.7 and 7.4.)

The triplet correlation in this system is quite complicated, even when S_{abc} is assumed to be independent of the conformation of the subunits. Nevertheless, we can make the following statements regarding the indirect triplet correlations:

1. There is no pairwise additivity neither in the sense $g_{\mathrm{abc}} = g_{ab}g_{bc}g_{ac}$ nor in the sense $g_{abc} = g_{ab}g_{bc}$, even when the direct correlation S_{abc} is strictly pairwise additive.

2. When *one* of h_α is unity, it makes two of the indirect pair correlations unity, but the indirect triplet correlation becomes equal to the third pair correlation. For example, when $h_a = 1$, we have

$$y_{ab} = y_{ac} = 1 \tag{5.6.10}$$

but

$$y_{abc} = y_{bc} \tag{5.6.11}$$

This follows directly from the definition of y_{abc}.

3. Whenever *two* of the hs are unity, there is neither pair nor triplet indirect correlation.

4. When only one of η_{ab} or η_{bc} is unity, the triplet indirect correlation may differ from unity. When $\eta_{ab} = \eta_{bc} = 1$, we have $y_{abc} = 1$.

5. Finally, as shown in the model in Section 5.5, it is easy to choose parameters such that the "short-range" correlations are negligible but the "long-range" correlation is large and, in addition, the triplet correlation is equal to the long-range correlation.

For instance, with $h_b \approx 1$, i.e., the binding to b does not discriminate between L and H while h_a and h_c are larger than unity, we have

$$y_{ab} \approx y_{bc} \approx 1 \tag{5.6.12}$$

but y_{ac} could be large and $y_{abc} \approx y_{ac}$.

This example should serve again as a warning signal when processing experimental data with a presumption regarding the negligibility of the "long-range" correlation (see also Section 5.10).

We conclude this section by noting that when the subunits become identical, and $h_a = h_b = h_c = h$ and $\eta_{ab} = \eta_{bc} = \eta$, the sites are, in general, not identical in either the weak or the strict sense. This is in contrast to the model discussed at the end of Section 5.5, where we found only one intrinsic constant. Here, in general, there are two different intrinsic binding constants, denoted by $k^{(1)}$ and $k^{(2)}$ for binding on the first (or third) site and on the second site, respectively. The general expression is complicated, but for $\eta = 1$ we have the following simple expressions:

$$k^{(1)} = \frac{Q_L\sqrt{Q_{LL}}q_L + Q_H\sqrt{Q_{HH}}q_H}{Q_L\sqrt{Q_{LL}} + Q_H\sqrt{Q_{HH}}} \tag{5.6.13}$$

and

$$k^{(2)} = \frac{Q_L Q_{LL} q_L + Q_H Q_{HH} q_H}{Q_L Q_{LL} + Q_H Q_{HH}} \tag{5.6.14}$$

Note the appearance of $\sqrt{Q_{\alpha\alpha}}$ in $k^{(1)}$, but $Q_{\alpha\alpha}$ in $k^{(2)}$. The reason is that a ligand approaching the first (or third) subunit sees an equilibrium concentration, of L and H, different from a ligand approaching the second subunit. These also differ from the equilibrium concentrations of L and H for an *isolated* subunit, for which the binding constant is

$$k^{(0)} = \frac{Q_L q_L + Q_H q_H}{Q_L + Q_H} \tag{5.6.15}$$

5.7. A SIMPLE SOLVABLE MODEL

We present here a simple model where long-range and nonadditivity of the correlations can be studied explicitly in terms of the ligand–ligand, and ligand–site interactions. With this model we can clearly see the different behavior of the three models discussed in previous sections and, by generalization, we shall see that the same mechanism applies for correlations between particles in the liquid state.

The three variants of this model are described in Fig. 5.9. The ligands are simple Lennard-Jones particles, at the center of which a point dipole of strength **d** is embedded. The orientation of the ligand on the site is always the same, say upward, as in Fig. 5.9. The adsorbent macromolecule consists of three subunits denoted by a, b, and c, each of which has one binding site to which we also refer as a, b, and c. Near each site the macromolecule has a dipole of strength D, which can be oriented either upward or downward. The orientation of the dipole D is determined

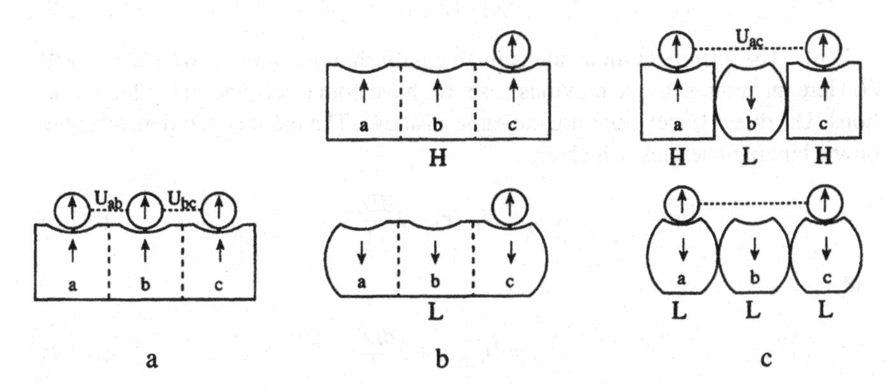

Figure 5.9. Three binding models for linearly arranged subunits: (a) direct interaction only (no conformational changes); (b) two conformations L and H for the entire molecule; (c) two conformations L and H for each subunit. The arrows indicate the direction of the dipoles embedded in the ligand and in each subunit.

by the state of the macromolecule, L or H. We assume that the orientation of D affects the binding energy (because of its proximity to the ligand), but has a minor effect on the total energy of the state of the macromolecule. We denote by R the ligand–ligand distance at the adjacent site, and by r the distance between the dipoles d and D. We now examine the different behavior of the three cases corresponding to Sections 5.3, 5.5, and 5.6.

(a) The adsorbent molecule is rigid, and binding of a ligand does not induce conformational changes: In this case, the correlation between the ligands is entirely due to direct interactions. These are:

$$U(1,2) = U(2,3) = U_{LJ}(R) + \frac{d^2}{R^3} \tag{5.7.1}$$

$$U(1,3) = U_{LJ}(2R) + \frac{d^2}{8R^3} \tag{5.7.2}$$

$$U(1,2,3) = U_{LJ}(2R) + 2U_{LJ}(R) + \frac{17d^2}{8R^3} \tag{5.7.3}$$

If the long-range interaction $U(1, 3)$ can be neglected compared with $U(1, 2)$, the (direct) triplet correlations can be written as

$$S(1,2,3) \approx S(1,2)S(2,3) \tag{5.7.4}$$

and

$$S(1,3) \approx 1 \tag{5.7.5}$$

(b) The adsorbent molecule is in an equilibrium mixture of two states L and H: Here, in contrast to the previous case, we have both direct and indirect correlations. The direct correlations are the same as above. The indirect part depends now on the binding energies, which are

$$U_H = U_{LJ}(r) - 2\frac{dD}{r^3} \tag{5.7.6}$$

and

$$U_L = U_{LJ}(r) + 2\frac{dD}{r^3} \tag{5.7.7}$$

The LJ part of the binding energies is of no importance for the indirect correlation; only the difference $U_H - U_H$ enters into the correlation. Note the different orientations of the dipoles D in the two states L and H (Fig. 5.9b).

If we again set

$$K = \frac{Q_H}{Q_L}, \qquad h = \exp[-\beta(U_H - U_L)] = \exp(4\beta dD/r^3) \tag{5.7.8}$$

then in terms of K and h we have

$$y(1, 2) = y(1, 3) = y(2, 3) = \frac{(1+K)(1+Kh^2)}{(1+Kh)^2} \tag{5.7.9}$$

and

$$y(1, 2, 3) = \frac{(1+K)^2(1+Kh^3)}{(1+Kh)^3} \tag{5.7.10}$$

Here, we see the fundamental differences between the direct and indirect correlations. First, all three indirect pair correlations are equal and *independent* of the distance R. They do depend on r (through h), but not on the ligand–ligand distance.

The triplet correlation can be shown to be always larger than unity. Here, we show that it is not pairwise additive, not in the sense $g(1, 2, 3) = g(1, 2)g(2, 3)$ nor in the sense $g(1, 2, 3) = g(1, 2)g(2, 3)g(1, 3)$. For this particular model we have

$$\frac{y(1, 2, 3)}{y(1, 2)y(2, 3)} = \frac{(1+Kh)(1+Kh^3)}{(1+Kh^2)^2} \tag{5.7.11}$$

and

$$\frac{y(1, 2, 3)}{y(1, 2)y(1, 3)y(2, 3)} = \frac{(1+Kh)^3(1+Kh^3)}{(1+K)(1+Kh^2)^3} \tag{5.7.12}$$

Clearly, there is no hint that any of these relations are close to unity. Incidentally, we note that for the case $h_a \neq h_b \neq h_c$ and $K = 1$, we have an exact relation of the form

$$g_{abc} = g_{ab} + g_{ac} + g_{bc} - 2 \tag{5.7.13}$$

which is clearly very different from the "additivities" mentioned before.

(c) Three subunits, each of which can be in either of the two states L or H: Perhaps the most interesting results are for the model of Section 5.6, shown schematically on the rhs of Fig. 5.9.

Here, we find a simple, explicit, and informative expression for all the indirect correlations in the system. In our model we again set $h_a = h_b = h_c = h$, $\eta_{ab} = \eta_{bc} = \eta$, and $K = 1$, and find the explicit results

$$y_{ab} = 1 - \frac{(\sqrt{\eta_{ab}} - 1)(h-1)^2}{(\sqrt{\eta_{ab}} + 1)(h+1)^2} \tag{5.7.14}$$

$$y_{bc} = 1 - \frac{(\sqrt{\eta_{bc}} - 1)(h-1)^2}{(\sqrt{\eta_{bc}} + 1)(h+1)^2} \tag{5.7.15}$$

$$y_{ac} = 1 + \frac{(\sqrt{\eta_{ab}} - 1)(\sqrt{\eta_{bc}} - 1)(h-1)^2}{(\sqrt{\eta_{ab}} + 1)(\sqrt{\eta_{bc}} + 1)(h+1)^2} \tag{5.7.16}$$

$$y_{abc} = 1 + \frac{(3 - \sqrt{\eta_{ab}} - \sqrt{\eta_{bc}} - \sqrt{\eta_{ab}\eta_{bc}})(h-1)^2}{(\sqrt{\eta_{ab}} + 1)(\sqrt{\eta_{bc}} + 1)(1+h)^2} \tag{5.7.17}$$

The above correlations are for three identical subunits, i.e., $\eta_{ab} = \eta_{bc} = \eta$. However, in Eqs. (5.7.14)–(5.7.17) we have left the different subscripts on η so as to stress the dependence of each correlation on the particular sequence of parameters η. Thus, when $\eta_{ab} = 1$, there is no transmission of information across the boundary between a and b, hence the circuit connecting a and b becomes short and $y_{ab} = 1$. Likewise, when $\eta_{bc} = 1$, we find that $y_{bc} = 1$. On the other hand, the communication between a and c depends on the transmission of information across the two boundaries ab and bc, if either η_{ab} or η_{bc} is unity, the circuit connecting a and c becomes short, and $y_{ac} = 1$. Note that the long-range correlation y_{ac} differs from the short-range correlations y_{ab} and y_{bc}. The difference is not due to the distance between the ligands, but to the number of boundaries across which the information must be transmitted. In Eq. (5.7.16), we have two boundaries to cross, and for $h_a = h_b = h_c = h$ and $\eta_{ab} = \eta_{bc} = \eta$ we have

$$y_{ac} = 1 + \frac{(\sqrt{\eta} - 1)^2(h-1)^2}{(\sqrt{\eta} + 1)^2(h+1)^2}$$

$$= 1 + \gamma\left(\frac{\sqrt{\eta} - 1}{\sqrt{\eta} + 1}\right)^2 \tag{5.7.18}$$

For any $h \neq 1$, the long-range correlation is always positive ($y_{ac} \geq 1$). It decreases from $1 + \gamma$ to 1 for $0 \leq \eta \leq 1$, and increases from 1 to $1 + \gamma$ for $1 \leq \eta < \infty$. We see that there exists a "distance dependence" of y_{ac}, which in this model depends on the number of boundaries between the subunits that connect the sites a and b. For a system of $l + 1$ subunits, there are l boundaries and the long-range correlation $y_{\alpha\omega}$ has the general form

$$y_{\alpha\omega} = 1 + \gamma\left(-\frac{\sqrt{\eta}-1}{\sqrt{\eta}+1}\right)^{l} \tag{5.7.19}$$

The dependence of $y_{\alpha\omega}$ on the "distance" l, the number of intervening boundaries, is different depending on η. When $0 < \eta \le 1$, $y_{\alpha\omega} \ge 1$, i.e., we always have positive cooperativity, and $y_{\alpha\omega}$ decreases monotonically with l (Fig. 5.10a). Note that as $\eta \to 0$, $y_{\alpha\omega}$ becomes infinitely long-range, i.e., $y_{\alpha\omega} = 1 + \gamma$. When $\eta > 1$, we see from Eq. (5.7.19) that $y_{\alpha\omega} > 1$ for even values of l, but $y_{\alpha\omega} < 1$ for odd values of l. Figure 5.10b shows $y_{\alpha\omega}$ as a function of l for two values of $\eta > 1$. Clearly, because of the factor $(-1)^l$ in Eq. (5.7.19), $y_{\alpha\omega}$ will oscillate above and below unity according to whether l is even or odd, respectively. In the limit $\eta \to \infty$, $y_{\alpha\omega}$ will be infinitely long-range, but now oscillating between $1 + \gamma$ and $1 - \gamma$ where $\gamma \ge 1$ is defined in Eq. (5.7.18). This behavior is reminiscent of the behavior of the indirect correlation function $y(R)$ in liquids. To see the connection between the two systems, replace the subunits by single molecules, allow for distance-dependent "subunit–subunit" interaction [$U(R)$ in the theory of liquids], and allow many different sequences of "subunits" to transmit the information; we then obtain the indirect correlation in liquids. Thus, the mechanism of transmission of information between ligands is essentially the same as that between any two particles in a condensed medium.

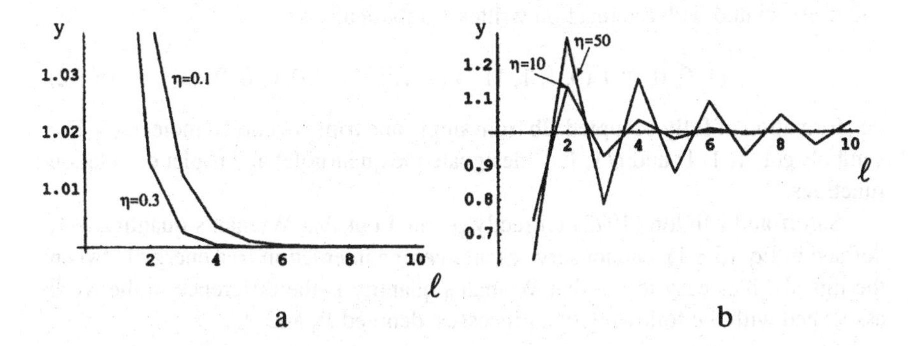

Figure 5.10. Dependence of the long-range correlation $y_{\alpha\omega}$ on l, the number of intervening boundaries, for different values of η, indicated next to each curve.

5.8. A MEASURE OF THE AVERAGE CORRELATION IN A BINDING SYSTEM

5.8.1. Introduction and Historical Background

In two-site systems, there is only one correlation function which characterizes the cooperativity of the system. In systems with more than two identical sites, for which additivity of the higher-order correlations is valid, it is also true that the pair correlation does characterize the cooperativity of the system. This is no longer valid when we have different sites or nonadditivity effects. In these cases there exists no single correlation that can be used to characterize the system, hence the need for a quantity that measures the average correlation between ligands in a general binding system. There have been several attempts to define such a quantity in the past. Unfortunately, these are valid only for additive systems, as will be shown below.

The first attempt was made by Wyman (1964). He suggested extracting from the Hill plot a quantity referred to as the "*total*" or "*overall measure*" of the "*free energy of interaction*" between the ligands. What he actually obtained is the ratio between the last and first binding constants which, for hemoglobin (Hb), is k_4/k_1 where k_1 is the *first* intrinsic binding constant and k_4 (in our notation, $k_{1/111}$) is the *last* (conditional) binding constant, i.e., binding to the fourth site given that the three sites are already occupied.

The corresponding work, translated into our notation, is

$$WY \equiv -RT\ln(k_4/k_1) = -RT\ln\frac{g(1,1,1,1)}{g(1,1,1)}$$

$$= -RT\ln\left[\frac{Q(1,1,1,1)Q(0,0,0,0)^3}{Q(1,0,0,0)^4}\right]\left[\frac{Q(1,0,0,0)^3}{Q(1,1,1,0)Q(0,0,0,0)^2}\right] \tag{5.8.1}$$

and is associated with the reaction written symbolically as

$$(1,0,0,0) + (0,1,1,1) \rightarrow (1,1,1,1) + (0,0,0,0) \tag{5.8.2}$$

i.e., formation of fully occupied Hb from singly and triply occupied molecules. The symbols $g(1, 1, 1, 1)$ and $g(1, 1, 1)$ designate the quadruplet and triplet correlation functions.

Saroff and Minton (1972) correctly pointed out that Wyman's quantity, *WY*, defined in Eq. (5.8.1), cannot serve as an *average* interaction free energy between the ligands. It is easy to see that Wyman's quantity is the difference in the work associated with the following two processes, denoted P_3 and P_4,

$$4\,(1,0,0,0) \xrightarrow{P_4} (1,1,1,1) + 3\,(0,0,0,0) \tag{5.8.3}$$

and

$$3\,(1,0,0,0) \xrightarrow{P_3} (1,1,1,0) + 2\,(0,0,0,0) \tag{5.8.4}$$

Minton and Sarof (1974) defined a new quantity, also referred to as the "*average Gibbs free energy of interaction*," by the integral

$$MS \equiv 4RT \int_0^1 [\ln x^{(r)} - \ln x^{(i)}]d\theta \tag{5.8.5}$$

where $x^{(r)}$ is the ligand concentration in equilibrium with the *real* cooperative system at fractional saturation θ, and $x^{(i)}$ is the ligand concentration in equilibrium with an *ideal*, or noncooperative, system at fractional saturation θ. As we shall see below, the quantity MS, defined in Eq. (5.8.5), is essentially the work associated with the process P_4 defined in Eq. (5.8.3), i.e., this is the work required to form the *fully occupied* molecule from four singly occupied molecules [see Eq. (5.8.21)]. In terms of a correlation function, this work is $-RT \ln g(1, 1, 1, 1)$, i.e., it measures the quadruplet correlation function. Dividing by 4, one obtains an "average" free energy of interaction, per ligand, in the *fully occupied molecule*. A related quantity also suggested by Wyman is essentially only the first term in Eq. (5.8.5),

$$4RT \int_0^1 \ln x^{(r)} d\theta = -4RT \ln k_1 - RT \ln g(1, 1, 1, 1) \tag{5.8.6}$$

which includes four times the binding work to the first site as well as the correlation work.

We shall see in the next subsection that if all the cooperativities in the system are pairwise additive, then either Wyman's or Minton and Saroff's definitions may be used as a measure of the overall cooperativity in the system. When the system is cooperative but nonadditive, then neither Wyman's nor Minton and Saroff's quantities can serve as the *average cooperativity* in the system. Furthermore, the average cooperativity in the system is, in general, dependent on ligand activity or concentration.* A simple example will illustrate this point. Suppose we have three sites arranged linearly, as in Fig. 5.11. Two ligands occupying neighboring sites repel each other, but at next-neighbor sites they attract each other. In this system, at relatively low ligand concentration, the system will manifest *positive* cooperativity. This is so because the occupation of next-nearest neighbors will be more likely than that of nearest neighbors. At very high ligand concentration, the third ligand is forced to occupy the central site, despite the repulsive forces. At this end, the system will manifest negative cooperativity. A quantitative numerical illustration of this behavior is shown in Subsection 5.8.3.

*Note that each correlation has been defined as the $\lambda \to 0$ limit and hence independent of λ or C. Here we show that the *average* correlation, constructed from the $\lambda \to 0$ limit of the correlations, is concentration-dependent.

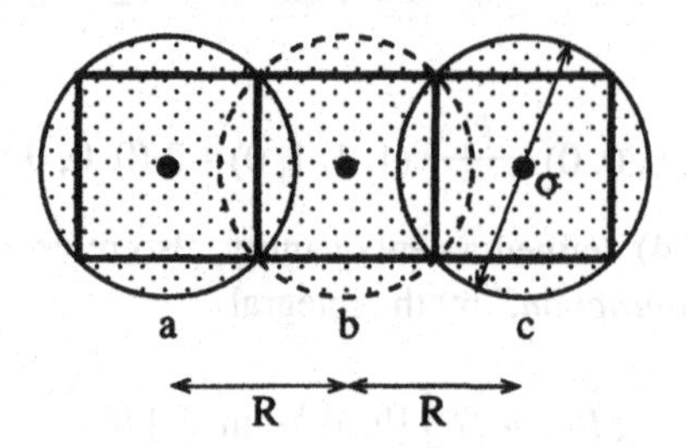

Figure 5.11. Schematic illustration of a linear three-site binding system. The ligands attract each other when bound at the sites a and c, but they repel each other when bound to adjacent sites a, b or b, c.

It is clear that neither Wyman's nor Minton and Saroff's measures "do justice" to all types of cooperativities in the system, and certainly cannot account for variation in cooperativity at different stages of the binding process. In the next subsection we define a new measure of the *average* correlation in any binding system, and show how to extract this quantity from experimental data.

5.8.2. Definition of the Average Correlation in Any Binding System

It is assumed that we have experimental data on the binding isotherm of a system known to consist of m binding sites. We denote by $\theta_T(C)$ the average number of ligands bound to the whole system when the ligand concentration is C (the ligand is assumed to be either in an ideal gas or in a dilute ideal solution, so that its absolute activity λ is proportional to its concentration, $\lambda = \lambda_0 C$).

We first determine the limiting slope of the BI, defined by

$$K_T = \lim_{C\to 0}\left(\frac{\partial\theta_T}{\partial C}\right) \tag{5.8.7}$$

If the system is known to have m identical sites, either in a strict or in a weak sense, then $K_T = mk$, where k is the *intrinsic* binding constant. On the other hand, if we know that the system has m *different* binding sites, each having a different intrinsic binding constant k_α, then we must determine each of these from the limiting slope of the corresponding individual BI, i.e.,

$$k_\alpha = \lim_{C\to 0}\left(\frac{\partial\theta_\alpha}{\partial C}\right) \tag{5.8.8}$$

where $\theta_T = \Sigma\theta_\alpha$ and $K_T = \Sigma k_\alpha$, the sum being over all m sites.

We define the two auxiliary functions

$$I(C) = \int_0^C \frac{\theta_T(C')}{C'} dC' \tag{5.8.9}$$

and

$$I_i(C) = \ln \prod_\alpha (1 + k_\alpha C) \tag{5.8.10}$$

The first, $I(C)$, can be determined from the experimental data for any finite value of C. The second, $I_i(C)$, may be interpreted as the same integral $I(C)$, but computed for a hypothetical system of independent sites. As we shall see below, this interpretation is somewhat risky and should be avoided. The function $I_i(C)$ is better viewed as *defined* in Eq. (5.8.10), with the binding constants determined in Eq. (5.8.8).

In terms of the two functions $I(C)$ and $I_i(C)$ and the intrinsic binding constants k_α, we define the quantity

$$g(C) = \frac{\exp[I(C)] - 1 - \Sigma_\alpha k_\alpha C}{\exp[I_i(C)] - 1 - \Sigma_\alpha k_\alpha C} \tag{5.8.11}$$

We now show that for any finite value of C, the quantity $\overline{g}(C)$ is a proper average of the correlations in the system. Note that $\overline{g}(C)$ can be determined directly from the experimental data. The units of the concentration C should be the same as those of $(k_\alpha)^{-1}$; hence Ck_α is a dimensionless quantity.

We first treat the most general case where all sites are different. The grand partition function of a single adsorbent molecule is

$$\xi = Q(0) + Q(1)\lambda + \sum_{l=2}^{m} Q(l)\lambda^l \tag{5.8.12}$$

where $Q(l)$ is the canonical PF of the molecule having exactly l $(0 \le l \le m)$ bound ligands.

For each l, the PF $Q(l)$ is a sum over $\binom{m}{l}$ different *specific* terms $Q_s(l)$, each of which is the PF of the adsorbent molecule having exactly l ligands occupying l *specific* sites. Hence the grand PF can be written as

$$\xi = Q(0) + \sum_\alpha Q_\alpha(1)\lambda + \sum_{l=2}^{m} \sum_{s(l)} Q_s(l)\lambda^l \tag{5.8.13}$$

The total BI is simply the average number of ligands in the system and is obtained from

$$\theta_T = n_T = \lambda \frac{\partial \ln \xi}{\partial \lambda} = \left[\sum_{\alpha} Q_\alpha(1)\lambda + \sum_{l=2}^{m} l \sum_{s(l)} Q_s(l)\lambda^l \right] \Big/ \xi \tag{5.8.14}$$

The subscript s stands for *specific*, while s(l) stands for a set of l *specific* indices, say $i_1, i_2, \ldots, i_l$, where i_k indicates a specific site. The sum over s(l) is the sum over $\binom{m}{l}$ specific configurations for each specific l. The sum over α in $\Sigma Q_\alpha(1)\lambda$ is simply over all the m sites. The statistical mechanical expression for k_α is

$$k_\alpha = \lim_{C \to 0} \frac{\partial \theta_\alpha}{\partial C} = \frac{Q_\alpha(1)}{Q(0)} \lambda_0 \tag{5.8.15}$$

where $Q_\alpha(1)$ is the PF of the system singly occupied at the specific site α. The correlation function of order l for any specific configuration of the ligands s(l) is defined by*

$$g_s(l) = \frac{Q_s(l)Q(0)^{l-1}}{\prod_\alpha Q_\alpha(1)} = S_s(l)y_s(l) \tag{5.8.16}$$

where the rhs of Eq. (5.8.16) is valid when $g_s(l)$ may be factorized into direct and indirect parts. With the definitions (5.8.15) and (5.8.16), we can rewrite the grand PF of the systems in the form

$$\xi = Q(0)\left[1 + \sum_\alpha k_\alpha C + \sum_{l=2}^{m} \sum_{s(l)} g_s(l) \prod_\alpha k_\alpha C^l \right] \tag{5.8.17}$$

The integral in Eq. (5.8.9) is immediate,

$$I(C) = \int_0^C \frac{\theta_T(C')}{C'} dC' = \ln[\xi(C)/Q(0)] \tag{5.8.18}$$

hence the quantity $\overline{g}(C)$ defined in Eq. (5.8.11) is

$$g(C) = \sum_{l=2}^{m} \sum_{s(l)} P_s(l, C) g_s(l) \tag{5.8.19}$$

*Note that this is already the $\lambda \to 0$ limit of the l-order correlation function.

where the probability distribution is defined, for any C, by

$$P_s(l, C) = \frac{\prod_{\alpha} k_\alpha C^l}{\prod_{\alpha} (1 + k_\alpha C) - 1 - \Sigma_\alpha k_\alpha C} \tag{5.8.20}$$

Since $1 \le P_s(l, C) \le 1$ and $\Sigma_{l\ge 2}\Sigma_{s(l)} P_s(l, C) = 1$ for any C, the quantity $\bar{g}(C)$ is a proper average over all the correlations in the system.

Two limiting cases should be noted when $C \to 0$, $P_s\,(l, C) = 1$ for $l = 2$ and $P_s\,(l, C) = 0$ for $l > 2$. Hence the average quantity $\bar{g}(C)$ reduces to the average *pair* correlations in the system. This is consistent with our expectation from $\bar{g}(C)$, since at this limit only the pair correlations are operative. On the other hand, for $C \to \infty$, we have $P_s(l, C) = 1$ for $l = m$ and $P_s(l, C) = 0$ for $2 \le l < m$; at this limit only the highest correlation $g(m)$ is represented by $\bar{g}(C)$. This is essentially the MS measure (provided the k_α are determined from Eq. (5.8.8), not chosen as the binding constant for a single subunit; see below), i.e.,

$$MS = -RT \ln g(m) = \lim_{C\to\infty} [-RT \ln g(C)] \tag{5.8.21}$$

Clearly, except for the two limits ($C \to 0$ and $C \to \infty$), $\bar{g}(C)$ is determined by *all* existing correlations in the systems.

Two special cases should also be noted. If all $g_s(l) = 1$, i.e., there are no correlations of any order, then also $\bar{g}(C) = 1$ for any C. On the other hand, in the case of a two-site system, $m = 2$, there exists only one correlation $g_{ab}(2)$; hence $\bar{g}(C) = g_{ab}(2)$. For all other cases, the weights (5.8.20) given to each correlation are proportional to the product of the corresponding binding constants. This is similar to the weights given to any configuration in a hypothetical reference system of independent sites but with the same binding constants. For instance, if $k_i k_j > k_r k_t$, then the occupation of the pair of sites i, j will be more probable than the pair r, t in this reference system, hence the weight given to the pair i, j will be larger than to the pair r, t. We stress, however, that the denominator in Eq. (5.8.11) is not the PF of the reference system, since we have subtracted the two terms $1 + \Sigma k_\alpha C$. One should also be careful in interpreting $\Pi(1 + k_\alpha C)$ as the PF of a system where all correlations have been "switched off." If all the correlations were due to *direct* interaction between the ligands, then it would be easy to visualize a real system for which all the interactions are "switched off." On the other hand, when the correlations are indirect (as in the case of most biochemical systems), then it is not clear what one must "switch off" to eliminate the correlations. Minton and Saroff suggested using k_0, the binding constant for a single isolated subunit. This is quite risky: first, because not every binding system has well-defined subunits; second, k_0 will in general depend on the equilibrium distribution of the states of the subunit. On the other hand, k_α (or k, if the subunits are identical) depends on the equilibrium distribution of the states of the *entire* adsorbent molecule. Hence, in general, k_0 is not related to k_α, and using k_0 instead of the intrinsic binding constant might not lead to a useful definition of an average correlation. (A simple example is given at the end of Sections 5.6 and 6.3, where $k^{(1)}$ and $k^{(2)}$ are the two *different* k_αs, and $k^{(0)}$ is the equivalent of k_0.)

For all the above reasons we have defined $\bar{g}(C)$ without reference to any hypothetical, independent-site system. One simply extracts both $I(C)$ and all k_α from the experimental data, and then constructs the quantity $\bar{g}(C)$. When the sites are identical in a weak sense, i.e., all $k_\alpha = k$, some of the correlations for a given l might differ. For example, four identical subunits arranged in a square will have only one intrinsic binding constant k, but two different pair correlation functions. For this particular example we have four nearest-neighbor pair correlations $g_{nn}(2)$, and two second-nearest-neighbor pair correlations $g_{sn}(2)$. The average correlation for this case is

$$g(C) = \frac{4k^2 g_{nn}(2)C^2 + 2k^2 g_{sn}(2)C^2 + 4k^3 g(3)C^3 + k^4 g(4)C^4}{6k^2C^2 + 4k^3C^3 + k^4C^4} \tag{5.8.22}$$

On the other hand, for a system of four identical sites in the strict sense, i.e., when all $k_\alpha = k$ *and* all the correlations of any given order are identical, such as the arrangement of four identical subunits at the vertices of a perfect tetrahedron, we have

$$\bar{g}(C) = \frac{6k^2 g(2)C^2 + 4k^3 g(3)C^3 + k^4 g(4)C^4}{6k^2C^2 + 4k^3C^3 + k^4C^4} \tag{5.8.23}$$

For the general case of m identical sites, in the strict sense we have

$$g(C) = \frac{\sum_{l=2}^{m} \binom{m}{l} k^l C^l g(l)}{\sum_{l=2}^{m} \binom{m}{l} k^l C^l} \tag{5.8.24}$$

We shall see in the next subsection that $g(C)$ can change dramatically as a function of concentration. However, if one insists on having a single measure of the average cooperativity, a convenient choice for the general case could be $C = 1$ [measured in the same units of $(k_\alpha)^{-1}$]. A more convenient choice for the case of identical sites (in either respect) is $C = k^{-1}$, in which case Eq. (5.8.24) reduces to

$$\bar{g}(C = k^{-1}) = \frac{\sum_{l=2}^{m} \binom{m}{l} g(l)}{\sum_{l=2}^{m} \binom{m}{l}} \tag{5.8.25}$$

This is an average with an almost binomial distribution, $\binom{m}{l}$. In the binomial distribution we count all 2^m occupation configurations of the system of m sites. In the denominator of Eq. (5.8.25), on the other hand, we count all the configurations for which a correlation function is definable, i.e., we exclude the empty and singly occupied configurations.

5.8.3. Some Numerical Illustrations

We present here two examples of the application of the measure $\overline{g}(C)$ for three-site systems. In all the plots below we draw $\overline{g}(C) - 1$ instead of $\overline{g}(C)$, as this is more convenient for observing the sign of the average cooperativity.

Example 1. Three Identical Sites with Direct and Indirect Correlations

For this demonstration we use a binding system of three identical subunits arranged in an equilateral triangle (i.e., the sites are identical in the strict sense). The corresponding GPF is

$$\xi = Q(0) + 3Q_s(1)\lambda + 3Q_s(2)\lambda^2 + Q(3)\lambda^3 \tag{5.8.26}$$

Here, s again stands for specific. We use a simple two-state model for each subunit (Section 5.6), with *negative direct* correlations $S(2) = 0.1$ and $S(3) = 0.1^3$. The indirect part of the correlation is due to conformational changes induced by the binding process. The molecular parameters chosen for this model are

$$\begin{aligned} &\text{(a)} \quad \eta = 0.1, \quad h = 70, \quad K = 0.2, \quad Q_{LL} = Q_{HH} \\ &\text{(b)} \quad \eta = 0.2, \quad h = 1000, \quad K = 0.02, \quad Q_{LL} = Q_{HH} \end{aligned} \tag{5.8.27}$$

Figure 5.12 shows the BI and the quantities $\overline{g}(C) - 1$ for this model. This illustration shows that although the binding isotherms "seem" to belong to a negative cooperative system, it is, in fact, meaningless in general to refer to the "cooperativity" of the system where there exists more than one type of cooperativity. In Fig. 5.12a, the curve starts with positive cooperativity, mainly due to the *indirect* part, i.e.,

$$S(2) = 0.1, \quad y(2) = 13.92, \quad \text{and} \quad g(2) = S(2)y(2) = 1.39 \tag{5.8.28}$$

but as the ligand concentration increases, the dominant contribution comes from the *direct* correlation, i.e.,

$$S(3) = 0.1^3, \quad y(3) = 519.7, \quad \text{and} \quad g(3) = S(3)y(3) = 0.5197 \tag{5.8.29}$$

Thus, the average cooperativity $[\overline{g}(C) - 1]$ changes from positive to negative.

The phenomenon is reversed in illustration (b) shown in Fig. 5.12b. Here, we initially have

$$S(2) = 0.1, \quad y(2) = 7.9, \quad \text{and} \quad g(2) = S(2)y(2) = 0.79 \tag{5.8.30}$$

i.e., the system starts initially with *negative* cooperativity. However, at high

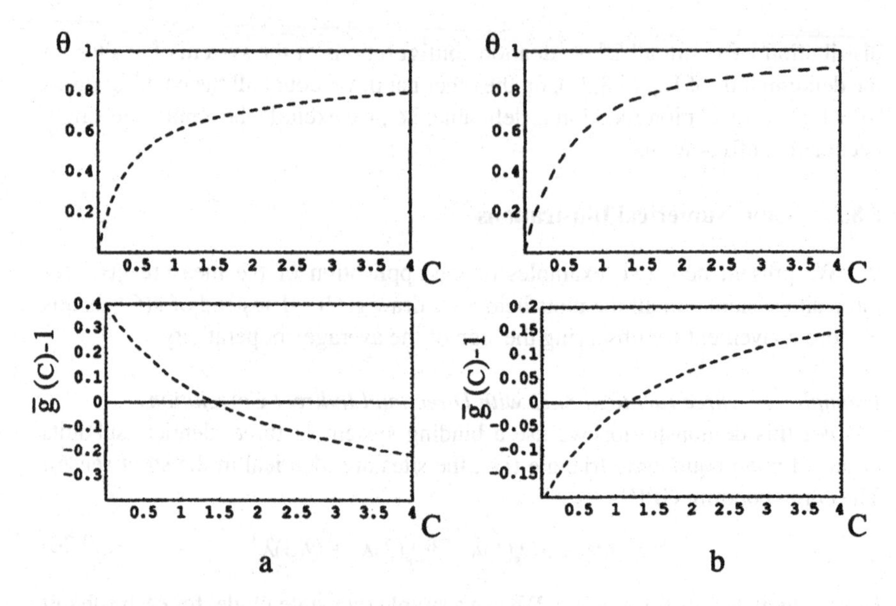

Figure 5.12. The BI and the average correlation, in the form $g(C) - 1$ for (a) the model (a) with parameters in Eq. (5.8.27) and (b) the model (b) with parameters in Eq. (5.8.27).

densities we find

$$S(3) = 0.1^3, \qquad y(3) = 1295, \qquad \text{and} \qquad g(3) = S(2)y(3) = 1.295 \tag{5.8.31}$$

and the average cooperativity is dominated by the positive indirect triplet correlation.

Example 2. The Linear Three-Site Model of Fig. 5.11

We return to the example shown in Fig. 5.11. This illustration shows how deceptive a measure of the average cooperativity could be such as the one suggested by Minton and Saroff.

We choose a Lennard-Jones interaction between ligands of the form

$$U_{LJ}(R) = 100\left[-\left(\frac{\sigma}{R}\right)^6 + \left(\frac{\sigma}{R}\right)^{12}\right] \tag{5.8.32}$$

where σ and R are in dimensionless units, and where the distance between adjacent sites is one unit and the distance between the first and third sites is two units. In this model, the binding constants for the three sites are chosen to be unity. There are, however, two different pair correlations. For the choice of $\sigma = 1$, 1.3, and 1.4 (in the same units of R), we find that initially the system shows positive cooperativity

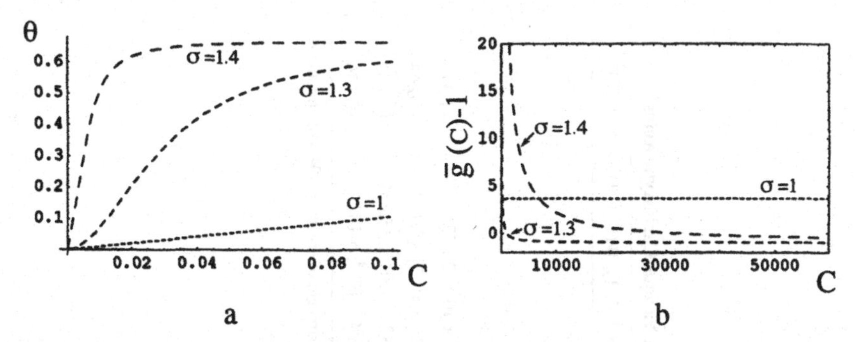

Figure 5.13. (a) Binding isotherms and (b) average correlations for the model shown in Fig. 5.11. The curves correspond to the three values $\sigma = 1$, 1.3, and 1.4 in Eq. (5.8.32).

(Fig. 5.13a). This is due to the fact that initially the attractive, next-nearest-neighbor interaction dominates the behavior of the system. However, at very high ligand activity a third ligand is forced to occupy the central site, producing an overall strong *negative* correlation. In Fig. 5.13b, we see that at very high concentration the $\sigma = 1.3$ and $\sigma = 1.4$ curves become negative. Clearly, if one uses the MS measure, it will pick up only the negative limiting value of $\bar{g}(C) - 1$, in spite of the fact that the system is positively cooperative in the major part of the binding process.

These two examples (and a few more in the following chapters) clearly demonstrate that there is no single number that conveys the cooperativity of the system. Sometimes, it is also meaningless to claim that a system is positively or negatively cooperative. The introduction of a concentration-dependent measure of the average correlation between ligands is useful because it "does justice" to all types of correlations in the system. When the binding process begins at low ligand concentration, the behavior of the binding isotherm is determined first by the intrinsic binding constants, then by the pair correlations, and, as the concentration increases, successively higher-order correlations come into play. Eventually, at $C \to \infty$, all the sites will be occupied. At this end only the highest-order correlation determines the behavior of the binding isotherm. It is only at this end that the MS measure correctly represents the correlation in the system.

5.9. CORRELATIONS BETWEEN TWO AND THREE PROTONS

Table 5.1 presents some values for the intrinsic binding constants, and also pair and triplet correlations for benzoic acid, benzene dicarboxylic acids, and benzene tricarboxylic acids.

Table 5.1.
Intrinsic Binding Constants, Correlation Functions, and Corresponding Free Energies (in kcal/mol) for Various Carboxylic Acids[a]

Acid	k_1	$k_2 = k_{1/1}$	$k_3 = k_{1/11}$	$g(1, 1), W(1, 1)$	$g(1, 1, 1), W(1, 1, 1)$	δW
Benzoic acid	1.58×10^4	—	—	—	—	—
Benzene-1,2-dicarboxylic acid	1.28×10^5	1.78×10^3	—	1.39×10^{-2}, 2.534	—	—
Benzene-1,3-dicarboxylic acid	1.99×10^4	8.34×10^3	—	4.19×10^{-1}, 0.516	—	—
Benzene-1,4-dicarboxylic acid	1.44×10^4	6.93×10^3	—	4.81×10^{-1}, 0.434	—	—
Benzene-1,3,5-tricarboxylic acid	1.67×10^4	7.76×10^3	3.95×10^2	4.65×10^{-1}, 0.454	1.1×10^{-2}, 2.672	+1.31
Benzene-1,2,3-tricarboxylic acid	[2.36×10^5]	[1.12×10^4]	[1.57×10^3]	[4.75×10^{-2}], [1.805]	[3.17×10^{-4}], [4.773]	[−0.64]
Benzene-1,2,4-tricarboxylic acid	[4.39×10^4]	[5.25×10^3]	[6.72×10^2]	[1.19×10^{-1}], [1.259]	[1.825×10^{-3}], [3.736]	[−0.041]
cis–cis-Cyclohexane-1,3,5-triamine	8.37×10^9	5.01×10^8	2.38×10^7	5.99×10^{-2}, 1.668	1.704×10^{-4}, 5.141	+0.136

[a]The experimental values are taken from Kortum, Vogel, and Andrussow (1961). Most of the measurements were carried out at an ionic strength of less than 0.04 mol/dm^3 and at 25 °C. Average quantities are included in square brackets.

The intrinsic binding constants were calculated from the experimental thermodynamic constants (see Section 2.3) as follows:

For the monocarboxylic acid,

$$k_1 = K_1 \tag{5.9.1}$$

For the dicarboxylic acid,

$$k_1 = K_1/2, \qquad k_2 = (k_{1/1}) = 2K_2 \tag{5.9.2}$$

For the tricarboxylic acid,

$$k_1 = K_1/3, \qquad k_2 = (k_{1/1}) = K_2, \qquad k_3 = (k_{1/11}) = 3K_3 \tag{5.9.3}$$

The pair and triplet correlations and the corresponding free energies are

$$g(1, 1) = \frac{k_1 k_2}{k_1^2} = \frac{k_2}{k_1}, \qquad g(1, 1, 1) = \frac{k_1 k_2 k_3}{k_1^3} \tag{5.9.4}$$

and

$$W = -k_B T \ln g \tag{5.9.5}$$

There are several points worth noting. First, the pair correlation in benzene-1,2-dicarboxylic acid (phthalic acid) is strongly negative, probably due to the strong repulsive electrostatic interaction between the two protons. The correlation becomes weaker as the proton–proton distance increases in the 1,3(meta)- and 1,4(para)-isomers. Second, the *pair* correlation in the benzene-1,3,5-tricarboxylic acid has a value closer to $g(1, 1)$ for the benzene-1,4-dicarboxylic acid, although the proton–proton distances are the same as in the 1,3-dicarboxylic acid. This is clearly due to the effect of the third carboxylic group on the pair correlation in the tricarboxylic acid, indicating a significant nonadditivity. We can confirm this by calculating the nonadditivity in $W(1, 1, 1)$ which, for the benzene-1,3,5-tricarboxylic acid, is

$$\delta W = W(1, 1, 1) - 3W(1, 1) = 2.672 - 3 \times 0.454 = 1.31 \text{ kcal/mol} \tag{5.9.6}$$

This is far from being negligible compared with the values of $W(1, 1, 1)$ or $W(1, 1)$. The source of this nonadditivity is not entirely clear. Since there is no major conformational change in the molecule, it has been attributed [e.g., by Saroff (1987)] to resonance energy and to solvent effects.

Before speculating on the origin of the nonaddivity in the tricarboxylic acid, we note that the *pair* correlation in the cyclohexane-triamine is much stronger than the corresponding pair correlation in benzene-tricarboxylic acid—probably due to the shorter distances between the protons in the former. Furthermore, the nonadditivity in $W(1, 1, 1)$ in this case is (see the last row in Table 5.1)

$$\delta W = W(1, 1, 1) - 3W(1, 1) = 5.141 - 3 \times 1.668 = 0.136 \text{ kcal/mol} \quad (5.9.7)$$

which is negligible compared with either $W(1, 1)$ or $W(1, 1, 1)$.

We recall that direct electrostatic interactions on their own are additive. In a solvent the interaction between three point charges could be nonadditive, due to correlation mediated through the solvent. However, we can rule out this effect in order to explain δW in the 1,3,5-tricarboxylic acid. The reason is that the nonadditivity of $W(1, 1, 1)$ in the cyclohexane-1,3,5-triamine, in spite of the shorter distances between the protons, is quite negligible. Therefore, we conclude that the correlation between protons mediated through the solvent is not the reason for the large nonadditivity effect in the tricarboxylic acid. The solvent can, however, affect the pair and triplet correlations through solvation of the benzene ring. It is well known that the charge distribution on the benzene ring changes upon protonation of benzoic acid. This, in turn, could change the solvation free energy of the acid. Thus, although there exists no major conformational change induced by the process of protonation, the free-energy levels (due to solvation; see Chapter 9) do change upon protonation. This can produce indirect correlations, which we already know to be in general nonadditive.

These indirect correlations (discussed in more detail in Section 9.3) also partially explain the relative larger negative cooperativities in the clyclohexane triamine compared with the corresponding cooperativities in the benzene tricarboxylic acid. As noted earlier, the proton–proton distances in the triamine are shorter than in the tricarboxylic acid, hence the *direct* cooperativities in the former are expected to be larger (more negative) than the latter. In addition, if solvation of the benzene ring also produces indirect cooperativities in the tricarboxylic acid (but not in the triamine), this, as we know, produces positive cooperativity. Hence this effect will tend to make the cooperativities in the tricarboxylic acid less negative, as is observed for the 1,3,5-isomer.

In Table 5.1 we also included some data on benzene-1,2,3- and benzene-1,2,4-tricarboxylic acid. Here, in contrast with the 1,3,5-isomer where all sites are identical in a strict sense, the three sites are not identical, not even in a weak sense. In the 1,2,3-isomer, there are two different binding constants and two different pair correlations. In the 1,2,4-isomer, all the sites are different and we have three intrinsic binding constants and three pair correlations. Therefore, the values reported in Table 5.1 [in square brackets] should be understood in an *average* sense (for details, see Appendix J). Nevertheless, it is of interest that the pair correlations are roughly correlated with the average distance between pairs of protons, being largest in the 1,3,5-isomer and shortest in the 1,2,3-isomer. Another, quite surprising, finding is the small nonadditivity effect in the 1,2,3- and 1,2,4-isomers compared with the large nonadditivity in the 1,3,5-isomer. This is probably due to a stronger solvation effect in the symmetrical 1,3,5-isomer.

5.10. BINDING OF PROTEINS TO DNA

5.10.1. Introduction

A central problem in molecular biology is the elucidation of the mechanism by which regulatory proteins control gene expression. It is known that the cooperative binding of repressor proteins to specific sites on the DNA, called the operator, is the key to understanding the mechanism of turning on and off the transcription of specific genes. The working model, developed mainly by Ptashne, has therefore been justifiably referred to as "a generic switch" [Ptashne (1992), Hochschild and Ptashne (1988), Johnson, Meyer, and Ptashne (1979)].

A quantitative model for gene regulation by λ-phage repressor has been developed by Ackers *et al.* [Ackers, Johnson, and Shea (1982); Senear, Brenowitz, Shea, and Ackers (1986)]. Their molecular model consists of three binding sites, on which the repressor can bind, with different binding constants. By analyzing the individual binding isotherms, obtained by footprinting titration, they found that there is a strong positive cooperativity between adjacent sites. More recently, the same experimental data were processed differently, arriving at different, sometimes opposite, conclusions ([Saroff (1993)].

The BI for a system with three different binding sites is given in Eq. (5.3.4) in terms of the intrinsic binding constants k_a, k_b, and k_c and the various *direct* correlations. The more general form for the total BI, per binding system, is

$$n = \lambda \frac{\partial \ln \xi}{\partial \lambda} = \frac{(k_a + k_b + k_c)C + 2(k_{ab} + k_{bc} + k_{ac})C^2 + 3k_{abc}C^3}{1 + (k_a + k_b + k_c)C + (k_{ab} + k_{bc} + k_{ac})C^2 + k_{abc}C^3} \tag{5.10.1}$$

Clearly, with this BI one cannot resolve all the seven intrinsic binding constants $k_a, k_b, k_c, k_{ab}, k_{ac}, k_{bc}$, and k_{abc}. These can be obtained, however, from the individual BIs which, for this case, are

$$\theta_a = \frac{k_a C + (k_{ab} + k_{ac})C^2 + k_{abc}C^3}{D} \tag{5.10.2}$$

$$\theta_b = \frac{k_b C + (k_{ab} + k_{bc})C^2 + k_{abc}C^3}{D} \tag{5.10.3}$$

$$\theta_c = \frac{k_c C + (k_{ac} + k_{bc})C^2 + k_{abc}C^3}{D} \tag{5.10.4}$$

where D is the denominator in Eq. (5.10.1) and $\bar{n} = \theta_a + \theta_b + \theta_c$.

It should be stressed that the above individual BIs must be measured under the same conditions used for the total Bl. This is not an easy experimental task, since one must follow the fractional saturation at a specific site, say a, while all other sites are free to bind ligands, as if we were measuring the total BI. These are different from the conditional BI, where one follows the fractional saturation at one site, *given* that other sites are constrained to be empty or occupied (see also Section 2.1).

The relevant individual BIs enable one to determine all the intrinsic binding constants k_a, k_b, k_c, k_{ab}, k_{ac}, k_{bc}, and k_{abc}.

Having determined all the intrinsic constants from the experimental data, one can calculate the corresponding correlations. The latter are of major importance in understanding the cooperative behavior of binding systems in general, and biochemical systems such as hemoglobin, regulatory enzymes, etc. in particular. However, in order to understand the cooperative behavior of the system it is indispensable to know the molecular content of the correlation functions without which the experimental data might be erroneously processed and therefore misinterpreted. This information cannot be obtained from the phenomenological description of the BI in terms of the measurable intrinsic binding constants. Instead, one must appeal to statistical mechanics to obtained the required information. The first step in analyzing the molecular content of the correlation function is to acknowledge the fact that the quantity $W(a, b) = -k_BT \ln g_{ab}$ is the *free-energy* change for the process, which we write symbolically as

$$(a, 0, 0) + (0, b, 0) \rightarrow (a, b, 0) + (0, 0, 0) \tag{5.10.5}$$

i.e., we take two singly-occupied systems and form one empty and one doubly-occupied system. Similarly, the quantity $W(a, b, c) = -k_BT \ln g_{abc}$ is the free energy associated with the process

$$(a, 0, 0) + (0, b, 0) + (0, 0, c) \rightarrow (a, b, c) + 2(0, 0, 0) \tag{5.10.6}$$

As shown in Section 5.3, under very special circumstances, $W(a, b)$ and $W(a, b, c)$ reduce to the corresponding *interaction energies* $U(a, b)$ and $U(a, b, c)$, respectively. If this is the case, and when the sites a, b, and c are arranged linearly, then one can approximately assume that

$$W(a, c) \approx U(a, c) \approx 0 \quad \text{and} \quad W(a, b, c) \approx U(a, b, c) \approx U(a, b) + U(b, c) \tag{5.10.7}$$

i.e., one may neglect the long-range interaction and assume pairwise additivity for the triplet interaction energy. Unfortunately, these two approximations do not apply, in general, to the pair and triplet correlations. It is very common to refer to W as a *free energy* of *interaction*, but treat it as if it were an *interaction energy*, hence attributing the properties (5.10.7) to $W(a, b)$ and $W(a, b, c)$. In the next subsection we explore several sources of long-range and nonadditivity of the cooperativity that are peculiar to the repressor–operator binding system.

5.10.2. Sources of Long-Range and Nonadditivity of the Correlation Functions

We briefly present here four different sources of long-range and nonadditivity of the cooperative binding of the λ repressor to the operator.

(a) Mass ratio effect. In most binding systems, ranging from simple dicarboxylic acids to hemoglobin, one can justifiably neglect the mass of the ligand compared with the mass of the adsorbent molecule (Appendix B). This is not the case for the λ repressor. The binding of the first ligand on any site significantly changes the total mass of the adsorbent molecule, and this in turn changes the translational PF of the adsorbent molecule, hence the binding constants for the second and third ligand will be affected. For classical particles with mass m_L, the translational PF is

$$q_{tr} = V/\Lambda^3, \qquad \Lambda = \frac{h}{\sqrt{2\pi m_L k_B T}} \tag{5.10.8}$$

where h and k_B are the Planck and Boltzmann constants, respectively.

Therefore, in the absence of any other effect, the free energies $W(a, b)$ and $W(a, b, c)$ associated with processes (5.10.5) and (5.10.6) due to translational effects only are

$$W_{tr}(a, b) = W_{tr}(a, c) = W_{tr}(b, c) = -k_B T \ln\left[\frac{q_{tr}(a, b, 0)q_{tr}(0, 0, 0)}{q_{tr}(a, 0, 0)q_{tr}(0, b, 0)}\right]$$

$$= -\frac{3}{2} k_B T \ln\left[\frac{1+2x}{(1+x)^2}\right] \tag{5.10.9}$$

where $x = m_L/m_O$ is the ratio of the mass of the ligand **L** to that of the operator **O**.

Similarly, for the triplet cooperativity we have

$$W_{tr}(a, b, c) = -k_B T \ln\left[\frac{q_{tr}(a, b, c)q_{tr}(0, 0, 0)^2}{q_{tr}(a, 0, 0)q_{tr}(0, b, 0)q_{tr}(0, 0, c)}\right]$$

$$= -\frac{3}{2} k_B T \ln\left[\frac{1+3x}{(1+x)^3}\right] \tag{5.10.10}$$

From Eqs. (5.10.9) and (5.10.10) it is clear that both the pair and triplet cooperativities are *negative*, i.e., $W \geq 0$ for any x.

Figure 5.14 shows $W_{tr}(a, b)$, $W_{tr}(a, b, c)$, and two different nonadditivities defined by

$$\delta_1 = W(a, b, c) - W(a, b) - W(b, c) \tag{5.10.11}$$

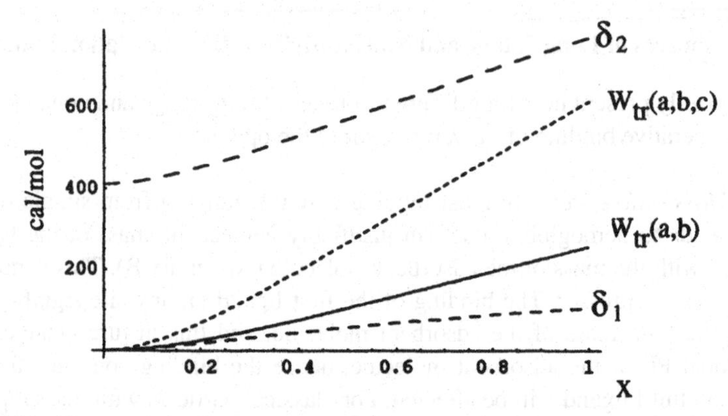

Figure 5.14. The mass ratio effect on the pair and triplet cooperativities and deviations from additivity, δ_1 and δ_2, defined in Eqs. (5.10.11) and (5.10.12), as a function of mass ratio $x = m_L/m_O$.

and

$$\delta_2 = -k_BT \ln\left[\frac{g(a, b, c)}{g(a, b) + g(b, c)}\right] \tag{5.10.12}$$

The first quantity, δ_1, is the deviation from additivity as expected from the interaction energy [Eq. (5.10.7)]. The second quantity, δ_2, is the deviation from additivity of the triplet correlation as used by Ackers *et al.* (1982, 1986) [see Eq. (5.10.23) below]. We see that even when there are no ligand–ligand interactions in the system, the mere fact that the mass ratio x is not negligibly small produced correlations that are both long-range and significantly nonadditive. It is interesting to note that δ_2, the nonadditivity with respect to Acker's assumption [i.e., that $g(a, b, c) = g(a, b) + g(b, c)$], is nonzero even for $x = 0$, in which case $\delta_2 = -k_BT \ln(1/2) = 410.6$ cal/mol. Both δ_1 and δ_2 increase with mass ratio x. Note also that since x is independent of T, W_{tr} is entirely due to the entropy effect.

(b) Moment of inertia effect. Again, we note that in most binding systems, since the mass ratio x is small, one can neglect the effect of ligands on the mass distribution in the adsorbent molecule, hence on its rotational PF. This is not so for the binding of the λ repressor. The exact effect is difficult to calculate (such a calculation requires knowledge of the coordinates of each atom in the system). However, to obtain a rough estimate of the order of magnitude of this effect, we assume that the system is strictly one-dimensional and that the mass of the entire operator is concentrated at three points roughly located at a distance d between the centers of the binding units (Fig. 5.15). The addition of one ligand will change the mass distribution in the adsorbent molecule, hence also its rotational PF. Assuming that this is the only effect of the binding of the ligand and neglecting symmetry

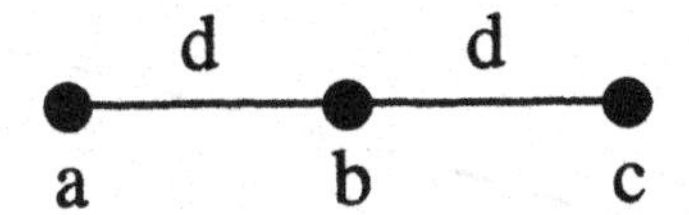

Figure 5.15. A simplified one-dimensional adsorbent molecule having three binding sites. The total mass of the molecule m_O is distributed equally at three points a, b, and c. The mass at any site is $\frac{1}{3}m_O$ when it is unoccupied, and $\frac{1}{3}m_O + m_L$ when occupied by a ligand of mass m_L.

factors, we find

$$\left.\begin{aligned} W_{rot}(a, b) \approx W_{rot}(b, c) &= -k_B T \ln\left[\frac{q_{rot}(a, b, 0)q_{rot}(0, 0, 0)}{q_{rot}(a, 0, 0)q_{rot}(0, b, 0)}\right] \\ &= -k_B T \ln\left[\frac{I(a, b, 0)I(0, 0, 0)}{I(a, 0, 0)I(0, b, 0)}\right] = 0 \\ W_{rot}(a, c) &= -k_B T \ln\left[\frac{4(1+3x)}{(2+3x)^2}\right] \end{aligned}\right\} \tag{5.10.13}$$

where I is its moment of inertia. The triplet cooperativity is

$$\begin{aligned} W_{rot}(a, b, c) &= -k_B T \ln\left[\frac{q_{rot}(a, b, c)q_{rot}(0, 0, 0)^2}{q_{rot}(a, 0, 0)q_{rot}(0, b, 0)q_{rot}(0, 0, c)}\right] \\ &= -k_B T \ln\left[\frac{4(1+3x)}{(2+3x)^2}\right] = W_{rot}(a, c) \end{aligned} \tag{5.10.14}$$

Figure 5.16 shows $W_{rot}(a, c) = W_{rot}(a, b, c)$ and δ_2 as a function of x. In this model, since $I(a, b, 0) = I(a, 0, 0)$, there is no correlation between nearest-neighbor sites, but there is "long-range" correlation, and the triplet correlation is the same as the long-range correlation. This is, of course, a result of the simplified model in which we have concentrated all the mass of the molecule at three points: the center of mass, and two equidistant points from the center of mass. Note again that δ_2 defined in Eq. (5.10.12) is nonzero even at $x = 0$ ($\delta_2 = 410.5$ cal/mol at $x = 0$), but $\delta_1 = W(a, c)$. In this case the nonadditivity of the triplet correlation is not negligible. Note also that all the quantities W_{rot} are entirely due to the entropy effect, e.g., $-\partial W_{rot}(a, b, c)/\partial T$ is equal to $T\Delta S(a, b, c)$ where $\Delta S(a, b, c)$ is the entropy change for the process (5.10.6).

(c) Effect of two (or more) configurations of the bound ligands. In most binding models it is assumed that the bound ligand is characterized by a single state. In the case of the λ repressor, there is evidence that the protein attains at least two orientations when bound to the site. In general, if the ligand (L) has two configurations or conformations on the site, there will be a contribution to the pair

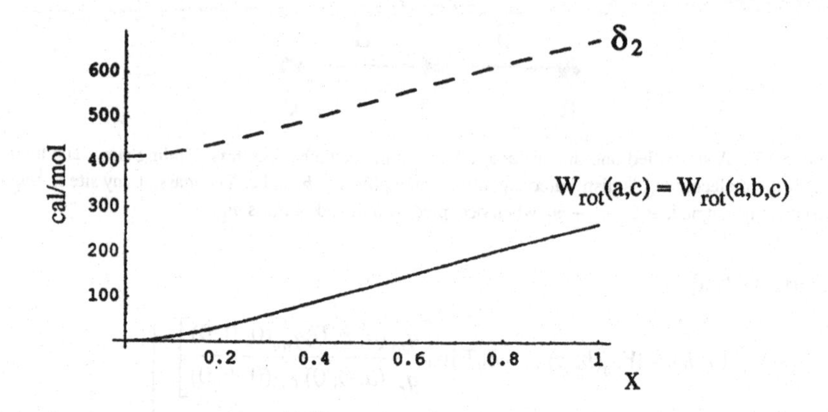

Figure 5.16. Effect of the moment of inertia on the pair and triplet cooperativities [here, $W_{rot}(a, c) = W_{rot}(a, b, c)$ and $W_{rot}(a, b) = W_{rot}(b, c) = 0$] and the deviation from additivity, δ_2 as a function of the mass ratio $x = m_L/m_O$.

cooperativity of the form

$$W_{L\text{-}conf}(a, b) = -k_BT \ln \left[\frac{\Sigma_{ij} \exp[-\beta(U_a^i + U_b^j + U_{ab}^{ij})]}{\Sigma_i \exp[-\beta U_a^i] \Sigma_i \exp[-\beta U_b^i]} \right] \tag{5.10.15}$$

where U_a^i is the binding energy to site a of a ligand in state i, and U_{ab}^{ij} is the corresponding pair *interaction energy*. A simplified form of this contribution is obtained when the binding energies are independent of the conformational state of the ligand, in which case Eq. (5.10.15) reduces to

$$W_{L\text{-}conf}(a, b) = -k_BT \ln [\sum_{ij} \exp(-\beta U_{ab}^{ij})] \tag{5.10.16}$$

A further simplification, relevant to the binding of a repressor to DNA, is that interaction exists only between two specific orientations, like that denoted by RL

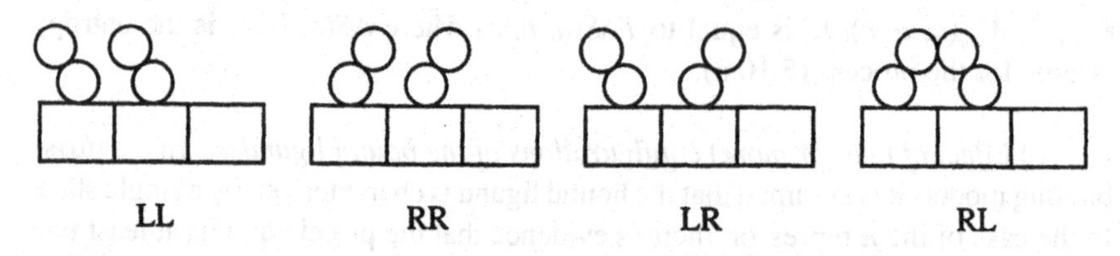

Figure 5.17. Four possible configurations of the ligands occupying two adjacent sites. Only one configuration (RL) contributes to the interaction energy.

in Fig. 5.17. The result in this case is

$$W_{\text{L-}conf}(a, b) = W_{\text{L},conf}(b, c) = -k_B T \ln\left[\frac{\exp(-\beta U_{ab}^{RL}) + 3}{4}\right] \quad (5.10.17)$$

and

$$W_{\text{L-}conf}(a, c) = 0 \quad (5.10.18)$$

It is noteworthy that $W_{\text{L-}conf}(a, b)$ is not equal to U_{ab}^{RL} (as is the case when the ligand has only one configuration). The difference between the *free energy* of interaction (W) and the energy of interaction (U) arises from counting all possible configurations of the pair of ligands (Fig. 5.17), *including* those which do not contribute to the interaction energy (denoted by LL, RR, and LR in Fig. 5.17).

The triplet cooperativity for the simplified case discussed above is

$$\begin{aligned} W_{\text{L-}conf}(a, b, c) &= -k_B T \ln\left[\sum_i \sum_j \sum_l \exp(-\beta U_{ab}^{ij} - \beta U_{bc}^{jl})\right] \\ &= -k_B T \ln\left[\frac{\exp(-\beta U_{ab}^{RL}) + \exp(-\beta U_{bc}^{RL}) + 2}{4}\right] \end{aligned} \quad (5.10.19)$$

Here, only four out of eight possible configurations (Fig. 5.18) contribute to the interaction energy. In terms of the correlation functions, this is equivalent to

$$g_{\text{L-}conf}(a, b, c) = g_{\text{L-}conf}(a, b) + g_{\text{L-}conf}(b, c) - 1 \quad (5.10.20)$$

Once again, we note that the triplet *cooperativity* is not simply related to the pair *interactions*—as expected from the triplet interaction energy, Eq. (5.10.7). It is neither a product of the pair correlation (as normally expected from nearest-neighbor interactions), nor the sum of pair correlations $g(a, b)$ and $g(b, c)$ as assumed by Ackers *et al.* (1982, 1986). Again, the difference between the *free energy* of

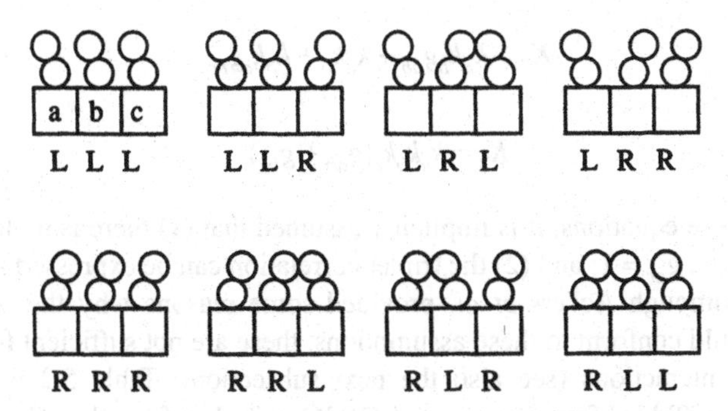

Figure 5.18. Eight possible configurations of ligands occupying three sites. Only four configurations contribute to the interaction energy.

interaction (i.e., the cooperativity) and the *energy* of interaction arises because one must count in the former case also configurations which do not contribute to the interaction *energy*. (These are the four configurations that do not include the nearest-neighbor pair *RL* in Fig. 5.18.) It is worth noting that this effect can produce negative triplet cooperativity in a system where there is one negative and one positive pair cooperativity [e.g., $g(a, b) = 0.5$, $g(b, c) = 1.2$, and $g(a, b, c) = 0.7$]. This is not possible when one assumes additivity in the form $g(a, b, c) = g(a, b) + g(b, c)$.

(d) Effect due to conformational changes induced in the adsorbent molecule. This effect has been discussed at great length in Sections 5.4–5.7. We have seen that if the ligand induces conformation changes in the adsorbent molecule, then long-range correlations can occur (and sometimes, even stronger than the short-range correlations), and the triplet correlations could be significantly nonadditive.

5.10.3. Processing the Experimental Data on Binding of the λ Repressor to the Operator

If we assume that the operator consists of three different binding sites, then the total BI is given by Eq. (5.10.1). Clearly, all seven constants cannot be obtained from this BI. However, these can be resolved from the three individual BI equations, (5.10.2)–(5.10.4).

Ackers *et al.* (1982, 1986) analyzed data on the binding of the λ repressor to the left and right operators obtained by footprinting titration. In order to obtain the correlations in this system, they wrote the following relations between the (overall) macroscopic binding constants K_i $(i = 1, 2, 3)$, defined in Section 2.3, and the intrinsic binding constants k_a, k_b, and k_c:

$$K_1 = k_a + k_b + k_c \tag{5.10.21}$$

$$K_2 = k_a k_b g_{ab} + k_a k_c + k_b k_c g_{bc} \tag{5.10.22}$$

$$K_3 = k_a k_b k_c (g_{ab} + g_{bc}) \tag{5.10.23}$$

In writing these equations, it is implicitly assumed that (1) there is no long-range correlation, i.e., $g_{ac} = 1$; and (2) the triplet correlation can be expressed as the sum $g_{ab} + g_{bc}$. Although Ackers *et al.* provided some reasons why the *interaction energies* should conform to these assumptions, these are not sufficient for the *free energies* of interactions (see also the next subsection). Table 5.2 is almost a reproduction of Table I from Senear *et al.* (1986). It is clear from the column headed "Operator configuration" and footnote *a* to Table 5.2 that the symbol ↔ denotes

Table 5.2

Microscopic Configurations and Associated Free Energies for the λ Operator-Repressor System[a,b]

Species	Operator configuration			Free-energy contribution	Total free energy	Free-energy contributions (our notation)
	Site 1	Site 2	Site 3			
1	O	O	O	Reference state	$\Delta G_{s1} = 0$	—
2	R_2	O	O	ΔG_1	ΔG_{s2}	ΔG_a
3	O	R_2	O	ΔG_2	ΔG_{s3}	ΔG_b
4	O	O	R_2	ΔG_3	ΔG_{s4}	ΔG_c
5	$R_2 \leftarrow$	$\rightarrow R_2$	O	$\Delta G_1 + \Delta G_2 + \Delta G_{12}$	ΔG_{s5}	$\Delta G_a + \Delta G_b + \Delta G(a, b)$
6	R_2	O	R_2	$\Delta G_1 + \Delta G_3$	ΔG_{s6}	$\Delta G_a + \Delta G_c + \Delta G(a, c)$
7	O	$R_2 \leftarrow$	$\rightarrow R_2$	$\Delta G_2 + \Delta G_3 + \Delta G_{23}$	ΔG_{s7}	$\Delta G_b + \Delta G_c + \Delta G(b, c)$
8	$R_2 \leftarrow$	$\rightarrow R_2$	R_2	$\Delta G_1 + \Delta G_2 + \Delta G_3 + \Delta G_{12}$	ΔG_{s8}	
9	R_2	$R_2 \leftarrow$	$\rightarrow R_2$	$\Delta G_1 + \Delta G_2 + \Delta G_3 + \Delta G_{23}$	ΔG_{s9}	$\Delta G_a + \Delta G_b + \Delta G_c + \Delta G(a, b, c)$

[a]Individual operator sites are denoted O if vacant, or R_2 if occupied by a repressor dimer. *Pairwise interactions* between *adjacent occupied sites* are denoted (↔); ΔG_{12} and ΔG_{23} are the free energies of *cooperative interaction* between *adjacent occupied sites*, defined as the difference between ΔG_s for any species and the sum of the intrinsic free energies of binding to occupied sites.

[b]The table, except for the last column, is the same as Table I from Senear *et al.* (1986). The numbers 1, 2, 3 correspond to sites *a*, *b*, and *c*. The underlines in footnote *a* are by the present author.

pairwise interactions between adjacent occupied sites. As noted earlier, assumptions (5.10.7) are good approximations for the interaction *energies*, but they are not valid approximations for the corresponding *free energies*. Since ΔG_{12} is *defined* for species 5 as the difference in free energies, $\Delta G_{s5} - \Delta G_{s2} - \Delta G_{s3}$, one cannot use the same quantity for species 8. A similar comment applies to ΔG_{23} defined for species 7 and used also for species 9. In the last column of Table 5.2 we have listed the free energies as used in the calculations reported below. Thus, by processing the experimental data according to Eqs. (5.10.21)–(5.10.23) Ackers *et al.* reduced the number of parameters from seven to five. [In earlier work (Ackers *et al.*, 1982), they also assumed that $g_{ab} = g_{bc}$, hence reducing the number of unknowns to four: three intrinsic binding constants and only *one* pair correlation.]

We have repeated the calculations of the intrinsic binding constants as well as the correlation functions without making the assumptions in Eqs. (5.10.21)–(5.10.23). Instead, we write

$$K_1 = k_a + k_b + k_c \tag{5.10.24}$$

$$K_2 = k_a k_b g_{ab} + k_a k_c g_{ac} + k_b k_c g_{bc} \tag{5.10.25}$$

$$K_3 = k_a k_b k_c g_{abc} \tag{5.10.26}$$

Here, we make no assumptions regarding the values of either g_{ac} or g_{abc}. The resulting intrinsic binding constants and correlation functions are shown in Table 5.3.*

We note that in spite of the large differences in the two sets of results reported in Table 5.3, the binding isotherms computed with these three sets of results were almost indistinguishable on the scale of Fig. 5.19. The most important differences between the calculated correlations and those reported by Senear *et al.* are, first, there is a large negative cooperativity between sites *a* and *c*, while Senear *et al.* assumed from the outset that no long-range cooperativity exists, and second, the triplet correlation is not additive, i.e., neither δ_1 nor δ_2 is zero, while Senear *et al.* assumed from the outset that $\delta_2 = 0$.

More recently, Saroff (1993) has processed the same experimental data in a different way. However, Saroff uses the same assumptions regarding $g(a, c)$ and $g(a, b, c)$ as made in Eqs. (5.10.21)–(5.10.23). In addition, Saroff introduces new quantities, referred to as *occupied–unoccupied* and *occupied–occupied* interactions, which, in my opinion, are neither well-defined nor necessary quantities. A critical discussion of this approach has been given elsewhere.†

*For details of the calculations, see Ben-Naim (1997, 1998).
†See Ben-Naim (1998) as well as Appendix I.

Table 5.3
Values of the Intrinsic Binding Constants (in lit/mol), Correlation Functions, and Corresponding Free Energies (in kcal/mol) for Binding of the λ Repressor to the Left Operator

	Senear *et al.* (1986)[a]	Computedb
k_a	1.945×10^{10}	2.18×10^{10}
k_b	1.051×10^{9}	1.95×10^{9}
k_c	1.759×10^{9}	5.66×10^{9}
g_{ab}	73.1 [–2.5]	46.53 [–2.23]
g_{ac}	1.0 [0]	9.03×10^{-5} [5.42]
g_{bc}	73.1 [–2.5]	3.48 [–0.73]
g_{abc}	146.2 [–2.9]	40.32 [–2.15]

[a]Computations based on Eqs. (5.10.21)–(5.10.23) (Senear *et al.*, 1986).
[b]Computations based on Eqs. (5.10.24)–(5.10.26) (Ben-Naim, 1998).

Given the arguments in Section 5.10.2; and knowing that (1) the mass of the repressor is not negligible compared with that of the DNA, (2) there are at least two possible orientations of the repressor on the binding sites, and (3) the DNA is far from being rigid; and that some conformational changes are likely to be induced by the binding process [Kondelka *et al.* (1988), Kondelka and Carlson (1992), Steitz (1990)], we conclude that indirect correlations (i.e., correlations not due to direct ligand–ligand interaction) are likely to be important in this system, hence the assumptions made in Eqs. (5.10.21)–(5.10.23), and the results obtained by both Ackers *et al.* and by Saroff are invalid. The main point of our criticism is that even if $g_{ac} = 1$ and $g_{abc} = g_{ab} + g_{bc}$ are approximately correct, this should be revealed as an *output* of processing of the experimental data, not as an *input*, as done by Ackers *et al.* and by Saroff.

5.10.4. Conclusions

There is a general tendency, especially in the biochemical literature, to introduce a *free energy* quantity but, though admitting that this is a free energy, treat it as if it were an *energy* quantity. This inevitably leads to misinterpretations of the experimental data. In the case of binding phenomena, it seems to me that this confusion arises from blurring the distinction between the phenomenological and molecular theories.

The phenomenological approach developed mainly by Wyman, and used by Ackers *et al.*, starts by defining all the binding constants as equilibrium constants for the binding process. Thus, k_a is defined in terms of the standard free-energy change for the process, written symbolically as

$$(0, 0, 0) + \mathbf{L} \rightarrow (a, 0, 0), \qquad k_a = \exp(-\beta \Delta G_a^0) \qquad (5.10.27)$$

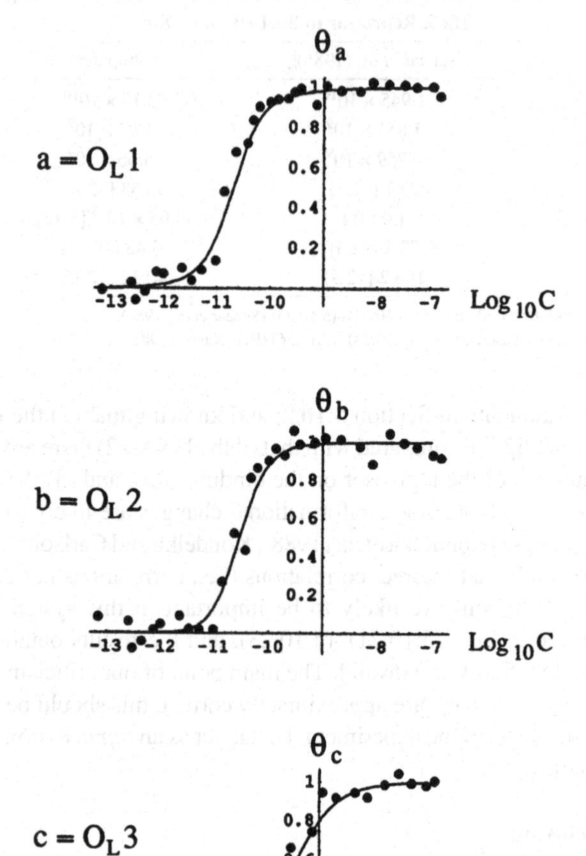

Figure 5.19. Binding isotherms for the λ repressor to the left operator. The points are experimental data taken from Figure 1 of Senear *et al.* (1986). Full curves are the least-squares fit to Eqs. (5.10.2)–(5.10.4). The theoretical curves obtained by the parameters calculated by Senear *et al.* are indistinguishable in this figure from the full curves.

Similarly, the (conditional) binding constant to site a, given that site b is already occupied, is defined in terms of the standard free-energy change for the process,

$$(0, b, 0) + \mathbf{L} \rightarrow (a, b, 0), \qquad k_{a/b} = \exp(-\beta\Delta G^0_{a/b}) \tag{5.10.28}$$

From Eqs. (5.10.27) and (5.10.28), it follows that the quantity referred to as the "interaction free energy" or "interaction constant" and denote by α_{ab} is given by

$$\alpha_{ab} = k_{a/b}/k_a = \exp(-\beta\Delta G^0_{a/b} + \beta\Delta G^0_a) \tag{5.10.29}$$

which is simply related to the standard free energy of the process,

$$(a, 0, 0) + (0, b, 0) \rightarrow (a, b, 0) + (0, 0, 0) \tag{5.10.30}$$

Thus, all the intrinsic binding constants as well as the interaction parameters are related to *free energy* changes for well-defined processes. With the phenomenological approach, one cannot proceed any further to interpret the quantities k_a, $k_{a/b}$, or α_{ab}. This is simply because the molecular *content* of these quantities, or their corresponding free energies, do not reveal themselves to the phenomenological theory. To do so, one must appeal to statistical mechanics in order to express these quantities in terms of the molecular properties of the system, such as the mass, moment of inertia, ligand–ligand, ligand–site interactions, etc. Of course, it is legitimate to call α_{ab}, defined in Eq. (5.10.29), "interaction free energy," "interaction energy," or "interaction coefficient." However, *naming* a quantity "interaction energy" does not *make* it an interaction energy. Doing that could lead to absurd consequences. We demonstrate this by the following example. Suppose we measured the quantities k_a, k_b, k_c, and k_{abc}, and define the quantity

$$\alpha_{abc} = \frac{k_{abc}}{k_a k_b k_c} \tag{5.10.31}$$

In the phenomenological approach there is no way of interpreting α_{abc} (to which we have referred as the triplet correlation function). However, if one treats α_{abc} as if it were an "*interaction energy*," then the natural consequence would be to write, for a linear arrangement of three sites,

$$\alpha_{abc} \approx \alpha_{ab} + \alpha_{bc} \tag{5.10.32}$$

i.e., neglecting the long-range "interaction energy" α_{ac}. This is effectively equivalent to Ackers' assumption, which is made explicitly in writing Eq. (5.10.23) while not explicitly admitting it.

Another, more intelligent guess would be to assume that α_{abc} has a Boltzmann-type dependence on U_{abc}. In this case the natural approximation would be

$$\alpha_{abc} \approx \exp(-\beta U_{abc}) \approx \exp(-\beta U_{ab} - \beta U_{bc}) = \alpha_{ab}\alpha_{bc} \tag{5.10.33}$$

i.e., α_{abc} is a *product* of the pair of "interaction parameters." This is the most common assumption made in the literature, especially in the case of hemoglobin. But, what if

$$\alpha_{abc} = A(a, b, c)\exp(-\beta U_{abc}) \tag{5.10.34}$$

where $A(a, b, c)$ depends on the sites a, b, and c but not on the interaction energy U_{abc}? An example of such behavior is given in Sections 5.5 and 5.6. Here, even when U_{abc} is strictly additive in the sense of (5.10.7), α_{abc} *cannot* be written either as a sum or as a product of α_{ab} and α_{bc}. In general, α_{abc} depends on the interaction energy in a more complicated manner than in Eq. (5.10.34), and there exists no simple relationship between α_{abc}, and α_{ab} and α_{bc}.

Thus, assuming either Eq. (5.10.32) or (5.10.33) in the processing of experimental data must lead to an erroneous conclusion regarding the various interaction parameters.

On the other hand, if one adopts the molecular approach a great deal can be learned about the dependence of the binding constants and free energies of interaction on the molecular properties of the system—as we have done in this and the previous chapters.

It is true that under very special cases, such as the model discussed in Section 5.3, the *interaction free energy* parameter α_{ab} depends only on the ligand–ligand *interaction energy* $U(a, b)$. In this case it is a good approximation to neglect the long-range interaction, i.e., $U(a, c) \approx 0$. Similarly, the triplet interaction free energy in this particular case can be assumed to be pairwise additive, i.e.,

$$W(a, b, c) = U(a, b, c) \approx U(a, b) + U(b, c) \tag{5.10.35}$$

We stress, however, that these two properties of the free energy of interaction are valid only in very special cases. We have shown in Sections 5.4–5.7, that, in general, the interaction free energies do not conform to this behavior; there could be long-range correlations, i.e., α_{ac} [or $g(a, c)$], that differ from unity, and triplet correlations that differ from the product $g(a, b)g(\mathbf{b}, c)$. [There is no known example where the triplet correlation may be written as a *sum* of pair correlations $g(a, b) + g(b, c)$, as assumed by Ackers *et al.*]

The molecular approach, adopted throughout this book, starts from the statistical mechanical formulation of the problem. The interaction free energies are identified as correlation functions in the probability sense. As such, there is no reason to assume that these correlations are either short-range or additive. The main difference between direct and indirect correlations is that the former depend only on the *interactions* between the ligands. The latter depend on the manner in which ligands affect the partition function of the adsorbent molecule (and, in general, of the solvent as well). The argument is essentially the same as that for the difference between the intermolecular potential and the potential of the mean force in liquids.

Therefore, when processing experimental data one must treat these correlations as unknown parameters to be determined by fitting the data to theoretical curves. As we have shown in Section 5.10.3, processing experimental data in this way produces quite different results from those obtained by Ackers *et al.*, as well as by Saroff, where the results are biased by the assumptions concerning the correlations. In Appendix H, we examine a simple model where all the parameters of the system are known exactly. We process a synthetic set of "experimental data" with and without the assumptions made by Ackers *et al.* The results obtained are quite instructive.

6

Four-Site Systems: Hemoglobin

6.1. INTRODUCTION

The extension of the theoretical framework required to deal with systems having four (or more) sites is straightforward. Formally, the extension involves the addition of one more summation to the GPF of the system; see Eq. (6.2.1) below. Conceptually, the only new quantity that appears in the PF is the quadruplet correlation. Like the triplet correlation it may, or may not, be pairwise additive, depending on the specific origin of the correlation. As in the three-site system, where we have compared the behavior of the linear and triangular arrangements, here we examine three different arrangements: the linear, the square, and the tetrahedral (Fig. 6.1a). We shall see that even when the *subunits* are identical (both chemically and structurally), the four *sites* may be either different, identical in a weak sense, or identical in a strict sense. The origin of these differences is discussed in Sections 6.3–6.5. We then examine the correlations in two experimental examples: the first are correlations between protons in benzene-tetracarboxylic acid; the second, the hemoglobin molecule—no doubt the most important and most studied cooperative biomolecule.

6.2. THE GENERAL THEORETICAL FRAMEWORK

We start with the analogue of the model discussed in Sections 4.7 and 5.6. The system consists of *four* identical subunits,* each having one binding site. Each subunit can be in one of two conformational states, L or H, having different

*In the case of hemoglobin discussed in Section 6.8, the four subunits are not identical, but in this and subsequent sections, we assume that the subunits are identical.

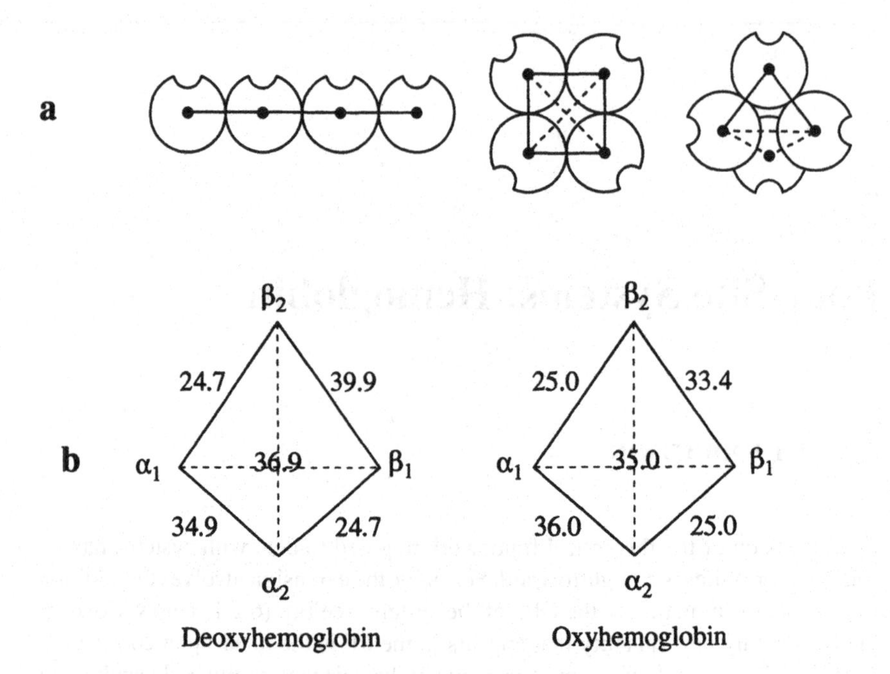

Figure 6.1. (a) Three different arrangements of four identical subunits; linear, square, and tetrahedral. Their corresponding PFs may be constructed from Table 6.1. (b) Schematic description of the distances (in Å) between the four subunits of hemoglobin, in two conformational states.

energies and different binding energies. The GPF for such a system is written as

$$\xi = \sum_{i=0}^{4} Q(i)\lambda^i = \sum_{i=0}^{4} \left[\sum_{\alpha\beta\gamma\delta} Q_{\alpha\beta\gamma\delta}(i) \right] \lambda^i \qquad (6.2.1)$$

Here, $Q(i)$ is the canonical PF of a system having i bound ligands. Each of these may be written as a sum over all possible conformations of the subunits. Since each of the indices α, β, γ, δ can be either L or H, we have altogether sixteen terms in the sum over these indices. The explicit form of each of the factors $Q_{\alpha\beta\gamma\delta}(i)$ depends on the specific arrangements of the subunits (i.e., linear, square, or tetrahedral) as well as the specific i sites on which the i ligands are bound. Each of these factors consists of a "common factor" depending only on the geometrical arrangement of the subunits, and a "variable factor" depending on the specific sites that are occupied. Table 6.1 presents the corresponding constituents of $Q_{\alpha\beta\gamma\delta}(i)$ for the three models: the linear, square, and tetrahedral. To construct $Q_{\alpha\beta\gamma\delta}(i)$ we multiply the

Table 6.1
The Constituents of the Coefficients $Q(i)$ in the Partition Function (6.2.1)[a]

Arrangement	Common factor	Variable factors				
		$Q^*(0)$	$Q^*(1)$	$Q^*(2)$	$Q^*(3)$	$Q^*(4)$
Linear[b]	$Q_\alpha Q_\beta Q_\gamma Q_\delta Q_{\alpha\beta} Q_{\beta\gamma} Q_{\gamma\delta}$	1	$2q_\alpha + 2q_\beta$	$2q_\alpha q_\beta S + 2q_\alpha q_\gamma + q_\alpha q_\delta + q_\beta q_\gamma S$	$2q_\alpha q_\beta q_\gamma S^2 + 2q_\alpha q_\beta q_\delta S$	$q_\alpha q_\beta q_\gamma q_\delta S^3$
Square	$Q_\alpha Q_\beta Q_\gamma Q_\delta Q_{\alpha\beta} Q_{\beta\gamma} Q_{\gamma\delta} Q_{\delta\alpha}$	1	$4q_\alpha$	$4q_\alpha q_\beta S + 2q_\alpha q_\gamma$	$4q_\alpha q_\beta q_\gamma S^2$	$q_\alpha q_\beta q_\gamma q_\delta S^4$
Tetrahedral	$Q_\alpha Q_\beta Q_\gamma Q_\delta Q_{\alpha\beta} Q_{\beta\gamma} Q_{\gamma\delta} Q_{\delta\alpha} Q_{\beta\delta} Q_{\alpha\gamma}$	1	$4q_\beta$	$6q_\alpha q_\beta S$	$4q_\alpha q_\beta q_\gamma S^3$	$q_\alpha q_\beta q_\gamma q_\delta S^6$

[a] To construct the required $Q(i)$ for any particular arrangements of the sites, multiply the common factor in the first column by the corresponding variable factor $Q^*(i)$ in the appropriate row and sum over $\alpha\beta\gamma\delta$.
[b] For the linear arrangement α, β, γ, and δ are the states of the first, second, third, and fourth subunits, respectively.

"common factor," relevant to the subunits' arrangement, by the "variable factor" $Q^*(i)$ in the relevant row. Summation over all the indices α, β, γ, δ (= L, H) gives the quantity $Q(i)$ in Eq. (6.2.1).

Note that direct correlation factors S are included in $Q^*(i)$ ($i \geq 2$) and are assumed to be additive.

As an example, we construct $Q(3)$ for the linear model. The common factor is $Q_\alpha Q_\beta Q_\gamma Q_\delta Q_{\alpha\beta} Q_{\beta\gamma} Q_{\gamma\delta}$ [which has one Q_α ($\alpha = L, H$) for each subunit in state α, and one $Q_{\alpha\beta}$ for each subunit–subunit interaction]. This is multiplied by the variable factor $[2q_\alpha q_\beta q_\gamma S^2 + 2q_\alpha q_\beta q_\delta S]$ and then summed over all the indices $\alpha\beta\gamma\delta$. (Note that in the linear arrangement, the order of the indices $\alpha\beta\gamma\delta$ corresponds to the spatial order of the subunits. Hence, neglecting long-range *direct* interactions, we have a factor S^2 for the occupation of three consecutive sites, but only a factor S for the occupation of the first, second, and fourth sites.) Thus,

$$Q(3) = \sum_{\alpha\beta\gamma\delta} Q_\alpha Q_\beta Q_\gamma Q_\delta Q_{\alpha\beta} Q_{\beta\gamma} Q_{\gamma\delta} [2q_\alpha q_\beta q_\gamma S^2 + 2q_\alpha q_\beta q_\delta S] \tag{6.2.2}$$

The four terms in square brackets correspond to the four configurations of placing three ligands at the four sites. In two of these there is only one direct correlation S, and in the other two there are two factors of direct correlation.

The form of the GPF as written in Eq. (6.2.1) applies to any system of four (different or identical) subunits. However, the construction of $Q_{\alpha\beta\gamma\delta}(i)$ from Table 6.1 is valid only when the subunits are *identical*. Henceforth, we shall always assume that the *subunits* are *identical*. Nevertheless, the properties of the *sites* are not necessarily *identical*, i.e., some of the intrinsic binding constants or correlations might be different. We now consider various cases.

In the tetrahedral model, which possesses the highest symmetry, all the sites are identical in the strict sense. This means that there is only *one* (first) intrinsic binding constant, only one pair correlation, one triplet and one quadruplet correlation.

In the square model, the sites are identical only in a weak sense. This means that there is only *one* (first) intrinsic binding constant, but we have *two* different pair correlations, which are denoted by $g^{(12)}$ and $g^{(13)}$ for the nearest and next-nearest neighbors. Here we also have one triplet and one quadruplet correlation.

In the linear model, which possesses the lowest symmetry, we find two different binding constants, denoted by $k^{(1)}$ and $k^{(2)}$ for binding to the first (or fourth) and second (or third) subunits, respectively. (Clearly, $k^{(1)} = k^{(2)}$ when there are only *direct* correlations in the system; see Section 6.3.) We have *four* different pair correlations denoted by $g^{(12)}$, $g^{(13)}$, $g^{(14)}$, and $g^{(23)}$, and two different triple correlations denoted by $g^{(123)}$ and $g^{(124)}$. Note that in Table 6.1 we assigned a direct correlation factor S only for nearest-neighbor pairs and assumed that S is independent of the state of the subunits.

The explicit expressions for either the binding constants or the correlation functions are, in general, quite complicated even when the subunits are identical. We present here some selected expressions obtained for special cases. As before, we use the reduced parameters

$$K = Q_H/Q_L,\quad h = q_H/q_L,\quad K' = \frac{Q_{HH}}{Q_{LL}},\quad \eta = \frac{Q_{LH}^2}{Q_{LL}Q_{HH}} \tag{6.2.3}$$

6.3. THE LINEAR MODEL

Perhaps the simplest way of understanding the behavior of the linear model is to examine the PF of the system when $\eta = 1$, but $Q_{LL} \neq Q_{HH}$. In this case the PF factorizes into four factors: two factors correspond to the edge subunits, the first and fourth; and two to the center subunits, the second and third. Thus,

$$\begin{aligned} \xi = \xi_{edge}^2 \xi_{center}^2 &= [Q_H\sqrt{Q}_{HH} + Q_L\sqrt{Q}_{LL} + (Q_H\sqrt{Q}_{HH}q_H + Q_L\sqrt{Q}_{LL}q_L)C]^2 \\ &\times [Q_HQ_{HH} + Q_LQ_{LL} + (Q_HQ_{HH}q_H + Q_LQ_{LL}q_L)C]^2 \end{aligned} \tag{6.3.1}$$

Note the difference in the two factors in Eq. (6.3.1). Since the edge subunits participate only in one subunit–subunit interaction but the center subunits are flanked by two subunits, $\sqrt{Q_{HH}}$ appears in ξ_{edge} but Q_{HH} in ξ_{center}.

The difference in behavior of the edge and center subunits is the reason for having two *different* intrinsic binding constants, denoted by $k^{(1)}$ and $k^{(2)}$, respectively. Recall that the subunits themselves are *identical*, hence q_L and q_H are the same for each subunit, which is also the same for a *separate* subunit. It is the different *averages* over q_L and q_H that make $k^{(1)}$ and $k^{(2)}$ different. The explicit expressions for the general case are complicated. They are relatively simple for the particular case $\eta = 1$:

$$k^{(1)} = \frac{Q_L\sqrt{Q}_{LL}q_L + Q_H\sqrt{Q}_{HH}q_H}{Q_L\sqrt{Q}_{LL} + Q_H\sqrt{Q}_{HH}}\lambda_0$$

and (6.3.2)

$$k^{(2)} = \frac{Q_LQ_{LL}q_L + Q_HQ_{HH}q_H}{Q_LQ_{LL} + Q_HQ_{HH}}\lambda_0$$

Note the different weights in forming the averages of q_L and q_H. The binding constant on a *separated* subunit (i.e., the isolated monomer) is obtained from Eq. (6.3.2) by substituting $Q_{LL} = Q_{HH}$, in which case

$$k^{(0)} = \frac{Q_Lq_L + Q_Hq_H}{Q_L + Q_H}\lambda_0 \tag{6.3.3}$$

This, again, is the average of q_L and q_H and, in general, differs from $k^{(1)}$ and $k^{(2)}$, and cannot serve as an approximate binding constant for the four-subunit system. Note that when $Q_{LL} = Q_{HH} = Q_{LH}$ and $Q_L = Q_H$, all three constants $k^{(0)}$, $k^{(1)}$, and $k^{(2)}$ become identical and equal to $(q_L + q_H)\lambda_0/2$.

We next turn to the indirect pair correlations. As noted before, we expect to have four different pair correlations that we denote by $g^{(12)}$, $g^{(13)}$, $g^{(14)}$, and $g^{(23)}$. Clearly, the difference between these arises from the different "routes of communication" between the different pairs of sites. Again, the explicit expression for these, in the general case, is very complicated. Nevertheless, for the particular case $K = 1$ and $K' = 1$, we have the following relatively simple but still informative expressions:

$$y^{(12)} = 1 - \frac{(\sqrt{\eta} - 1)(q_H - q_L)^2}{(\sqrt{\eta} + 1)(q_H + q_L)^2}, \qquad g^{(12)} = y^{(12)}S \tag{6.3.4}$$

$$y^{(13)} = 1 + \frac{(\sqrt{\eta} - 1)^2(q_H - q_L)^2}{(\sqrt{\eta} + 1)^2(q_H + q_L)^2}, \qquad g^{(13)} = y^{(13)} \tag{6.3.5}$$

$$y^{(14)} = 1 - \frac{(\sqrt{\eta} - 1)^3(q_H - q_L)^2}{(\sqrt{\eta} + 1)^3(q_H + q_L)^2}, \qquad g^{(14)} = y^{(14)} \tag{6.3.6}$$

$$y^{(23)} = 1 - \frac{(\sqrt{\eta} - 1)(q_H - q_L)^2}{(\sqrt{\eta} + 1)(q_H + q_L)^2}, \qquad g^{(23)} = y^{(23)}S \tag{6.3.7}$$

Note that a factor S appears only for nearest-neighbor correlations. In all cases when either $q_L = q_H$ or $\eta = 1$, there will be no indirect correlations (both pair and higher order). The direct interactions contribute a factor S for $g^{(12)}$ and $g^{(23)}$, but $S \approx 1$ for the cases $g^{(13)}$ and $g^{(14)}$.

The signs of the indirect correlations are determined by the factor $(-1)^l(\sqrt{\eta} - 1)^l$, where l is the number of boundaries between the communicating subunits. Thus, for $y^{(12)}$ we have only one boundary. The indirect cooperativity is positive ($y^{(12)} > 1$) when $\eta < 1$ and negative when $\eta > 1$.

For $y^{(13)}$, we have two boundaries to cross in order to transmit information (on occupation) between the first and third sites, hence $(-1)^2(\sqrt{\eta} - 1)^2$ is positive, i.e., this pair cooperativity is always positive. For $y^{(14)}$, we have three boundaries to cross, hence $(-1)^3(\sqrt{\eta} - 1)^3$ is positive or negative according to whether $\eta < 1$ or $\eta > 1$, respectively. Note, however, that the magnitude of $y^{(14)}$ differs from $y^{(12)}$. This produces a "distance" dependence of the pair correlation that will be further discussed in Chapter 8. Finally, $y^{(23)}$ is in general different from the other pair correlations, and is equal to $y^{(12)}$ for the particular choice $K = 1$ and $K' = 1$, as assumed in Eq. (6.3.7).

The triplet and quadruplet correlations are, in general, very complicated for the general cases, and may be shown to have the general form

$$y^{(123)} = 1 - (\sqrt{\eta} - 1)(q_H - q_H)^2 P, \qquad g^{(123)} = y^{(123)} S^2 \tag{6.3.8}$$

and

$$y^{(1234)} = 1 - (\sqrt{\eta} - 1)(q_H - q_H)^2 P', \qquad g^{(1234)} = y^{(1234)} S^3 \tag{6.3.9}$$

where P and P' are positive complicated expressions. Again, we find that either $\eta = 1$ or $h = 1$ is a necessary and sufficient condition for $y = 1$ (y being any indirect correlation).

6.4. THE SQUARE MODEL

The square model is simpler than the linear model. Since all the sites are identical, in a weak sense, we have only *one* (first) intrinsic binding constant defined by

$$k = \frac{Q(1, 0, 0, 0)}{Q(0, 0, 0, 0)} \lambda_0 \tag{6.4.1}$$

This has a particularly simple form for $\eta = 1$ and $Q_{LL} \neq Q_{HH}$,

$$k = \frac{Q_L Q_{LL} q_L + Q_H Q_{HH} q_H}{Q_L Q_{LL} + Q_H Q_{HH}} \lambda_0 \tag{6.4.2}$$

which is identical in form to $k^{(2)}$ in Eq. (6.3.2). The reason is that in both cases the binding subunit is flanked by two subunits. When $Q_{LL} = Q_{HH}$, we have again the binding constant for a single *separated* subunit, i.e.,

$$k^{(0)} = \frac{Q_L q_L + Q_H q_H}{Q_L + Q_H} \lambda_0 \tag{6.4.3}$$

which is the same as Eq. (6.3.3).

In the square model we have two different pair correlations: the nearest and second-nearest neighbors. The corresponding direct correlations are S and S'. In general, S' is neglected compared with S. Neglecting altogether the direct ligand–ligand correlation, we have for the indirect correlations, in the particular case $K = 1$

and $K' = 1$,

$$y_{nn}(2) = 1 - \frac{(\eta - 1)(\eta + 1)(q_H - q_L)^2}{(1 + 6\eta + \eta^2)(q_H + q_L)^2} \tag{6.4.4}$$

$$y_{sn}(2) = 1 + \frac{(\eta - 1)^2(q_H - q_L)^2}{(1 + 6\eta + \eta^2)(q_H + q_L)^2} \tag{6.4.5}$$

$$y(3) = 1 - \frac{(\eta - 1)(q_H - q_L)^2}{(1 + 6\eta + \eta^2)(q_H + q_L)^2} \tag{6.4.6}$$

$$y(4) = 1 - (\eta - 1)(q_H - q_L)^2 P \tag{6.4.7}$$

Note that the sign of the nearest-neighbor pair correlation depends on whether $\eta > 1$ or $\eta < 1$. This is similar to $g^{(12)}$, in Eq. (6.3.4). On the other hand, the second-nearest-neighbor pair correlation is always positive, as in $g^{(13)}$; see Eq. (6.3.5). The reason for this difference is again due to having one intervening boundary in the former case, but two boundaries in the latter. As before, either $\eta = 1$ or $q_H = q_L$ is a necessary and sufficient condition for $y = 1$ (any order). We note also that even in this particular case ($K = 1$, $K' = 1$) there exists no additivity either of the form $y(3) = y_{nn}(2)y_{nn}(2)y_{sn}(2)$ or $y(3) = y_{nn}(2)y_{nn}(2)$. A similar comment applies for $y(4)$. P is again a complicated positive expression.

6.5. THE TETRAHEDRAL MODEL

This is the model with the highest symmetry, hence there is only one (first) intrinsic binding constant and one correlation of each order. As always, the intrinsic binding constant is given by

$$k = \frac{Q(1, 0, 0, 0)}{Q(0, 0, 0, 0)} \lambda_0 = (X_L^0 q_L + X_H^0 q_H)\lambda_0 \tag{6.5.1}$$

where X_L^0 and X_H^0 are the mole fractions of the L and H states of a subunit, connected to all the other subunits. This is

$$X_L^0 = \frac{1 + 3\eta K^2 + \eta^{3/2} K(3 + K)^3}{1 + 6\eta^2 K^2 + K^4 + 4\eta^{3/2}(K + K^3)} \tag{6.5.2}$$

where $K = K(K')^{3/2}$. This is fairly complicated. The reason is that the $L \rightleftharpoons H$

equilibrium in a single subunit is determined not only by K (as in the isolated subunit), but by the interactions among *all* the other subunits. When $K' = 1$ and $\eta = 1$, the intrinsic binding constant reduces to the simpler form

$$k = \frac{Q_L Q_{LL}^{3/2} q_L + Q_H Q_{HH}^{3/2} q_H}{Q_L Q_{LL}^{3/2} + Q_H Q_{HH}^{3/2}} \lambda_0 \tag{6.5.3}$$

Compare with Eq. (6.4.2). The difference arises because here each subunit is in contact with *three* subunits, while in the square model it is in contact with *two* subunits. Compare also with Eq. (6.3.2). The direct correlations in this system are simple: S, S^3, and S^6.

The indirect correlations for the particular case $K = 1$, $K' = 1$ all have the form

$$y = 1 - (\sqrt{\eta} - 1)(q_H - q_L)^2 P \tag{6.5.4}$$

where P is always positive and differs for the various orders. Again, we stress that the indirect correlations are not additive, even when the direct correlations are additive.

Owing to the importance of the tetrahedral model for hemoglobin, we present also the form of the correlation functions in the limit $\eta \to 0$. Recall (Section 4.7) that in this limit all the subunits change simultaneously from H to L. Thus, we have a two-state adsorbing system. In this case the intrinsic binding constant is

$$k = \frac{Q_L^4 Q_{LL}^6 q_L + Q_H^4 Q_{HH}^6 q_H}{Q_L^4 Q_{LL}^6 + Q_H^4 Q_{HH}^6} \lambda_0 \tag{6.5.5}$$

as expected from a two-state system with "energy levels" $4E_L + 6E_{LL}$ and $4E_H + 6E_{HH}$. The correlation functions for this case are

$$\left.\begin{aligned} g(2) &= 1 + \frac{K^4(h-1)^2}{(1+hK^4)^2} \\ g(3) &= 1 + \frac{K^4(h-1)^2(K^4 + 2hK^4 + h + 2)}{(1+hK^4)^3} \\ y(4) &= 1 + \frac{K^4(h-1)^2[(1+3K^4+3K^8)(h^2+2h) + K^8 + 3K^4 + 3]}{(1+hK^4)^4} \end{aligned}\right\} \tag{6.5.6}$$

Thus, all cooperativities in this limit are positive, as expected from a two-state binding system.

6.6. THE AVERAGE COOPERATIVITY OF THE LINEAR, SQUARE, AND TETRAHEDRAL MODELS: THE "DENSITY OF INTERACTION" ARGUMENT

We compare here the average correlation in the three models of the four-site system. In the case of *direct* interactions only, it is intuitively clear and easily proven that the average correlation depends only on the sign of $S(2) - 1$ (assuming the subunits are identical, that direct interactions are pairwise additive, and neglecting long-range interactions). Hence, when $S(2) > 1$ (positive direct correlation), we always have

$$\bar{g}_T(C) \geq \bar{g}_S(C) \geq \bar{g}_L(C) \tag{6.6.1}$$

and

$$\theta_T(C) \geq \theta_S(C) \geq \theta_L(C) \tag{6.6.2}$$

where the subscripts T, S, and L denote the tetrahedral, square, and linear models, respectively. The reverse is true for $S(2) < 1$ (negative direct correlation). This is demonstrated in Fig. 6.2. For $S(2) > 1$, the curves for L, S, and T show increasing

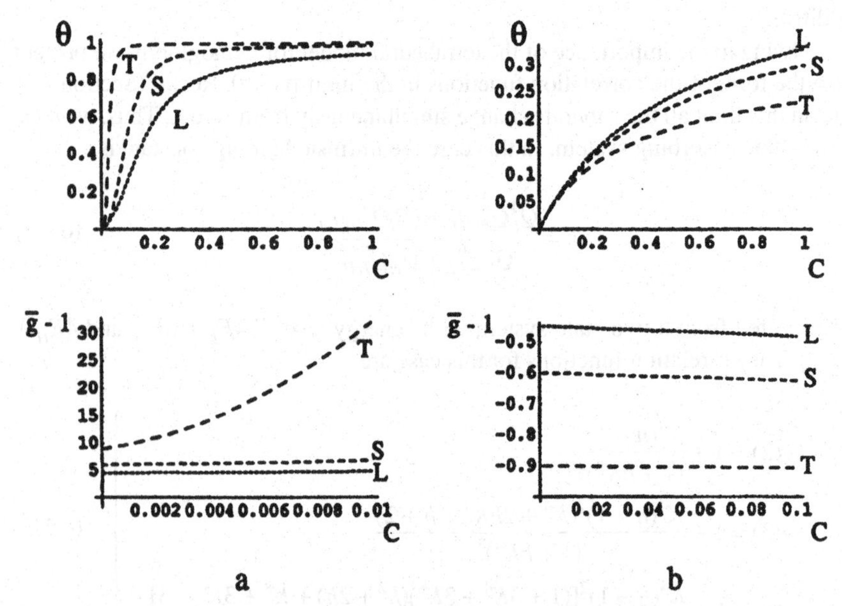

Figure 6.2. Binding isotherms and the average correlation, $\bar{g}(C) - 1$ for the tetrahedral (T), square (S), and linear (L) models. The sites are identical and all correlations are due to *direct* ligand–ligand pairwise additive interactions. (a) Curves for positive cooperativity, $S(2) = 10$; (b) curves for negative cooperativity, $S(2) = 0.1$. Note that in these systems the cooperativity increases in absolute magnitude from L to S to T.

cooperativity. This is clear from both the BIs and the average correlations, plotted as $\bar{g}(C) - 1$. The proof of inequalities (6.6.1) and (6.6.2) is easily obtained from the definitions of $\bar{g}(C)$ and $\theta(C)$, i.e.,

$$g_T(C) - g_s(C) = [S(2) - 1]P_1, \qquad g_S(C) - g_L(C) = [S(2) - 1]P_2 \quad (6.6.3)$$

and

$$\theta_T(C) - \theta_s(C) = [S(2) - 1]P_3, \qquad \theta_S(C) - \theta_L(C) = [S(2) - 1]P_4 \quad (6.6.4)$$

where P_i are positive numbers. This result has been ascribed to the effect of the "density of interaction" by T. L. Hill (1985). Clearly, the tetrahedral model has a higher "density of interaction" than the square model, and the latter higher than the linear model.

The "density of interaction" argument does not hold for the case of indirect correlations. We recall that when there are indirect correlations we have three different correlations and, in general, it is meaningless to refer to *the correlation* or *the cooperativity* of the system. As we have seen in Section 5.8, the average cooperativity is, in general, dependent on the ligand concentration and may even change sign during the binding process.

In Fig. 6.3 we compare the square and tetrahedral models for four identical subunits with parameters

$$h = 0.01, \qquad K = 1, \qquad Q_{HH} = 4Q_{LL}, \qquad \eta = 0.01 \quad (6.6.5)$$

Judging from the shapes of the BIs, we should conclude that the square model is *more* cooperative than the tetrahedral model. Indeed, all the cooperativities in this system are larger for the square model. We have computed all the correlations for these two models, using the parameters in (6.6.5). These yield: for the square model

$$g_{nn}(2) = 15.3, \qquad g_{sn}(2) = 13.5, \qquad g(3) = 698, \qquad g(4) = 37348 \quad (6.6.6)$$

and for the tetrahedral model

$$g(2) = 3.3, \qquad g(3) = 221, \qquad g(4) = 21058 \quad (6.6.7)$$

The average correlation, plotted as $\bar{g}(C) - 1$, shows that the square model starts initially with a small positive value and increases monotonously to the very large value of 37,348 at $C \to \infty$. On the other hand, the $\bar{g}(C) - 1$ curve for the tetrahedral model starts from a very small value and reaches the value of about 21,058 at very high concentrations. Clearly, both of the BIs appear as positive cooperative, but with much stronger cooperativity for the *square* model, in apparent defiance of the "density of interaction" argument.

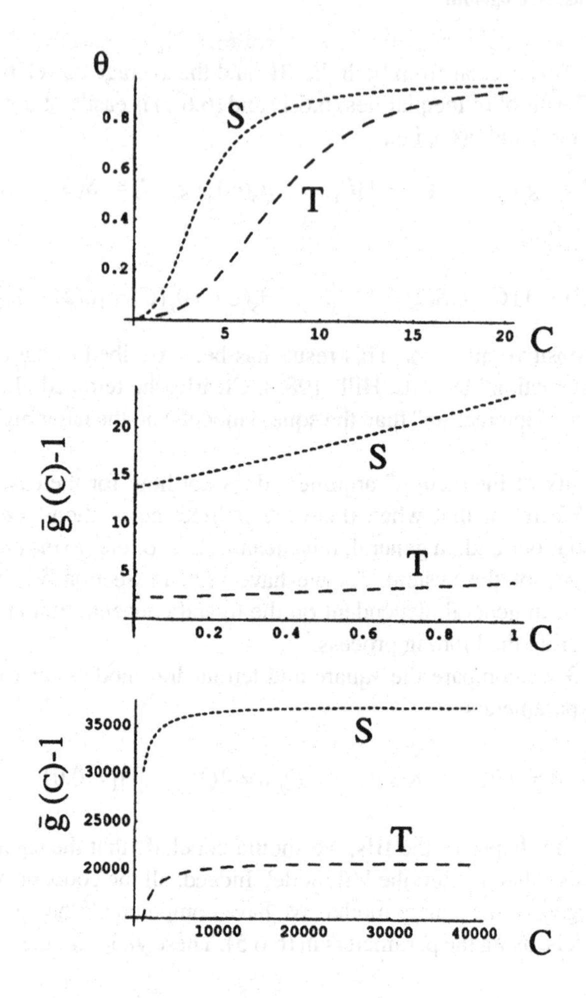

Figure 6.3. The binding isotherm and the average correlation [as $\bar{g}(C)-1$] for the square and tetrahedral models discussed in Section 6.6. No direct ligand–ligand interactions are assumed. The parameters for the indirect correlations are $h = 0.01$, $K = 1$, $Q_{HH} = 4Q_{LL}$, and $\eta = 0.01$. The lower two figures show the behavior of $\bar{g}(C)-1$ in the region $0 \le C \le 1$ and in the limit of large C. Note that in this model, with these particular parameters, the average cooperativity in the square model is everywhere *stronger* than in the tetrahedral model.

6.7. BENZENE-TETRACARBOXYLIC ACIDS

Figure 6.4 shows the structures of benzene-1,2,3,4- and 1,2,4,5- and 1,2,3,5-tetracarboxylic acids. The first is the analogue of the linear model discussed in Section 6.3. Here, we expect two different intrinsic binding constants, $k^{(1)}$ (for the

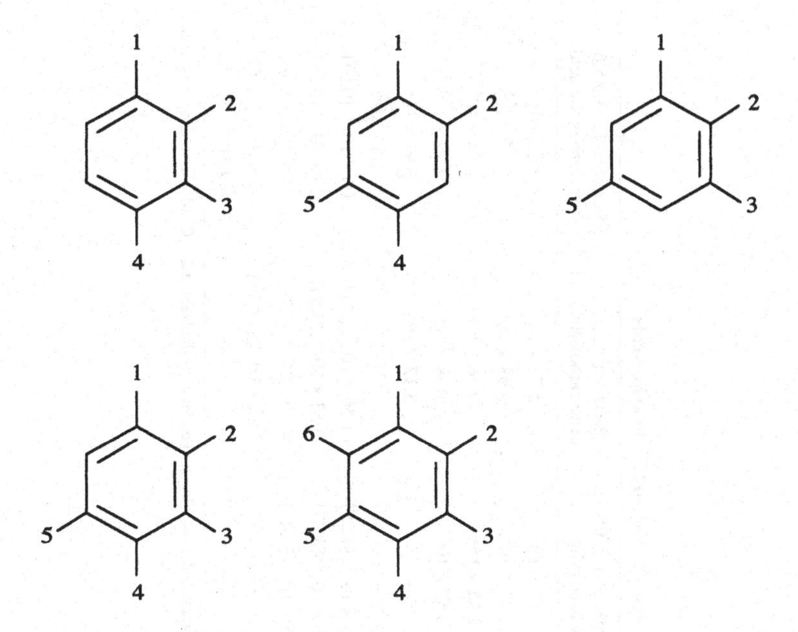

Figure 6.4. The various carboxylic acids referred to in Section 6.7. Three different isomers of benzene tetracarboxylic acid are drawn in the first row. The penta- and hexacarboxylic acids are drawn in the second row.

edge sites, 1 and 4) and $k^{(2)}$ (for the center sites, 2 and 3). The second conforms neither to the square nor to the tetrahedral model. Here, the four sites are equivalent, as in the square model, and although it has a lower symmetry than the square model the sites are identical only in the weak sense. Therefore, we expect to have only one intrinsic binding constant k. However, contrary to the square model where we have only two pair correlations, $g^{(12)}$ and $g^{(13)}$ (see Section 6.4), here we have, in general, *three* different pair correlations $g^{(12)}$, $g^{(14)}$, and $g^{(15)}$ (the numbering is as in Fig. 6.4). There is one triplet and one quadruplet correlation. The third isomer, 1,2,3,5-tetra-carboxylic acid, will have three different intrinsic binding constants $k^{(1)}$ (at positions 1 and 3), $k^{(2)}$ (at position 2), and $k^{(3)}$ (at position 5). There will be four different pair correlations $g^{(12)} = g^{(23)}$, $g^{(13)}$, $g^{(35)} = g^{(15)}$, and $g^{(25)}$, and three different triplet correlations $g^{(123)}$, $g^{(135)}$, and $g^{(125)} = g^{(235)}$.

One cannot resolve all the correlation functions from the experimental (thermodynamic) data. However, by processing the data in the same way that we processed the data as if they were strictly identical sites (see Section 5.9 and Appendix J), we obtain quantities that should be understood only in an average sense, as discussed in Appendix J. The results for benzene-tetracarboxylic acids are reported in Table 6.2. We recall the value of $k_1 = 1.58 \times 10^4$ for benzoic acid. We

Table 6.2
Intrinsic Binding Constants and Correlation Functions for Various Carboxylic Acids[a]

	Benzene-1,2,3,4-tetracarboxylic acid	Benzene-1,2,4,5-tetracarboxylic acid	Benzene-1,2,3,5-tetracarboxylic acid	Benzene-1,2,3,4,5-pentacarboxylic acid	Benzene-1,2,3,4,5,6-hexacarboxylic acid
k_1	4.1×10^5	1.06×10^5	1.63×10^5	5.71×10^5	1.51×10^6
k_2	3.56×10^4	2.08×10^4	1.85×10^4	8.93×10^4	3.12×10^5
k_3	2.68×10^3	1.12×10^3	4.84×10^3	9.26×10^3	5.54×10^4
k_4	4.54×10^2	3.33×10^2	9.52×10^2	1.08×10^3	2.72×10^3
k_5	—	—	—	3.12×10^2	3.91×10^2
k_6	—	—	—	—	1.5×10^2
$g(1, 1)$	8.7×10^{-2}, [1.45]	1.96×10^{-1}, [0.96]	1.13×10^{-1}, [1.29]	1.56×10^{-1}, [1.1]	2.06×10^{-1}, [0.93]
$g(1, 1, 1)$	5.68×10^{-4}, [4.42]	2.06×10^{-3}, [3.66]	3.36×10^{-3}, [3.37]	2.53×10^{-3}, [3.54]	6.19×10^{-3}, [3.01]
$g(1, 1, 1, 1)$	6.30×10^{-7}, [8.46]	6.46×10^{-6}, [7.08]	1.96×10^{-5}, [6.42]	4.79×10^{-6}, [7.26]	1.11×10^{-5}, [6.76]
$g(1, 1, 1, 1, 1)$	—	—	—	2.62×10^{-9}, [11.71]	2.86×10^{-9}, [11.65]
$g(1, 1, 1, 1, 1, 1)$	—	—	—	—	2.84×10^{-13}, [17.12]

[a]The experimental values are taken from Kortum, Vogel, and Andrussow (1961). Values in square brackets are $-k_B T \ln g$ (in kcal/mole) calculated at 25 °C. All the values are average quantities in the sense discussed in Appendix J.

have already seen in Section 5.9 (Table 5.1) that k_1 (except for 1,4-dicarboxylic acid) becomes larger upon the addition of more carboxylate groups. The same is true for the tetra-, penta-, and hexacarboxylic acids, as shown in the first row of Table 6.2. The reason is probably the additional attraction exerted on the proton by the neighboring carboxylate groups. Note that, on the average, k_1 is largest when the groups are close together, i.e., in the 1,2,3,4-isomer, and smallest when the groups are spread as far as possible, i.e., in the 1,2,4,5-isomer. The value of k_1 increases for the penta- and further increases for the hexacarboxylic acid.

The (average) pair cooperativities are all negative; they are largest [i.e., smallest $g(1, 1)$ or largest $W(1, 1)$] for the 1,2,3,4-isomer and smallest for the 1,2,4,5-isomer where, on average, the groups are spread as far as possible.

It is of interest that the nonadditivities of the triplet correlations are quite different for the three isomers (see also Table 5.1). The nonadditivities, computed as

$$\delta_3 = W(1, 1, 1) - 3W(1, 1) \tag{6.7.1}$$

are 0.067, 0.76, and −0.49 (in kcal/mol) for the 1,2,3,4-, 1,2,4,5-, and 1,2,3,5-isomers, respectively. Thus, when the groups are close to each other the triplet correlation is almost additive—indicating that the correlation is mainly due to direct proton–proton interactions. This is similar to the behavior in the tricarboxylic acid series reported in Table 5.1.

It is not clear how to estimate nonadditivities of the quadruplet correlations. Since $W(1, 1)$ is an average over all pairs, the simplest nonadditivity measure would be

$$\delta_4 = W(1, 1, 1, 1) - 6W(1, 1) \tag{6.7.2}$$

These values are −0.22, 1.29, and −1.317 (all in kcal/mol) for the three isomers. The trend is similar to that in the nonadditivity of the triplet cooperativities, i.e., smallest for the 1,2,3,4-isomers and quite large, with different signs, for the other two isomers.

Finally, we note that the nonadditivity in the hexacarboxylic acid, computed as

$$\delta_6 = W(1, 1, 1, 1, 1, 1) - 15W(1, 1) \approx 3.1 \text{ kcal/mol} \tag{6.7.3}$$

is quite large relative to $W(1, 1, 1, 1, 1, 1)$. This is similar to the most symmetric of the tricarboxylic acids, i.e., the 1,3,5-isomer (Table 5.1).

6.8. HEMOGLOBIN—THE EFFICIENT CARRIER OF OXYGEN

6.8.1. Introduction and a Brief Historical Overview

Hemoglin (Hb) is no doubt the most extensively studied protein, in general, and as an allosteric binding system, in particular. Although the details of its structure and function are well known, there are still some gaps in our understanding of the precise molecular mechanism that makes this molecule such an efficient carrier of oxygen from the lungs to the tissues.*

In this section we shall focus only on a few aspects of the function of this remarkable molecule. A more detailed treatment may be found in Antonini *et al.* (1971) and in Imai (1982).

The Hb molecule consists of four subunits: two α-subunits (each with 141 amino acid residues) and two β-subunits (each with 146 amino acid residues).† The distances between the subunits in the two conformations of the Hb are shown schematically in Fig. 6.1b. The tetramer as a whole has a roughly spherical shape. Since the four *subunits* are not identical, one cannot expect that the four binding *sites* will be strictly identical. Nevertheless, in most of the theoretical treatments of the binding properties of Hb, one assumes that the sites are nearly identical, hence all of the intrinsic binding constants, as well as the correlation functions, must be understood only in an average sense, as discussed in Appendix J.

Each of the subunits, α and β (as well as the closely related myoglobin molecule), has a prosthetic heme group to which the oxygen molecule binds. There are no covalent bonds between the subunits of Hb. The aggregate is maintained by a combination of weak direct subunit–subunit interactions as well as by indirect interactions mediated by the solvent.

Perhaps the earliest attempt to explain the sigmoidal shape of the BI of oxygen to Hb was made by A. V. Hill (1910). Hill assumed that a solution of Hb contains different aggregates of binding molecules. If n is the *average* size of the aggregate, then the "average" binding reaction is written as

$$\mathrm{Hb} + n\mathrm{O}_2 \rightarrow \mathrm{Hb(O_2)}_n \tag{6.8.1}$$

and the corresponding equilibrium constant K_H is given by

$$K_H = \frac{[\mathrm{Hb(O_2)}_n]}{[\mathrm{Hb}][\mathrm{O_2}]^n} = \frac{\theta}{1-\theta}[\mathrm{O_2}]^{-n} \tag{6.8.2}$$

where θ is the fractional saturation of Hb defined by

$$\theta = \frac{[\mathrm{Hb(O_2)}_n]}{[\mathrm{Hb}] + [\mathrm{Hb(O_2)}_n]} \tag{6.8.3}$$

The square brackets indicate either concentration or, for oxygen, partial pressure.

*Such an elaborate transport mechanism is not necessary for small organisms, where oxygen can diffuse between different parts of their bodies.

†The α- and β-subunits are structurally and evolutionarily related to each other and to myoglobin (Mb), the monomeric oxygen-binding protein in the muscles.

If K_H is a constant in the sense of being independent of oxygen concentration (or partial pressure), then one can transform Eq. (6.8.2) into

$$\ln \frac{\theta}{1-\theta} = n \ln[O_2] - \ln K_H \tag{6.8.4}$$

A plot of the measurable quantity $\ln\{\theta/(1-\theta)\}$ as a function of $\ln[O_2]$ should give a straight line with slope n. Such a plot is called the Hill plot and is widely used in reporting and interpreting experimental binding data.

The actual Hill plot for Hb is far from a linear line with constant slope. The actual curve has a varying slope between one to three. Thermodynamically, Eq. (6.8.2) implies that all n ligands bind simultaneously to Hb. There is no provision in this model for intermediary occupancy states. Therefore, this model is thermodynamically unacceptable. This is true *a fortiori* when n, obtained by fitting the experimental data, turns out to be a nonintegral number.*

Adair (1925) introduced the stepwise binding model which, for Hb, produces the so-called Adair equation (see Section 2.3). This has the form

$$\theta = \frac{A_1P + 2A_2P^2 + 3A_3P^3 + 4A_4P^4}{4(1 + A_1P + A_2P^2 + A_3P^3 + A_4P^4)} \tag{6.8.5}$$

where P is the partial pressure of oxygen and A_i are known as the Adair constants.

The Adair equation is equivalent to the BI written in terms of the thermodynamic binding constants K_i (see Section 2.3), namely.

$$\theta = \frac{K_1P + 2K_1K_2P^2 + 3K_1K_2K_3P^3 + K_1K_2K_3K_4P^4}{4(1 + K_1P + K_1K_2P^2 + K_1K_2K_3P^3 + K_1K_2K_3K_4P^4)} \tag{6.8.6}$$

where the relations between the A_i's and K_i's and the successive intrinsic binding constants are

$$\begin{aligned} A_1 &= K_1\ (= 4k_1), \qquad A_2 = K_1K_2\ (= 6k_1k_{1/1}) \\ A_3 &= K_1K_2K_3\ (= 4k_1k_{1/1}k_{1/11}), \qquad A_4 = K_1K_2K_3K_4\ (= k_1k_{1/1}k_{1/11}k_{1/111}) \end{aligned} \tag{6.8.7}$$

Here, the equalities in parentheses are strictly valid whenever the sites are identical in the strict sense. Otherwise, one should interpret these intrinsic binding constants as average quantities, as discussed in Appendix J. Since Eq. (6.8.5) has the correct functional form of $\theta(P)$, it is easy to obtain a good fit of the experimental data with the four Adair constants A_i, $i = 1, 2, 3, 4$.

Pauling (1935) attempted to rewrite the BI in terms of two parameters: an intrinsic binding constant k and an interaction parameter α. The latter was assigned

*Regrettably, the slope of the Hill plot is still widely used as a "convenient measure of cooperativity." See also Sections 4.3 and 4.6.

to each nearest-neighbor pair. Pauling considered two "configurations" of the hemes, the square and the tetrahedral, for which, in our notation, the BI is written, respectively, as

$$\theta_S = \frac{kP + (2\alpha + 1)k^2P^2 + 3\alpha^2k^3P^3 + \alpha^4k^4P^4}{1 + 4kP + (4\alpha + 2)k^2P^2 + 4\alpha^2k^3P^3 + \alpha^4k^4P^4} \tag{6.8.8}$$

and

$$\theta_T = \frac{kP + 3\alpha k^2P^2 + 3\alpha^3k^3P^3 + \alpha^6k^4P^4}{1 + 4kP + 6\alpha k^2P^2 + 4\alpha^3k^3P^3 + \alpha^6k^4P^4} \tag{6.8.9}$$

Note that in the square model only nearest-neighbor pairs are assigned an energy parameter α, while in the tetrahedral model each pair of ligands contributes a factor α. Both of these functions gave a good fit to experimental data with $k = 0.033$, and $\alpha = 12$ for the square model and $\alpha = 12^{2/3} = 5.2$ for the tetrahedral model.

Although at that time the structure of Hb was unknown and the allosteric model, based on the induced conformational changes, was not even suspected, Pauling was obviously puzzled when he wrote "it is difficult to imagine a connection between two hemes this far [referring to the tetrahedral arrangement] which would lead to an interaction energy as large as 1000 cal/mol." On the basis of this argument Pauling rejected the tetrahedral model in favor of the square model.

Referring to $-RT \ln \alpha$ as "interaction energy," it was only natural to assume that each "connected pair" would contribute a factor α. This is equivalent to the assumption of pairwise additivity of both the triplet and quadruplet correlation. In our language, these assumptions are equivalent for the square model to

$$g_{nn}(2) = \alpha, \qquad g_{sn}(2) = 1, \qquad g(3) = \alpha^2, \qquad g(4) = \alpha^4 \tag{6.8.10}$$

and for the tetrahedral model to

$$g(2) = \alpha, \qquad g(3) = \alpha^3, \qquad g(4) = \alpha^6 \tag{6.8.11}$$

These (explicit or implicit) assumptions were made by many researchers studying binding phenomena. We have seen in several places in this book (see Chapter 5, in particular Section 5.10) that the assumption of pairwise additivity is approximately valid only when the cooperativity originates from ligand–ligand *direct* interaction. It is not a valid approximation when the ligands interact indirectly. Pauling was clearly thinking in terms of (direct) interaction energies when he assigned "interaction energy" parameter to each "connected pair." Although he did express his puzzlement over the apparent "long-range" interaction, he did not suspect any possible source of indirect correlations.

A major development in our understanding of the mechanism of indirect correlation has been advanced in two classical papers by Monod, Changeux, and Jacob (1963) and Monod, Wyman, and Changeux (1965). In these papers the principle of transmitting information between ligands by means of conformational changes induced in the adsorbent molecule was clearly and convincingly described. This principle became known as the *allosteric effect*. It was first meant to describe the behavior of regulatory enzymes (see Chapter 8), but was also applied to the working mechanism of hemoglobin (originally referred to as the homotropic effect, i.e., correlation between two or more identical ligands; the term "heterotropic effect" was used for correlation between different ligands).

The Monod–Wyman–Changeux (MWC) model consists of the following three elements: (1) The protein (or adsorbent molecule) is assumed to possess at least two stereospecifically different binding sites. In regulatory enzymes one of these sites is the active site, the other is the regulatory site. The term *allosteric* refers here to the *different locations* of the two sites. (2) The protein can attain one of two conformations. These were originally denoted by R (for relaxed) and T (for tense). The transition between T and R was referred to as the *allosteric transition*. This is the second meaning attributed to the term allosteric. (3) The binding of a ligand on one site induces an *allosteric* transition in the protein. Since the *allosteric change* involves the entire protein, it affects also the local conformation on the *allosteric site*. Hence, a second ligand approaching the protein "will know" whether or not the first site is occupied. In the case of Hb, the two or more ligands are the same oxygen molecule. In the general case, the two *allosteric sites* are designed to accept two *different* ligands. In the first case the two ligands are isosteric (i.e., having the same structures) while the second case usually applies to a substrate and an effector having different structures—hence the term *allosteric* refers here to the different structures of the ligands. This is a third meaning applied to the term *allosteric*.

In this book we use the term *allosteric* only in its first meaning, i.e., for two different *locations* of the sites. It should be noted that the third assumption, made in the MWC model, was partially based upon a previous notion developed by Koshland (1958)—the so-called induced-fit model. Both Koshland's idea of the *induced fit* and the MWC model of allosteric induced transition require that the conformation of the protein responds to the binding process. The "purpose" of this response is different in the two models, however. The induced-fit idea was suggested to extend the old idea of the lock and key model proposed by Emil Fisher. It focuses on the *local* change in the conformation of the protein in such a way as to improve the *fit* or the binding affinity between the substrate and the protein. The MWC theory focuses on both the *local* and *global* changes in the conformation of the protein. The binding of a ligand on one site not only changes the local conformation of the protein at the binding site, but also at the allosteric site. This is the essence of the allosteric effect which we have referred to as the indirect correlation, or indirect cooperativity. The induced conformational change on one site is transmitted to all the other allosteric sites.

The MWC model is presently known as the *concerted* model, since the entire protein changes its conformation concertedly. The induced-fit model was later developed by Koshland, Nemethy, and Filmer (KNF) and is presently known as the *sequential* model. These two models were shown to be limiting cases of a more general scheme by Eigen (1967). These two limiting cases were discussed for a two-subunit molecule in Section 4.7. Since the sixties much structural, kinetic, and thermodynamic data have been accumulated. The main idea of the allosteric mechanism of cooperativity is now universally accepted.

In the next section we describe some experimental data on the binding of oxygen to Hb. We shall focus on the efficiency of the loading and unloading of oxygen at two not-too-different partial pressures.

6.8.2. A Sample of Experimental Data

Experimental data on binding oxygen to hemoglobin (Hb) are abundant. There are different types of Hb (human, horse, sheep, etc.), and different conditions under which the binding of different ligands are measured (such as temperature, pressure, and solvent composition). Also, the concentration of Hb in solution could affect its binding properties. Our theoretical treatment has been confined to that range of concentrations at which there is neither dissociation nor association of the Hb tetramer. We present here a very small sample of data illustrating the function of Hb as an oxygen carrier. A comprehensive treatment of binding data of Hb can be found in Imai's monograph (1982).

The most important point to bear in mind when examining binding data of Hb is that the Hb molecule does not operate in vacuum, nor in a pure one-component solvent. The molecular composition of the environment of the Hb molecule can have a decisive role in determining the efficiency of its function. In particular, the utility function, as determined by the pressure difference under which Hb operates, can change substantially when we vary the temperature, the pH, or the composition of the solvent (see the next subsection).

Figure 6.5 shows the BI $\theta(P_{O_2})$ of human adult Hb at different temperatures. [In this particular set of data, based on Imai and Yonetani (1975), the system was held in a fixed buffer solution with no addition of DPG or IHP; see below.]

From the general shape of the curves one gets the impression that the "overall" cooperativity of the Hb (under these specific conditions) *decreases* with *increase* in temperature. This impression is correct in general. However, if we look more closely at the various correlations in the system we see some temperature dependence that does not conform to this "general" trend.

Table 6.3 shows the pair, triplet, and quadruplet correlations at six temperatures.* It is quite clear that the correlations do not change monotonically with temperature. For instance, the pair correlation initially increases with temperature (which is indicative of the dominance of indirect correlation), then decreases up to 30 °C, and again increases. This is also true for the corresponding free energies,

*These should be understood only as *average* quantities, in the sense discussed in Appendix J.

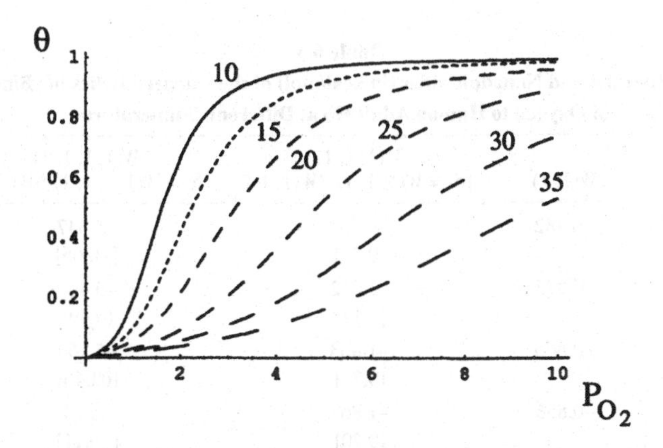

Figure 6.5. The BI, $\theta(P_{O_2})$, of human adult Hb at different temperatures (pressure in torr and temperature in °C). The temperatures are indicated next to each curve. Based on Table I from Imai and Yonetani (1975).

defined by $-k_BT \ln g$, and shown in Table 6.4 and Fig. 6.6. It is clear that these free energies do not change monotonically with temperature. Note that if all correlations were due to *direct* interactions, then the quantities W should be equal to the interaction energies U, and these are presumed to be temperature-independent.

Table 6.4 shows the free energies W corresponding to the correlation functions of Table 6.3. We have also indicated, in square brackets, the nonadditivities (with respect to pairs) defined by

$$\delta_3 = W(1, 1, 1) - 3W(1, 1), \qquad \delta_4 = W(1, 1, 1, 1) - 6W(1, 1) \quad (6.8.12)$$

No general trend is observed in these nonadditivities. It is noteworthy, however,

Table 6.3
Correlation Functions for Human Adult Hb at Different Temperatures[a]

t (°C)	$g(1, 1)$	$g(1, 1, 1)$	$g(1, 1, 1, 1)$
10	3.98	23.10	1.27×10^4
15	5.42	23.24	1.18×10^4
20	4.38	24.02	6.55×10^3
25	4.20	22.95	6.61×10^3
30	2.82	14.86	2.65×10^3
35	4.25	19.49	3.11×10^3

[a]Based on data from Table I of Imai and Yonetani (1975).

Table 6.4
Free Energies and Nonadditivities (in kcal/mol) of the Cooperativities for Binding of Oxygen to Human Adult Hb at Different Temperatures[a]

t (°C)	$W(1, 1)$	$W(1, 1, 1)$ [$\delta_3 = W(1, 1, 1)$-$3W(1, 1)$]	$W(1, 1, 1, 1)$ [$\delta_4 = W(1, 1, 1, 1)$-$6W(1, 1)$]
10	−0.782	−1.777 [0.57]	−5.347 [−0.65]
15	−0.973	−1.812 [1.11]	−5.400 [0.49]
20	−0.866	−1.863 [0.73]	−5.150 [0.046]
25	−0.855	−1.867 [0.70]	−5.24 [−0.11]
30	−0.629	−1.635 [0.25]	−4.78 [−1.00]
35	−0.891	−1.829 [0.84]	−4.95 [0.39]

[a]Based on data from Table I in Imai and Yonetani (1975).

that δ_3 are relatively large [compared with $W(1, 1, 1)$], but δ_4 are relatively small [compared with $W(1, 1, 1, 1)$]. Note also the changes in sign. It is unclear to what extent the latter quantities are significant or within experimental error.

Figure 6.7 shows the average correlation $g(P_{O_2})$ as a function of oxygen partial pressure. At the low-pressure limit (determined by the pair correlation only) there is no clear-cut monotonic dependence on temperature. At the high-pressure limit (as determined by the quadruplet correlations, see Section 5.8), we see that the

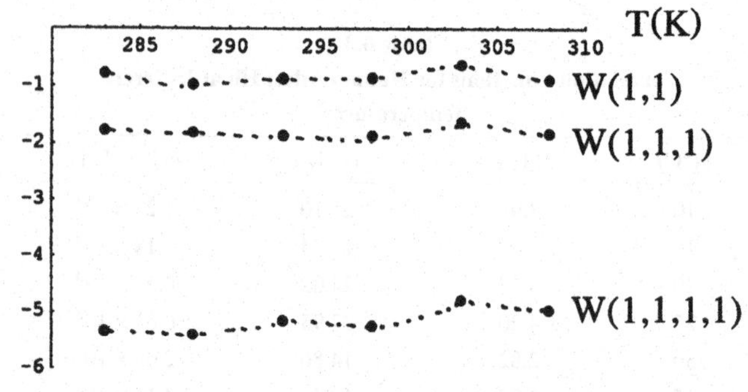

Figure 6.6. Values of $W(1, 1)$, $W(1, 1, 1)$, and $W(1, 1, 1, 1)$ (in kcal/mol) as a function of absolute temperature for human adult Hb. Based on data from Table I in Imai and Yonetani (1975).

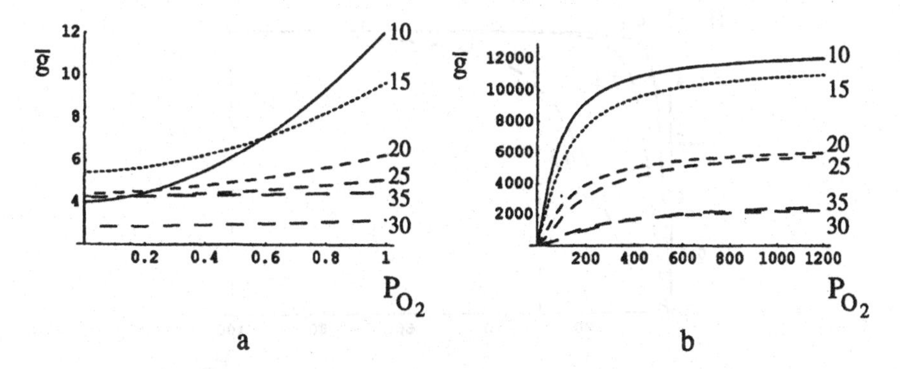

Figure 6.7. Average correlation $\overline{g}(P_{O_2})$ as a function of the partial pressure of oxygen P_{O_2} (in torrs), for the same system as in Fig. 6.5. (a) Low-pressure limit; (b) high-pressure limit. The temperatures are indicated next to each curve.

general trend (except for the two curves at 30 °C and 35 °C, which seems to be almost identical within the experimental error) is a decrease in the average correlation with temperature.

Before examining the effect of changing the composition of the solvent, it is important to note that the utility function (see the next subsection and Appendix K) as determined by the partial pressures of oxygen—about 100 torr in arterial blood and 30 torr in veinous blood—is almost zero for the system described in Fig. 6.5. In fact, the two pressures, 30 and 100 torr, do not even appear in this figure. it is also clear from this figure that the more cooperative system does not necessarily mean more efficient transport of oxygen. It is the utility function, as determined by the two *fixed* pressures, that is the important quantity. For instance, if the oxygen was to be transported between 4–10 torr, then it is the 30 °C curve which will have the highest utility value.

Next, we examine the effect of changing the solution composition on the BI and on the correlation functions.

Figure 6.8 shows three BIs of Hb [based on Table 6.2 from Imai (1982) for the same buffer conditions at 25 °C] at three different pH values of 9.1, 7.4, and 6.5. The overall appearance of these curves indicates that increasing the pH strengthens the cooperativity of the system. On the other hand, judging from the utility point of view, we see that the curve with the lowest pH has the highest utility value (which is about 0.2; this is quite small compared with values of the same system with added BPG and IHP, see below).

Figure 6.9 shows the average correlation $\overline{g}(P_{O_2})$ for Hb at 25 °C and at three different pH values. At very low pressures $\overline{g}$ is 0.92 for pH = 6.5; it increases to 2.84 at pH = 7.4 and then slightly decreases to 2.68 at pH = 9.1. On the other hand, at the high-pressure limit $\overline{g}$ has the lowest value of about 558 at pH = 9.1; it increases to 6,193 at pH = 7.4 and decreases again to 3,794 at pH = 6.5. Also the behavior of

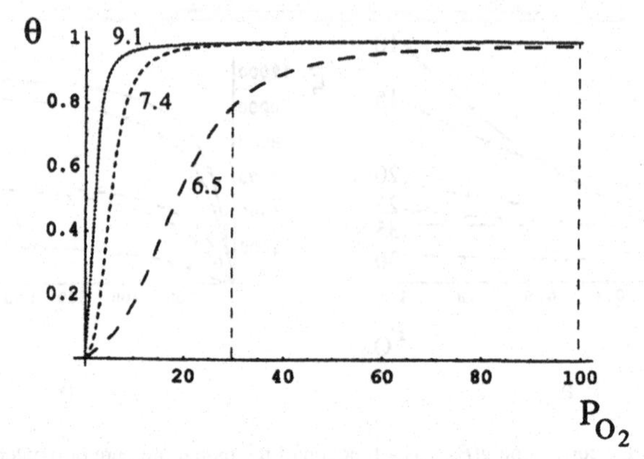

Figure 6.8. Binding isotherms for Hb at 25 °C at three different pH values (indicated next to each curve). The utility value is the difference θ(100) – θ(30).

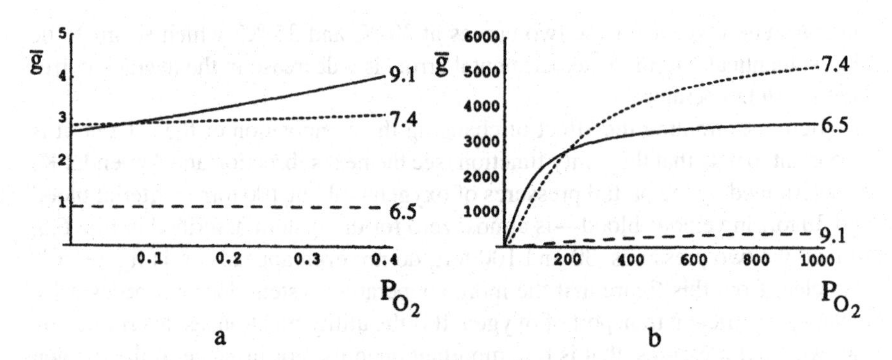

Figure 6.9. Average correlation $\bar{g}(P_{O_2})$ for Hb at three different pH values [at 25 °C based on Table 6.2 from Imai (1982)]. (a) Low-pressure limit; (b) high-pressure limit.

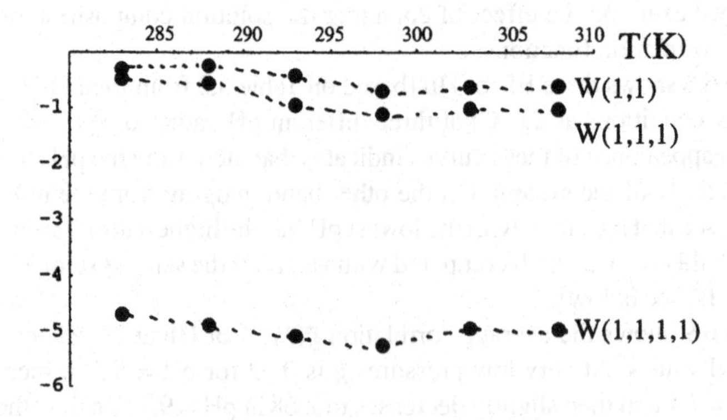

Figure 6.10. Values of $W(1, 1)$, $W(1, 1, 1)$, and $W(1, 1, 1, 1)$ (in kcal/mol) as a function of temperature, for human adult Hb with added 2 mM of BPG. Based on data from Imai and Yonetani (1975).

Table 6.5
Correlation Functions and Corresponding Free Energies (in kcal/mol at 25 °C) for Three Different pH Values[a]

pH	$g(1, 1)$ [$W(1, 1)$]	$g(1, 1, 1)$ [$W(1, 1, 1)$]	$g(1, 1, 1, 1)$ [$W(1, 1, 1, 1)$]
6.5	0.924 [0.047]	5.36 [−1.00]	558.5 [−3.77]
7.4	2.84 [−0.622]	39.14 [−2.18]	6193.9 [−5.20]
9.1	2.69 [−0.59]	67.79 [−2.51]	3794 [−4.91]

each of the correlations (Table 6.5) does not indicate a clear trend. We see that the pair correlation first slightly increases, then becomes negative upon lowering the pH. The triplet correlation decreases from 67.8 to 39.2 to 5.3. The quadruplet correlation again first increases, and then decreases considerably upon lowering the pH, i.e., increasing the concentration of protons in the solution.

We next turn to the effect of adding D-2,3-bisphosphoglycerate (BPG, previously known as DPG) and inositol hexaphosphate (IHP).

Figure 6.10 shows the free energies corresponding to the three correlations for the system where 2 mM of BPG was added. Similarly, Fig. 6.11 shows the free energies for the same system but with 2 mM of IHP added (instead of BPG). Figures 6.10 and 6.11 show immediately that replacing BPG by IHP moves all the points upward, i.e., the correlations become smaller and even become *negative* (the corresponding free energies become positive). The physiological significance of these changes is shown in Fig. 6.12, where we depict the BIs for the three systems in the same range of partial pressures between 0 to 100 torr.

It is clearly seen that within each group of curves the overall cooperativity became weaker as the temperature increases. This can be judged qualitatively from the steepness of the BIs, as well as quantitatively from the average correlations (shown on the lower panel of Fig. 6.12, in the range of 30–100 torr). The utility function, computed for $P_2 = 100$ torr and $P_1 = 30$ torr, is nearly zero for the system with no added solutes (besides the buffer solution). It is small for the system with 2 mM of BPG added and becomes relatively large for the system with 2 mM of IHP added. (Note, however, that the utility values do not increase monotonically with temperature; this is clear for the BI on the lhs of Fig. 6.12.) Thus, the change in utility values upon addition of BPG and IHP is much more dramatic than the change due to temperature.

Table 6.6 shows utility values for the three systems at the six temperatures. It is clear that the highest value is obtained for the system with 2 mM of IHP at a temperature between 25–30 °C.

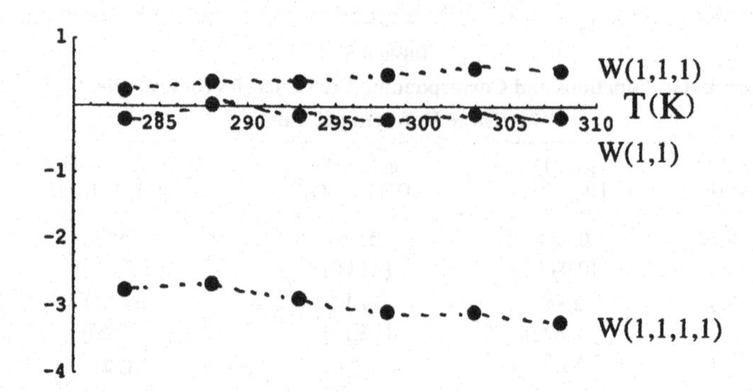

Figure 6.11. Same as Fig. 6.10, but with 2 mM of IHP instead of BPG.

6.8.3. Utility Function under Physiological Conditions

We have seen that there exists no simple correlation between the cooperativities in the Hb system and the utility values, as determined by the two partial pressures of 30 and 100 torr of oxygen at the loading and unloading ends. We have also seen that the utility function changes dramatically with temperature, pH, or addition of BPG and IHP. In all our considerations so far we have examined the change in the

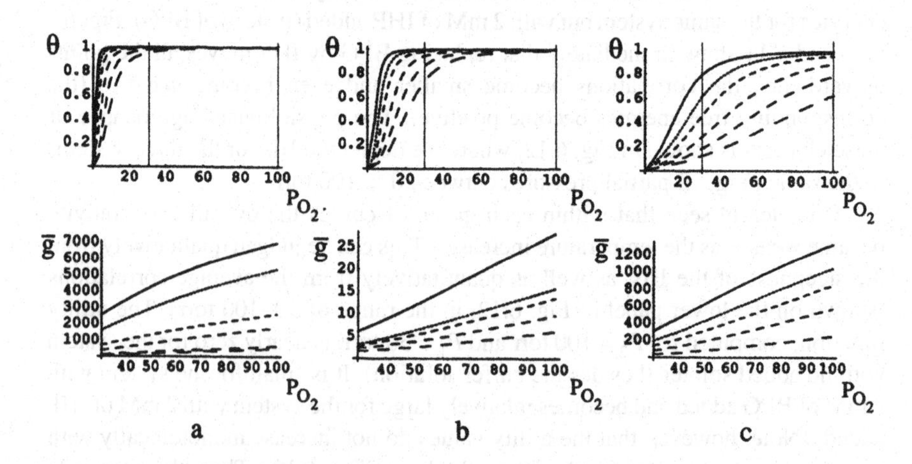

Figure 6.12. BI (upper panel) and average correlation $\bar{g}(P)$ (lower panel), for human adult Hb at temperatures of 10, 15, 20, 25, 30, 35 °C. The upper curves correspond to 10 °C and the lowest to 35 °C. All systems were in a fixed buffer solution [for details, see Imai and Yonetani (1975)]. (a) As in Fig. 6.5, no added BPG or IHP; (b) 2 mM of BPG added; (c) 2 mM of IHP added. The BIs were drawn in the same range of 0–100 torr, but the average correlations are drawn between 30–100 torr. The two vertical lines drawn at 30 and 100 torr are used to estimate the utility value for each system.

Table 6.6
Utility Values for the Three Systems Described in Fig. 6.12, Computed for $P_2 = 100$ torr and $P_1 = 30$ torr[a]

t (°C)	None	2 mM BPG	2 mM IHP
10	0	0.00	0.2
15	0	0.01	0.3
20	0	0.02	0.5
25	0	0.06	0.6
30	0.01	0.12	0.6
35	0.02	0.24	0.4

[a]The values were approximately estimated from the curves in Fig. 6.12.

utility function, as measured in a single BI, i.e., we have drawn the BI and examined the difference $\theta(P_2) - \theta(P_1)$ (where P_2 and P_1 are the loading and unloading partial pressures of oxygen). The situation is, of course, much more complicated in the real-life system, where loading and unloading of oxygen occurs. Here, the environment at the loading terminal (the lungs) might be quite different from the environment at the unloading terminal (the tissues). This requires consideration of the BI in a multiple space, i.e., θ as a function of P_{O_2} as well as a function of the concentrations of various added solutes. We shall demonstrate this point by the following two examples:

(a) BI as a function of partial pressure *and* pH: Consider the two-dimensional function $\theta(P, C_H)$, where P is the oxygen partial pressure and C_H the concentration H^+ in solution ($pH = -\log_{10} C_H$). In the previous subsection we discussed the utility function at *each* pH (i.e., at a *fixed* value of C_H). Suppose the two partial pressures for loading and unloading of oxygen were $P_2 = 10$ torr and $P_1 = 5$ torr (which are far from the real values; here, we use these values for demonstration purposes only).

Figure 6.13a is the same as Fig. 6.8, but drawn in the range of 0–30 torr. Measuring the utility value at $P_2 = 10$ and $P_1 = 5$ for each curve yields

$$\left.\begin{aligned} U_t(pH=6.5) &= \theta(P_2=10, pH=6.5) - \theta(P_1=5, pH=6.5) \approx 0.1 \\ U_t(pH=7.4) &= \theta(P_2=10, pH=7.4) - \theta(P_1=5, pH=7.4) \approx 0.4 \\ U_t(pH=9.1) &= \theta(P_2=10, pH=9.1) - \theta(P_1=5, pH=9.1) \approx 0.05 \end{aligned}\right\} \quad (6.8.13)$$

We see that between $P_2 = 10$ and $P_1 = 5$ the utility value is largest for pH = 7.4. At the same pressure range, the average correlation is largest at pH = 9.1 (Fig. 6.13b).*

*This shows that the utility function is, in general, not a linear function of the cooperativity (see also Appendix K).

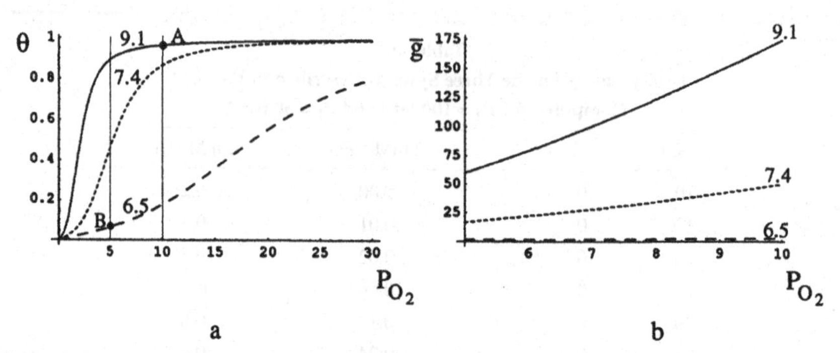

Figure 6.13. BI and average correlation for Hb at three different pH values. The data are the same as in Fig. 6.8, but here the BI is drawn in the pressure range of 0–30 torr. The utility function is computed for the arbitrarily selected pair of pressures $P_1 = 5$ and $P_2 = 10$ torr. The average correlations are drawn only in this range of pressures.

In each of the differences computed in Eq. (6.8.13) the pH has been *fixed* (pH = 6.5, pH = 7.4, and pH = 9.1), i.e., we have computed the utility value for each *single* BI (at a fixed pH).

In reality, the pH values are not the same at the loading and unloading terminals. In fact, the pH at the tissue is *lower* (i.e., more acidic) than the pH at the lungs. Thus, the utility function must be calculated not as differences of θ for each *fixed* pH as we have done in Eqs. (6.8.13), but at two different pH values as in the real loading and unloading locations. As an example (which is far from the real values for both the pH values and partial pressures), suppose the pH is 9.1 at the loading end but 6.5 at the unloading end. In this case, again using the two fictitious pressures $P_2 = 10$ and $P_1 = 5$, we would have loaded oxygen on the pH = 9.1-curve (point A in Fig. 6.13a), but unloaded the oxygen on a *different* pH = 6.5-curve (point B in Fig. 6.13a). The resulting utility function would be

$$U_t = \theta(P_2 = 10, pH = 9.1) - \theta(P_1 = 5, pH = 6.5) \approx 0.9 \qquad (6.8.14)$$

which is dramatically larger than the biggest value calculated in Eqs. (6.8.13). We see that by switching from one BI curve to another, the utility function has become almost unity, i.e., the oxygen is transmitted at about 90% efficiency.

(b) BI as a function of partial pressure *and* solute concentration: In the previous example we used fictitious pressures P_1 and P_2 to demonstrate the increase in efficiency of transporting oxygen between two pH values. In the second example we use the more realistic pressure differences of $P_2 = 100$ and $P_1 = 30$ torr, but fictitious pairs of solute concentrations. Figure 6.14 shows the BIs for Hb at one temperature $t = 35$ °C. One curve, denoted "none," is the same as the corresponding curve in Fig. 6.12a. The second, denoted IHP, is the same as the 35 °C curve of Fig. 6.12c.

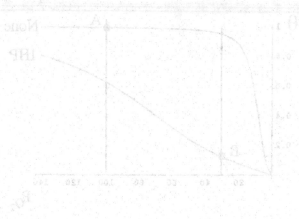

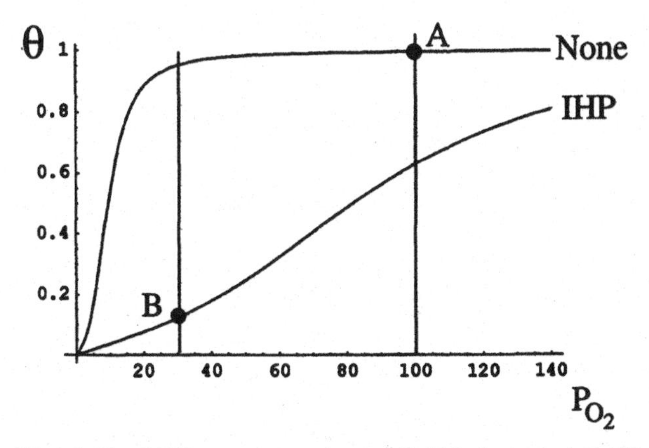

Figure 6.14. BIs as in Fig. 6.12, but at one temperature of 35 °C. One curve is for Hb with no IHP added (indicated "none"); the second curve is for Hb with 2 mM IHP added.

The utility functions calculated for these two curves are

$$U_t = \theta(P_2 = 100, none) - \theta(P_1 = 30, none) \approx 0.04$$
$$U_t = \theta(P_2 = 100, IHP) - \theta(P_1 = 30, IHP) \approx 0.5 \tag{6.8.15}$$

Clearly, at 35 °C the utility function is much larger for the system with 2 mM of IHP added. Now, suppose the Hb loads oxygen in an environment containing a negligible amount of IHP (point A in Fig. 6.14) but unloads the oxygen in a different environment, containing 2 mM of IHP (point B). In this case, operating between the real range of pressures, the actual utility value will be

$$U_t = \theta(P_2 = 100, none) - \theta(P_1 = 30, IHP) \approx 0.82 \tag{6.8.16}$$

Clearly, operating between two *different* environments considerably increases the utility value for transporting oxygen.

We have presented above two examples showing how the efficiency of transporting oxygen between two given pressures could be improved by changing a concentration of one solute (H^+ in the first and IHP in the second example). In a real-life system, the environment at the loading and unloading points of oxygen could differ with respect to several different solutes. The actual utility function should be calculated from the multidimensional BI, $\theta(P, C_1, C_2, \ldots, C_n)$, where $C_1, C_2, \ldots, C_n$ are the concentrations of the various solutes in the system. It is possible that the precise "tuning" of the various solute concentrations was selected by evolution to satisfy the specific needs of utility values of different living organisms.

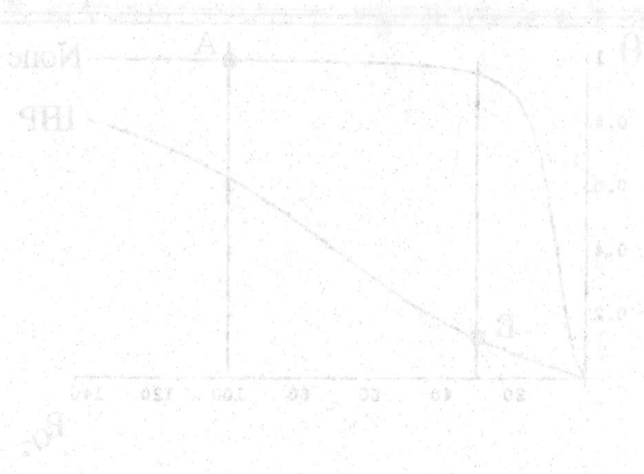

7

Large Linear Systems of Binding Sites

7.1. THE MATRIX METHOD

In this section we introduce the matrix method to rewrite the GPF of a linear system of m sites in a more convenient form. This is both an elegant and a powerful method for studying such systems. We start by presenting the so-called Ising model for the simplest system. We assume that each unit can be in one of two occupational states: empty or occupied. Also, we assume only nearest-neighbor (*nn*) interactions.* Both of these assumptions may be removed. In subsequent sections and in Chapter 8 we shall discuss four and eight states for each subunit. We shall not discuss the extension of the theory with respect to interactions beyond the *nn*. Such an extension is used, for example, in the theory of helix-coil transition.

The system is an extension of the models treated in Sections 4.3 and 5.3. It consists of m sites arranged in a linear sequence on the adsorbent molecule (Fig. 7.1a). The two states of each unit are "empty" and "occupied." The canonical PF of a single system having n ligands on the m sites ($n \le m$) is

$$Q(T, m, n) = \sum_{\substack{\text{all } \mathbf{s} \text{ with} \\ \text{fixed } m, n}} \exp[-\beta E(\mathbf{s})] \tag{7.1.1}$$

Here, the vector $\mathbf{s} = (s_1, s_2, \ldots, s_m)$ specifies the *configuration* of the system, i.e., $s_i = 0$ and $s_i = 1$ stand for site i being empty and occupied, respectively. Clearly, the total number of ligands is

$$n = \sum_{i=1}^{m} s_i \tag{7.1.2}$$

Each vector $\mathbf{s}$ specifies a configuration, or a specific distribution of the n ligands

*These include direct ligand–ligand and subunit–subunit interaction. The exclusion of long-range interaction does not preclude the occurrence of long-range correlation.

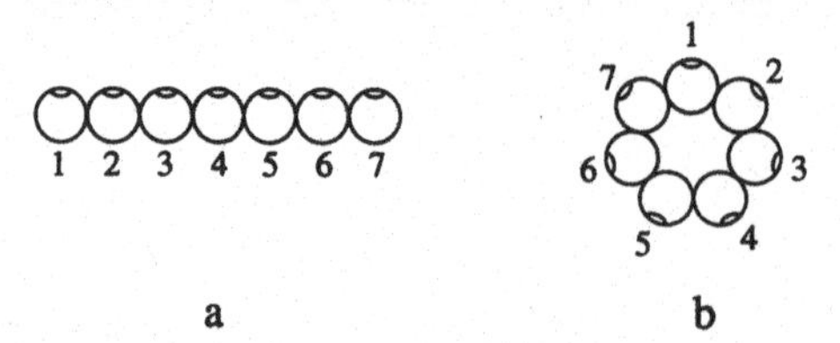

Figure 7.1. A system of $m = 7$ identical subunits arranged linearly: (a) open chain; (b) closed to a cycle.

on the m sites. There are altogether $\binom{m}{n}$ different configurations, each of which has a different energy, denoted by $E(\mathbf{s})$, and given by

$$E(\mathbf{s}) = \sum_{i=1}^{m-1} s_i s_{i+1} U_{11} + \sum_{i=1}^{m} s_i U = n_{11}(\mathbf{s})U_{11} + nU \tag{7.1.3}$$

where U is the binding energy, i.e., the interaction of the ligand with the system when it is bound to site i*; U_{11} is the *direct* interaction energy between two ligands on neighboring sites.

Since $s_i = 0, 1$, the total energy of the system consists of the total binding energy, nU, and a total of $n_{11}(\mathbf{s})U_{11}$ for the ligand–ligand interactions, where $n_{11}(\mathbf{s})$ is the number of nearest-neighbor pairs of ligands for the specific configuration $\mathbf{s}$.

The sum over $\mathbf{s}$ in Eq. (7.1.1) is over all the $\binom{m}{n}$ possible configurations $\mathbf{s}$. This is a sum over all the *states* of the system with a given n. Some of the terms in Eq. (7.1.1) might have the same energy. For instance, the configurations ($s_1 = s_2 = 1$, $s_3 = s_4 = \cdots = s_m = 0$) and ($s_1 = s_2 = 0$, $s_3 = s_4 = 1$, $s_5 = s_6 = \cdots = s_m = 0$) have the same energy, which is simply $U_{11} + 2U$. Therefore, one can change the order of summation and sum over all *energy levels* to obtain

$$Q(T, m, n) = \sum_{\substack{energy\\levels}} \Omega(E) \exp(-\beta E) \exp(-\beta n U) = \sum_{n_{11}} \Omega(n_{11}) \exp(-\beta n_{11} U_{11}) \tag{7.1.4}$$

Since the energy levels are determined by the number of nearest-neighbor pairs n_{11}, the sum over all energy levels is the same as the sum over all possible values of n_{11}, $\Omega(n_{11})$ being the number of configurations (or states) having the same value of n_{11}, hence of the same energy. Note that n_{11} in Eq. (7.1.4) is a variable and can take any possible value for a fixed n and m; $n_{11}(\mathbf{s})$ in Eq. (7.1.3) is the value of n_{11} for a specific configuration $\mathbf{s}$. Thus, for the specific configuration $\mathbf{s} = (1, 1, 1, 0, 0)$ of a system with $m = 5$ and $n = 3$ we have $n_{11}(\mathbf{s}) = 2$ (two nearest neighbors,) but n_{11} in

*We stress again that U is the interaction *energy* of the ligand with the *entire* adsorbent molecule—not only with the site itself; see also Appendix I. In this particular model U is independent of i. Also, the intrinsic binding constant is independent of i, i.e., the system has m weakly identical sites. In Section 7.2 we deal with a system where U is again independent of i, but there are, in general, different intrinsic binding constants.

Eq. (7.1.4) runs over $n_{11} = 0, 1, 2$ with $\Omega(0) = 1$, $\Omega(1) = 6$, and $\Omega(2) = 3$, the total number of configurations being $\Sigma\Omega(n_{11}) = \binom{5}{3} = 10$.

As in previous chapters, we now open the system with respect to the ligand. This is equivalent to letting n vary, with $0 \le n \le m$. The relevant GPF is now

$$\xi(T, m, \lambda) = \sum_{n=0}^{m} Q(T, m, n)\lambda^n = \sum_{n=0}^{m} \lambda^n \sum_{\binom{m}{n}} \exp[-\beta E(\mathbf{s})] \tag{7.1.5}$$

where $Q(T, m, n)$ is given in Eq. (7.1.1), but we stress the fact that this is a sum over $\binom{m}{n}$ configurations. Once we sum over all possible n, we extend the summation over *all* possible 2^m configurations, which is simply obtained from the identity

$$2^m = \sum_{n=0}^{m} \binom{m}{n} = (1 + 1)^m \tag{7.1.6}$$

The passage to the open system allows us to rewrite the GPF of the system in a condensed form, for any m. Thus, by summing over n in Eq. (7.1.5), we effectively remove the restriction on a *fixed n* (m is still fixed) that we had in the sum (7.1.1). The GPF can be rewritten in matrix form, as follows: We first write the sum (7.1.5) as a single sum over all possible 2^m configurations, namely,

$$\begin{aligned}
\xi &= \sum_{\substack{\text{all } \mathbf{s} \text{ with} \\ \text{fixed } m}} \lambda^{\Sigma s_i} \exp\left(-\beta \sum_{i=1}^{m-1} s_i s_{i+1} U_{11} - \beta \sum_{i=1}^{m} s_i U\right) \\
&= \sum_{\mathbf{s}} \lambda^{s_1} \exp(-\beta s_1 U - \beta s_1 s_2 U_{11}) \lambda^{s_2} \exp(-\beta s_2 U - \beta s_2 s_3 U_{11}) \cdots \\
&= \sum_{\mathbf{s}} (\lambda q)^{1/2 s_1} (\lambda q)^{1/2 s_1} S^{s_1 s_2} (\lambda q)^{1/2 s_2} (\lambda q)^{1/2 s_2} S^{s_2 s_3} (\lambda q)^{1/2 s_3} \\
&\qquad \cdots (\lambda q)^{1/2 s_{m-1}} S^{s_{m-1} s_m} (\lambda q)^{1/2 s_m} (\lambda q)^{1/2 s_m}
\end{aligned} \tag{7.1.7}$$

where in the first form on the rhs we sum over all possible configurations **s** with fixed m (but not n). In the second form we write explicitly the beginning of a typical term in this sum. In the third form we used the notations

$$q = \exp(-\beta U), \quad S = \exp(-\beta U_{11}) \tag{7.1.8}$$

to rewrite the same sum, but now we separate factors that "belong" to each consecutive pair of units. Thus, except for the first, $(\lambda q)^{1/2 s_1}$, and the last, $(\lambda q)^{1/2 s_m}$, factors, each of the factors in Eq. (7.1.7) is assigned to the pair $(i, i + 1)$ of neighboring sites. This allows us to rewrite the GPF in a very concise form, as follows:

We define the matrix **M** by

$$\mathbf{M} = \begin{pmatrix} 1 & (\lambda q)^{1/2} \\ (\lambda q)^{1/2} & \lambda q S \end{pmatrix} \tag{7.1.9}$$

This consists of all possible factors that are contributed by a *nn* pair. Since there are four possible states of a pair of units, we have the following assignments:

$$\left.\begin{aligned} (s_1 = 0, s_2 = 0) &\leftrightarrow M_{11} = 1 \\ (s_1 = 0, s_2 = 1) &\leftrightarrow M_{12} = (\lambda q)^{1/2} \\ (s_1 = 1, s_2 = 0) &\leftrightarrow M_{21} = (\lambda q)^{1/2} \\ (s_1 = 1, s_2 = 1) &\leftrightarrow M_{22} = \lambda q S \end{aligned}\right\} \tag{7.1.10}$$

It will be convenient, especially when the units can be in more than two states, to introduce the two unit row vectors

$$\langle 0| = (1, 0), \quad \langle 1| = (0, 1) \tag{7.1.11}$$

and the corresponding column vectors

$$|0\rangle = \begin{pmatrix} 1 \\ 0 \end{pmatrix}, \quad |1\rangle = \begin{pmatrix} 0 \\ 1 \end{pmatrix} \tag{7.1.12}$$

With this notation the elements of the matrix **M** are identified as

$$\left.\begin{aligned} M_{11} &= \langle 0|\mathbf{M}|0\rangle = 1 \\ M_{12} = M_{21} &= \langle 0|\mathbf{M}|1\rangle = \langle 1|\mathbf{M}|0\rangle = (\lambda q)^{1/2} \\ M_{22} &= \langle 1|\mathbf{M}|1\rangle = \lambda q S \end{aligned}\right\} \tag{7.1.13}$$

This notation is more convenient, since in $\langle s_i|\mathbf{M}|s_j\rangle$ the *states* of the pair *i* and *j* are specified while the subscripts in M_{lk} specify the location of the element in the *l*th row and *k*th column. The symbol $\langle s_i|\mathbf{M}|s_{i+1}\rangle$ is the factor contributed to the GPF by the *nn* pair *i* and $i + 1$, being in states s_i and s_{i+1}, respectively. This matrix element is obtained by multiplying the matrix **M** by a row vector on the left and by a column vector on the right. Since $\langle s_i|$ and $|s_{i+1}\rangle$ are unit vectors, this multiplication produces the element of the matrix that corresponds to the state of the *nn* pair *i* and $i + 1$.

To avoid confusion, we stress that the symbols $\langle a|$ and $|a\rangle$ denote *any* two-dimensional row and column vector,* respectively. In particular, whenever we

*In quantum mechanics these are referred to as "bra" and "ket" vectors. Multiplying the two forms the "bracket," which is a scalar product of $\langle a|$ and $|a\rangle$.

specifically introduce the *state* of a single unit within this symbol, as in Eqs. (7.1.11) and (7.1.12), we obtain a unit vector. In our case, we have only two unit vectors, $\langle 0|$ and $\langle 1|$. The scalar product, denoted by $\langle 0|a\rangle$, is therefore the *first* component of the vector $|a\rangle$. Similarly, $\langle 1|a\rangle$ is its second component.

With the notation in Eq. (7.1.13), we rewrite the GPF in Eq. (7.1.7) as

$$\xi = \sum_{\mathbf{s}} (\lambda q)^{1/2s_1} \langle s_1|\mathbf{M}|s_2\rangle\langle s_2|\mathbf{M}|s_3\rangle \cdots \langle s_{m-1}|\mathbf{M}|s_m\rangle(\lambda q)^{1/2s_m} \tag{7.1.14}$$

The summations are over all the indices s_i, each of which represents the state of the ith unit. Thus, except for the first and last factors, this sum produces $m-1$ matrix multiplications; hence,

$$\xi = \sum_{s_1}\sum_{s_m} (\lambda q)^{1/2s_1} \langle s_1|\mathbf{M}^{m-1}|s_m\rangle(\lambda q)^{1/2s_m} \tag{7.1.15}$$

If we define the vector

$$\langle v| = [1, (\lambda q)^{1/2}] \tag{7.1.16}$$

we may rewrite ξ in matrix notation as

$$\xi_{\mathbf{O}} = \langle v|\mathbf{M}^{m-1}|v\rangle \tag{7.1.17}$$

The subscript **O** was added to refer to an "open" chain.

A particularly simple form of the GPF is obtained for a cyclic system, i.e:, when the end units ($i = 1, i = m$) are closed (Fig. 7.1b), in which case one more interaction factor $S^{s_m s_1}$ is introduced in the PF. Hence Eq. (7.1.15) is repalced by

$$\xi_{\mathbf{C}} = \sum_{s_1}\sum_{s_m} \langle s_1|\mathbf{M}^{m-1}|s_m\rangle\langle s_m|\mathbf{M}|s_1\rangle$$

$$= \sum_{s_1} \langle s_1|\mathbf{M}^m|s_1\rangle = \mathrm{Tr}(\mathbf{M}^m) \tag{7.1.18}$$

where now the GPF of the cyclic system, $\xi_{\mathbf{C}}$, is the trace of the matrix $\mathbf{M}^m$. This is far easier to use than the open-system GPF in Eq. (7.1.17). We shall use $\xi_{\mathbf{C}}$ either when our system is genuinely cyclic, as in Section 7.2, or for very long open-chain systems. In the latter case, $\xi_{\mathbf{C}}$ is only an approximate PF for $\xi_{\mathbf{O}}$. However, in the thermodynamic limit of $m \to \infty$, the addition of the factor $S^{s_m s_1}$ has a negligible effect on the thermodynamic properties of the system.

The form of the GPF for the cyclic system is very convenient since as we shall see below, all we have to do is find the eigenvalues of the 2×2 matrix **M** in order to express the PF in terms of the molecular parameters of the system. To proceed, we need some theorems from matrix algebra, which we cite without proof.*

A matrix $\mathbf{A}$ is said to diagonalize matrix $\mathbf{M}$ if

$$\mathbf{AMA}^{-1} = \mathbf{D} \tag{7.1.19}$$

where $\mathbf{A}^{-1}$ is the inverse matrix of $\mathbf{A}$ and $\mathbf{D}$ is a diagonalized matrix; i.e., for a 2×2 matrix $\mathbf{D}$ has the form

$$\mathbf{D} = \begin{pmatrix} d_1 & 0 \\ 0 & d_2 \end{pmatrix} \tag{7.1.20}$$

Since our matrix $\mathbf{M}$ is symmetric and has real entries, there always exists an orthogonal matrix $\mathbf{A}$ which diagonalizes $\mathbf{M}$.

A matrix $\mathbf{A}$ that diagonalizes $\mathbf{M}$ will also diagonalize $\mathbf{M}^m$. Thus, for any m, we can write

$$\mathbf{AM}^m\mathbf{A}^{-1} = \mathbf{AMMM} \cdots \mathbf{A}^{-1} = \mathbf{AMA}^{-1}\mathbf{AMA}^{-1}\mathbf{AM} \cdots \mathbf{AMA}^{-1}$$

$$= \mathbf{D}^m = \begin{pmatrix} d_1^m & 0 \\ 0 & d_l^m \end{pmatrix} \tag{7.1.21}$$

where the unit matrix $\mathbf{I} = \mathbf{AA}^{-1}$ has been added between each pair of consecutive matrices $\mathbf{M}$. The last form on the rhs of Eq. (7.1.21) highlights the substantial simplification achieved by diagonalizing the matrix $\mathbf{M}$. Instead of computing the mth power of a *matrix*, all we need is the mth power of two *numbers*, the eigenvalues of matrix $\mathbf{M}$.

It is easy to verify that the trace of any product of matrices is invariant to cyclic permutations of the matrices, for instance,

$$\begin{aligned} \mathrm{Tr}(\mathbf{AB}) &= \mathrm{Tr}(\mathbf{BA}) \\ \mathrm{Tr}(\mathbf{ABC}) = \mathrm{Tr}(\mathbf{CAB}) &= \mathrm{Tr}(\mathbf{BCA}) \end{aligned} \tag{7.1.22}$$

This follows directly from the definition of the trace of any matrix $\mathbf{A}$ of order $m \times m$,

$$\mathrm{Tr}\mathbf{A} = \sum_{i=1}^{m} A_{ii} \tag{7.1.23}$$

i.e., $\mathrm{Tr}\mathbf{A}$ is simply the sum over all diagonal elements of $\mathbf{A}$.

Applying this rule to Eq. (7.1.19), we have

$$\mathrm{Tr}(\mathbf{AMA}^{-1}) = \mathrm{Tr}(\mathbf{A}^{-1}\mathbf{AM}) = \mathrm{Tr}(\mathbf{IM}) = \mathrm{Tr}(\mathbf{M}) \tag{7.1.24}$$

i.e., the trace of $\mathbf{M}$ is the same as the trace of $\mathbf{AMA}^{-1}$.

*For proof, the reader is referred to any elementary text in algebra.

Returning to the GPF in Eq. (7.1.18) and using Eq. (7.1.21), we write

$$\xi_C = \mathrm{Tr}(\mathbf{M})^m = \mathrm{Tr}(\mathbf{A}\mathbf{M}^m\mathbf{A}^{-1}) = \mathrm{Tr}(\mathbf{D}^m) = d_1^m + d_2^m \tag{7.1.25}$$

The eigenvalues are obtained by solving the characteristic equation (or the secular equation)

$$|\mathbf{M} - d\mathbf{I}| = 0 \tag{7.1.26}$$

where $\mathbf{I}$ is the identity matrix and the determinant $|\mathbf{M} - d\mathbf{I}|$ is the characteristic polynomial of the matrix $\mathbf{M}$. Thus, calculation of the GPF reduces to calculation of the two roots of a polynomial of second degree.

When applying the matrix method to very large values of m, it is only the largest root of the characteristic equation that is significant for calculating the thermodynamic properties of the system. For the particular matrix defined in Eq. (7.1.9), the characteristic equation is

$$\begin{vmatrix} 1-d & (\lambda q)^{1/2} \\ (\lambda q)^{1/2} & \lambda q S - d \end{vmatrix} = 0 \tag{7.1.27}$$

The two solutions for d are

$$d_{1,2} = \frac{1}{2}\left[1 + \lambda q S \pm \sqrt{4\lambda q + (1 - \lambda q S)^2}\right] \tag{7.1.28}$$

This is an explicit expression for $d_2 = d_+$ and $d_1 = d_-$ $(d_1 < d_2)$ in terms of the molecular parameters. Raising these to the mth power yields the final expression for the GPF in Eq. (7.1.25). In particular, for $d_1 < d_2$ we have

$$\xi = d_2^m\left[1 + \left(\frac{d_1}{d_2}\right)^m\right] \xrightarrow{\text{large } m} d_2^m \tag{7.1.29}$$

and

$$-k_B T \ln \xi = -m k_B T \ln d_2 \tag{7.1.30}$$

This shows that in the limit of large systems, the thermodynamics of the system is determined by the largest root of the secular equation.* In particular, $-k_B T \ln \xi$, as seen in Eq. (7.1.30) is an extensive property, i.e., it is proportional to the number of units.

*We have also dropped the subscript C to ξ. In the limit of very large m, the properties of the open and closed systems become identical.

We have developed above the specific case of a system with two states for each unit. However, most of the formal results are valid for any number of states. We simply reinterpret any sum over each s_i to run over all possible states of the unit. Let f be the number of states, or the number of degrees of freedom for a single unit. Then the general result in Eq. (7.1.18) is still valid, i.e.,

$$\xi = \mathrm{Tr}(\mathbf{M}^m) \tag{7.1.31}$$

where now **M** is an $f \times f$ instead of a 2×2 matrix. We shall discuss the cases $f = 4$ and $f = 8$ in Sections 7.4 and 8.8, respectively. Clearly, in this case there are f roots to the matrix **M**, but the final result for the GPF is still

$$\xi = d_{max}^m \tag{7.1.32}$$

where d_{max} is the largest root of **M**.

7.2. CORRELATION FUNCTIONS

From now on we use only the GPF for the *cyclic* system and drop the subscript C. Since our system has m identical units, the sites will always be identical in the weak sense. There is always one intrinsic constant for the first site but, in general, we have more than one pair correlation, triplet correlation, etc. As in Section 7.1 we develop, for simplicity, the case of two states $f = 2$, but most of the results are quite general.

The intrinsic binding constant for the particular model of Section 7.1 is obtained from

$$k = \frac{Q(1)\lambda_0}{mQ(0)} = \frac{\lambda_0}{m} \lim_{\lambda \to 0} \frac{\partial \ln \xi}{\partial \lambda} = q\lambda_0 \tag{7.2.1}$$

Note that $Q(1)$ is the canonical PF of the system having one ligand. Since there are m identical singly-occupied systems, we have divided by m to obtain the intrinsic binding constant k. Incidentally, for this particular model, also the *open* linear chain has only one k, equal to $q\lambda_0$. This is not true for more complicated models, such as that discussed in Sections 5.6, 6.3, and 7.4.

In order to obtain the correlation between any two (or more) events, we must collect all the relevant terms in the PF and form the correlation between the events. Here, as always in this book, we are interested in the *cooperativity* between the *ligands*. This was defined as the correlation between two particular events, such as "site i is occupied" and "site j is occupied." For instance, the nearest-neighbor (*nn*) correlation is defined by

$$g(i, i+1) = \frac{P(s_i = 1, s_{i+1} = 1)}{P(s_i = 1)P(s_{i+1} = 1)} \tag{7.2.2}$$

In general, one can define other correlations, say between two unoccupied sites, or one occupied and one unoccupied site, etc.* We shall not require these here. Also, as always, we shall need only the $\lambda \to 0$ limit of the correlations, since only these enter into the binding isotherm.

To obtain the singlet probability $P(s_1 = 1)$, we recall that, in general, each term in the GPF is proportional to the probability of finding the specific configuration **s**. For instance, in a cyclic system of $m = 3$, the probability of finding the configuration $\mathbf{s} = (s_1 = 1, s_2 = 0, s_3 = 1)$ is

$$P(\mathbf{s}) = \frac{\langle 1|\mathbf{M}|0\rangle\langle 0|\mathbf{M}|1\rangle\langle 1|\mathbf{M}|1\rangle}{\xi} \tag{7.2.3}$$

where ξ is the GPF of the *cyclic* system of three units.

To obtain $P(s_1 = 1)$, we simply sum over all possible states of $i = 2, 3, \ldots, m$ while holding the state of unit $i = 1$ at $s_i = 1$ (i.e., occupied). Thus, in general, we have

$$P(s_1 = 1) = \sum_{s_2,\ldots,s_m} P(s_1 = 1, s_2, s_3, \ldots, s_m)$$

$$= \sum_{s_1, s_2,\ldots,s_m} P(s_1, s_2, \ldots, s_m)\delta_{s_1,1} \tag{7.2.4}$$

where, in the second form on the rhs of Eq. (7.2.4), we have introduced the Kronecker delta function defined by

$$\delta_{s_1,1} = \begin{cases} 1 \text{ for } s_1 = 1 \\ 0 \text{ for } s_1 \neq 1 \end{cases} \tag{7.2.5}$$

and added also a summation over the index s_1. We can repeat similar steps, as in Section 7.1, to write this sum in terms of the matrix elements, namely,

$$P(s_1 = 1) = \frac{1}{\xi} \sum_{s_1, s_2,\cdots,s_m} \langle s_1|\mathbf{M}|s_2\rangle\langle s_2|\mathbf{M}|s_3\rangle \cdots \langle s_{m-1}|\mathbf{M}|s_m\rangle\delta_{s_1,1}$$

$$= \frac{1}{\xi}\sum_{s_1}\langle s_1|\mathbf{M}^m|s_1\rangle\delta_{s_1,1} = \frac{1}{\xi}\langle 1|\mathbf{M}^m|1\rangle \tag{7.2.6}$$

*For examples of such correlations in 1-D models, see Chapter 4 in Ben-Naim (1992).

Recall that $\langle 1|$ is a unit vector defined in Eq. (7.1.11). Hence $\langle 1|\mathbf{M}^m|1\rangle$ is the element at position 2,2 of the matrix $\mathbf{M}^m$, i.e., the element $(\mathbf{M}^m)_{2,2}$.

In order to simplify the form of Eq. (7.2.6), we need the eigenvectors of matrix **M**. The eigenvector corresponding to the eigenvalue d_i is defined by

$$\mathbf{M}|a_i\rangle = d_i|a_i\rangle \tag{7.2.7}$$

The eigenvectors are the rows or the column of an orthogonal matrix* **A** that diagonalizes $\mathbf{M}^*$. They fulfill the orthogonality condition

$$\langle a_i|a_j\rangle = \delta_{i,j} \tag{7.2.8}$$

The following identity exists for the unit matrix **I**,

$$\mathbf{I} = \sum_i |a_i\rangle\langle a_i| \tag{7.2.9}$$

Here, the sum is over all the states of one unit of the system.

With the help of the latter identity, Eq. (7.2.6) can be expressed in the form

$$P(s_1 = 1) = \frac{1}{\xi}\langle 1|\mathbf{M}^m|1\rangle = \frac{1}{\xi}\sum_{i,j}\langle 1|a_i\rangle\langle a_i|\mathbf{M}^m|a_j\rangle\langle a_j|1\rangle$$

$$= \frac{1}{\xi}\sum_{i,j}\langle 1|a_i\rangle d_j^m \delta_{i,j}\langle a_j|1\rangle$$

$$= \frac{1}{\xi}\sum_i \langle 1|a_i\rangle^2\, d_i^m = \frac{\Sigma_i \langle 1|a_i\rangle^2\, d_i^m}{\Sigma_i\, d_i^m} \tag{7.2.10}$$

In this equation we first insert the identity matrix **I** before and after $\mathbf{M}^m$, and then apply m times the operation (7.2.7) to produce d_j^m. We next apply the orthogonality condition (7.2.8), and the fact that all our vectors and matrices are real, hence $\langle 1|a_i\rangle = \langle a_i|1\rangle$. Finally, we use the expression for the GPF obtained in Eq. (7.1.25).

Since we have already solved for d_i in terms of the molecular parameters [see Eq. (7.1.28)], we must solve for the components of the eigenvectors $\langle 1|a_i\rangle$. These may be obtained by solving the two linear equations (7.2.7):

$$\begin{aligned} M_{11}\langle 0|a_1\rangle + M_{12}\langle 1|a_1\rangle &= d_1\langle 0|a_1\rangle \\ M_{21}\langle 0|a_1\rangle + M_{22}\langle 1|a_1\rangle &= d_1\langle 1|a_1\rangle \end{aligned} \tag{7.2.11}$$

and similar equations for $\langle 0|a_2\rangle$ and $\langle 1|a_2\rangle$. Since we are interested in cooperativities, we need only the components $\langle 1|a_i\rangle$ corresponding to the state "1," i.e., occupied. Hence, we must solve only for $\langle 1|a_1\rangle$ and $\langle 1|a_2\rangle$. Doing so produces, not surprisingly,

*The existence of such an orthogonal matrix is guaranteed for a real and symmetric matrix **M**.

the corresponding BI for a system of m (weakly) identical sites. This is so because the probability of finding a specific site occupied is equal to θ (see Section 2.1). For a subsequent evaluation of the various correlations we shall use the form (7.2.10). Note that in the limit $m \to \infty$, only the largest root is important. Since $d_1 < d_2$, we have

$$\lim_{m\to\infty} P(s_1 = 1) = \langle 1|a_2\rangle^2 \tag{7.2.12}$$

The singlet distribution in this limit is simply the *second* component of the *second* eigenvector.

To obtain the *nn*-pair correlation we need the probability of the joint event ($s_i = 1$, $s_{i+1} = 1$) i.e., both site i and $i+1$ being occupied. This may be obtained by following steps similar to those for $P(s_1 = 1)$. Since our system is cyclic, the *nn*-pair distribution is independent of index i, hence we write

$$\begin{aligned} P(s_1 = 1, s_2 = 1) &= \sum_{s_3,\ldots,s_m} P(s_1 = 1, s_2 = 1, s_3, s_4, \ldots, s_m) \\ &= \sum_{s_1,s_2,\ldots,s_m} P(s_1, s_2, \ldots, s_m)\delta_{s_1,1}\delta_{s_2,1} \\ &= \frac{1}{\xi}\sum_{s_1,s_2} \langle s_1|\mathbf{M}|s_2\rangle\langle s_2|\mathbf{M}^{m-1}|s_1\rangle\delta_{s_1,1}\delta_{s_2,1} \\ &= \frac{1}{\xi}\langle 1|\mathbf{M}|1\rangle\langle 1|\mathbf{M}^{m-1}|1\rangle \end{aligned} \tag{7.2.13}$$

By treating both $\langle 1|\mathbf{M}|1\rangle$ and $\langle 1|\mathbf{M}^{m-1}|1\rangle$ in the same manner as earlier in Eq. (7.2.10), we obtain

$$\begin{aligned} P(s_1 = 1, s_2 = 1) &= \frac{1}{\xi}\left[\sum_i \langle 1|a_i\rangle^2 d_i \sum_j \langle 1|a_j\rangle^2 d_j^{m-1}\right] \\ &= \frac{1}{\xi}\sum_{i,j} \langle 1|a_i\rangle^2\langle 1|a_j\rangle^2 d_i d_j^{m-1} \end{aligned} \tag{7.2.14}$$

Since $d_2 > d_1$ [see Eq. (7.1.28)], then for very large m, $(d_1/d_2)^m << 1$, so we obtain for the *nn*-pair correlation

$$g_{nn}(2) = g_{12}(2) = \frac{P(s_1 = 1, s_2 = 1)}{P(s_1 = 1)^2} = 1 + \frac{\langle 1|a_1\rangle^2}{\langle 1|a_2\rangle^2}\frac{d_1}{d_2} \tag{7.2.15}$$

The generalization for the pair correlation between two ligands, $l - 1$ units apart, is straightforward:

$$P(s_1 = 1, s_l = 1) = \sum_{\mathbf{s}} P(s_1, s_2, \ldots, s_m)\delta_{s_1,1}\delta_{s_l,1}$$

$$= \frac{1}{\xi} \sum_{s_1, s_l} \langle s_1 | \mathbf{M}^{l-1} | s_l \rangle \langle s_l | \mathbf{M}^{m-l+1} | s_1 \rangle \delta_{s_1,1}\delta_{s_l,1}$$

$$= \frac{1}{\xi} \langle 1 | \mathbf{M}^{l-1} | 1 \rangle \langle 1 | \mathbf{M}^{m-l+1} | 1 \rangle$$

$$= \frac{1}{\xi} \left[\sum_i \langle 1 | a_i \rangle^2 d_i^{l-1} \sum_j \langle 1 | a_j \rangle^2 d_j^{m-l+1} \right] \tag{7.2.16}$$

The correlation function for the two ligands at the first and lth sites is

$$g_{1l}(2) = \frac{P(s_1 = 1, s_l = 1)}{[P(s_1 = 1)]^2}$$

$$= \xi \frac{\sum_i \sum_j \langle 1 | a_i \rangle^2 \langle 1 | a_j \rangle^2 d_i^{l-1} d_j^{m-l+1}}{\sum_i \sum_j \langle 1 | a_i \rangle^2 \langle 1 | a_j \rangle^2 d_i^m d_j^m} \tag{7.2.17}$$

Again, for $d_1 < d_2$ and very large m we have

$$g_{1l}(2) = 1 + \frac{\langle 1 | a_1 \rangle^2}{\langle 1 | a_2 \rangle^2} \left(\frac{d_1}{d_2} \right)^{l-1} \tag{7.2.18}$$

This is the generalization of Eq. (7.2.15).

Using a similar procedure we can obtain correlations of any order. For example, in order to obtain the 1, 2, 3 triplet probability, we write

$$P(s_1 = 1, s_2 = 1, s_3 = 1) = \sum_{\mathbf{s}} P(s_1, s_2, s_3, \ldots, s_m)\delta_{s_1,1}\delta_{s_2,1}\delta_{s_3,1}$$

$$= \frac{1}{\xi} \sum_{s_1, s_2, s_3} \langle s_1 | \mathbf{M} | s_2 \rangle \langle s_2 | \mathbf{M} | s_3 \rangle \langle s_3 | \mathbf{M}^{m-2} | s_1 \rangle \, \delta_{s_1,1}\delta_{s_2,1}\delta_{s_3,1}$$

$$= \frac{1}{\xi} \langle 1 | \mathbf{M} | 1 \rangle \langle 1 | \mathbf{M} | 1 \rangle \langle 1 | \mathbf{M}^{m-2} | 1 \rangle$$

$$= \frac{1}{\xi} \sum_i \sum_j \sum_k \langle 1 | a_i \rangle^2 \langle 1 | a_j \rangle^2 d_i \langle 1 | a_k \rangle^2 d_k^{m-2} \tag{7.2.19}$$

and the corresponding triplet correlation is, in the limit $m \to \infty$,

$$g_{1,2,3} = \frac{P(s_1 = 1, s_2 = 1, s_3 = 1)}{P(s_1 = 1)^3}$$

$$= 1 + 2\frac{\langle 1|a_1\rangle^2}{\langle 1|a_2\rangle^2}\frac{d_1}{d_2} + \frac{\langle 1|a_1\rangle^4}{\langle 1|a_2\rangle^4}\left(\frac{d_1}{d_2}\right)^2 \tag{7.2.20}$$

Other correlations can be obtained similarly.

We now summarize some general results valid for any number of degrees of freedom f. The general form is presented for the cyclic system with a fixed m, and also for its $m \to \infty$ limit.

We recall that the GPF is

$$\xi = \mathrm{Tr}(\mathbf{M}^m) = \sum_{i=1}^{f} d_i^m \to d_{mx}^m \tag{7.2.21}$$

where d_{mx} is the largest eigenvalue of the $f \times f$ matrix $\mathbf{M}$. The singlet distribution is

$$P(s_1 = \alpha) = \frac{\langle \alpha|\mathbf{M}|\alpha\rangle}{\xi}$$

$$= \frac{\sum_{i=1}^{f} \langle \alpha|a_i\rangle^2 d_i^m}{\sum_i d_i^m} \to \langle \alpha|a_{mx}\rangle^2 \tag{7.2.22}$$

where $s_1 = \alpha$ means that the state of the unit numbered one is α. Since the system is cyclic and the subunits identical, it does not matter which *specific* unit we choose to number as the first. $|a_{mx}\rangle$ is the eigenvector corresponding to the largest eigenvalue, i.e.,

$$\mathbf{M}|a_{mx}\rangle = d_{mx}|a_{mx}\rangle \tag{7.2.23}$$

Note that $|a_i\rangle$ is the ith eigenvector, but $|\alpha\rangle$ is a *unit* vector corresponding to the α state. We shall use Greek letters for a *general unit* vector. For a *specific unit* vector we explicitly indicate the state, such as in Eqs. (7.1.11). In Section 7.4 we use the notation $|L1\rangle$ for the state "occupied and conformation L," and in Section 8.5 we use $|L, A, R\rangle$ for "occupied by A and by R and conformation L."

The *nn* distribution is [see Eqs. (7.2.13) and (7.2.14)]

$$P(s_1 = \alpha, s_2 = \beta) = \frac{\langle\alpha|\mathbf{M}|\beta\rangle\langle\beta|\mathbf{M}^{m-1}|\alpha\rangle}{\xi}$$

$$= \frac{1}{\xi}\left[\sum_{i=1}^{f}\langle\alpha|a_i\rangle\langle\beta|a_i\rangle d_i\right]\left[\sum_{j}\langle\alpha|a_j\rangle\langle\beta|a_j\rangle d_j^{m-1}\right]$$

$$\to \left[\sum_{i=1}^{f}\langle\alpha|a_i\rangle\langle\beta|a_i\rangle\left(\frac{d_i}{d_{mx}}\right)\right]\langle\alpha|a_{mx}\rangle\langle\beta|a_{mx}\rangle$$

$$= \langle\alpha|a_{mx}\rangle^2\langle\beta|a_{mx}\rangle^2\left[1 + \sum_{i\neq f}\frac{\langle\alpha|a_i\rangle\langle\beta|a_i\rangle}{\langle\alpha|a_{mx}\rangle\langle\beta|a_{mx}\rangle}\left(\frac{d_i}{d_{mx}}\right)\right]$$

$$= P(s_1 = \alpha)\, P(s_2 = \beta) g_{12}(s_1 = \alpha, s_2 = \beta) \tag{7.2.24}$$

where in the last form on the rhs of Eq. (7.2.24) we have written the pair distribution as a product of the two singlet distributions and a correlation function. The latter is defined, as usual, by

$$g_{12}(s_1 = \alpha, s_2 = \beta) = \frac{P(s_1 = \alpha, s_2 = \beta)}{P(s_1 = \alpha)P(s_2 = \beta)} = 1 + \sum_{i\neq f}\frac{\langle\alpha|a_i\rangle\langle\beta|a_i\rangle}{\langle\alpha|a_{mx}\rangle\langle\beta|a_{mx}\rangle}\left(\frac{d_i}{d_{mx}}\right) \tag{7.2.25}$$

Note that the sum over $i \neq f$ means over all states except that indexed $i = f$, which we choose to denote as the state having the largest eigenvalue.

For the 1, l pair distribution [see Eq. (7.2.16)], we have

$$P(s_1 = \alpha, s_l = \beta) = \frac{\langle\alpha|\mathbf{M}^{l-1}|\beta\rangle\langle\beta|\mathbf{M}^{m-l+1}|\alpha\rangle}{\xi}$$

$$\to \left[\sum_{i}\langle\alpha|a_i\rangle\langle\beta|a_i\rangle\left(\frac{d_i}{d_{mx}}\right)^{l-1}\right]\langle\alpha|a_{mx}\rangle\langle\beta|a_{mx}\rangle \tag{7.2.26}$$

The corresponding correlation function is [see Eq. (7.2.17)]

$$g_{1l}(s_1 = \alpha, s_l = \beta) = \frac{P(s_1 = \alpha, s_l = \beta)}{P(s_1 = \alpha)P(s_l = \beta)} = 1 + \sum_{i\neq f}\frac{\langle\alpha|a_i\rangle\langle\beta|a_i\rangle}{\langle\alpha|a_{mx}\rangle\langle\beta|a_{mx}\rangle}\left(\frac{d_i}{d_{mx}}\right)^{l-1} \tag{7.2.27}$$

By a straightforward generalization we can write the triplet and higher-order correlations. For instance, the triplet correlation $g_{1,2,3}(s_1 = \alpha, s_2 = \beta, s_3 = \gamma)$ is obtained from the corresponding triplet probability

$$P(s_1 = \alpha, s_2 = \beta, s_3 = \gamma) = \frac{\langle\alpha|\mathbf{M}|\beta\rangle\langle\beta|\mathbf{M}|\gamma\rangle\langle\gamma|\mathbf{M}^{m-2}|\alpha\rangle}{\xi}$$

$$= \frac{1}{\xi}\left[\sum_i \langle\alpha|a_i\rangle\langle\beta|a_i\rangle d_i\right]\left[\sum_j \langle\gamma|a_j\rangle\langle\beta|a_j\rangle d_j\right]\left[\sum_k \langle\alpha|a_k\rangle\langle\gamma|a_k\rangle\, d_k^{m-2}\right]$$

$$\to \left[\sum_i \langle\alpha|a_i\rangle\langle\beta|a_i\rangle \frac{d_i}{d_{mx}}\right]\left[\sum_j \langle\gamma|a_j\rangle\langle\beta|a_j\rangle \frac{d_j}{d_{mx}}\right]\langle\alpha|a_{mx}\rangle\langle\gamma|a_{mx}\rangle \quad (7.2.28)$$

The last form on the rhs of Eq. (7.2.28) is for $m \to \infty$.

The triplet correlation is thus [compare with Eq. (7.2.20)]

$$g(s_1 = \alpha, s_2 = \beta, s_3 = \gamma) = \frac{P(s_1 = \alpha, s_2 = \beta, s_3 = \gamma)}{P(s_1 = \alpha)P(s_2 = \beta)P(s_3 = \gamma)}$$

$$= 1 + [F(\alpha, \beta) + F(\beta, \gamma)] + F(\alpha, \beta)F(\beta, \gamma) \quad (7.2.29)$$

where we have introduced the notation F defined by

$$F(\varepsilon, \nu) = \sum_{i \neq f} \frac{\langle\varepsilon|a_i\rangle\langle\nu|a_i\rangle}{\langle\varepsilon|a_{mx}\rangle\langle\nu|a_{mx}\rangle}\left(\frac{d_i}{d_{mx}}\right) \quad (7.2.30)$$

We see that in order to calculate the thermodynamic properties of a long ($m \to \infty$) one-dimensional system, all that is required is the largest eigenvalue of the corresponding matrix. For the singlet distribution, we need the eigenvector that corresponds to the largest eigenvalue. For the correlation functions we need, in general, all the eigenvalues and the corresponding eigenvectors.

There is an important characteristic property of the distributions, hence of the correlations in the infinite one-dimensional system, namely, that all the distributions can be expressed in terms of the *nn*-pair distribution and the singlet distribution. As an example we write the triplet distribution (7.2.28) as

$$P(s_1 = \alpha, s_2 = \beta, s_3 = \gamma) = \frac{\langle\alpha|\mathbf{M}|\beta\rangle\langle\beta|\mathbf{M}|\gamma\rangle\langle\gamma|\mathbf{M}^{m-2}|\alpha\rangle}{\xi}$$

$$\xrightarrow{m\to\infty} \langle\alpha|\mathbf{M}|\beta\rangle\langle\beta|\mathbf{M}|\gamma\rangle\langle\alpha|a_{mx}\rangle\langle\gamma|a_{mx}\rangle d_{mx}^{-2} \quad (7.2.31)$$

On the other hand,

$$\frac{P(s_1=\alpha, s_2=\beta)P(s_2=\beta, s_3=\gamma)}{P(s_2=\beta)}=\frac{\langle\alpha|\mathbf{M}|\beta\rangle\langle\beta|\mathbf{M}|\gamma\rangle\langle\beta|\mathbf{M}^{m-1}|\alpha\rangle\langle\gamma|\mathbf{M}^{m-1}|\beta\rangle}{\xi\langle\beta|\mathbf{M}^{m}|\beta\rangle}$$

$$\to\langle\alpha|\mathbf{M}|\beta\rangle\langle\beta|\mathbf{M}|\gamma\rangle\langle\alpha|a_{mx}\rangle\langle\gamma|a_{mx}\rangle d_{mx}^2 \tag{7.2.32}$$

Hence we have the equality (in the limit $m\to\infty$)

$$P(s_1=\alpha, s_2=\beta, s_3=\gamma)=\frac{P(s_1=\alpha, s_2=\beta)\,P(s_2=\beta, s_3=\gamma)}{P(s_3=\beta)} \tag{7.2.33}$$

In terms of the corresponding correlations, Eq. (7.2.33) is transformed into

$$g(s_1=\alpha, s_2=\beta, s_3=\gamma)=g(s_1=\alpha, s_2=\beta)g(s_2=\beta, s_3=\gamma) \tag{7.2.34}$$

This is sometimes referred to as the "superposition approximation." It is not, however, the superposition approximation used in the theory of liquids, first because Eq. (7.2.34) is exact (in the limit $m\to\infty$), and second because the superposition approximation [as introduced by Kirkwood (1935) and used extensively in the theory of liquids] has the form

$$g(s_1=\alpha, s_2=\beta, s_3=\gamma)\approx g(s_1=\alpha, s_2=\beta)g(s_2=\beta, s_3=\gamma)g(s_3=\gamma, s_1=\alpha) \tag{7.2.35}$$

Using arguments similar to the above, we can easily see that the second nearest-neighbor correlation may assume the form

$$\begin{aligned} g(s_1=\alpha, s_3=\gamma)&=\frac{P(s_1=\alpha, s_3=\gamma)}{P(s_1=\alpha)P(s_3=\gamma)}\\ &=\frac{\sum_{s_2}P(s_1=\alpha, s_2, s_3=\gamma)}{P(s_1=\alpha)P(s_3=\gamma)}\\ &=\sum_{s_2}g(s_1=\alpha, s_2)P(s_2)g(s_2, s_3=\gamma) \end{aligned} \tag{7.2.36}$$

where the definition of the marginal distribution has been used in the second equality. Here, summing over the index s_2, the triplet distribution produces the pair distribution. In the third equality we used the result (7.2.33).

We stress again that both results (7.2.34) and (7.2.36) are valid in the limit of large m, strictly at $m\to\infty$. They are not valid for finite m, as we shall see again in the next section. We also note that the correlations mentioned above are defined for any events of the system and for any λ. In our study of correlations that appear in binding isotherms, we require only correlations between events of the form "site i is occupied," "site j is occupied," etc., and we need only the $\lambda\to 0$ limits of these correlations.

7.3. 1-D SYSTEM WITH DIRECT CORRELATIONS ONLY

The description of the two-state system, $f = 2$, was introduced earlier in Sections 7.1 and 7.2. Here, we present some quite obvious results for systems with *nn direct* interactions only. Since we discuss only a restricted group of events, we use a simpler notation for the correlations. Thus, instead of $g(s_1 = \alpha, s_2 = \beta)$, we simply use $g_{nn}(2)$ or $g_{12}(2)$ to denote pair correlations (between the event "site i occupied" and "site $i + 1$ occupied"). Also, we shall always refer to the $\lambda \to 0$ limit as the correlation and omit specific notation for this limit.

For the pair correlations in the $m \to \infty$ limit we have, as expected,

$$\begin{aligned} g_{nn}(2) &= g_{12}(2) = S \\ g_{1l}(2) &= 1 \text{ for } l \geq 2 \end{aligned} \tag{7.3.1}$$

where $S = \exp(-\beta U_{11})$. Clearly, the *nn* pair correlation is the *direct* correlation S and there exists no long-range correlation beyond the *nn*.

For higher-order correlations, the particular correlation depends on m and on whether the chain is open or closed. For instance, when $m = 3$

$$\begin{aligned} g_{123}(3) &= S^2 \text{ (open chain)} \\ g_{123}(3) &= S^3 \text{ (closed chain)} \end{aligned} \tag{7.3.2}$$

but when $m \geq 4$

$$g_{123}(3) = S^2 \text{ (either open or closed)} \tag{7.3.3}$$

All these can be read directly from the number of "bonds" between the ligands. The situation is far more complicated when indirect correlations are operative, as we shall see in the next section.

Figure 7.2 shows the BI for the open (a) and closed (b) systems, for $q = 1$ and $S = 100$, and $m = 4, 7, 10$.

We recall that for the open linear case $m = 2$, the value of $\lambda_{1/2}$ for which $\theta(\lambda) = 1/2$ is

$$\lambda_{1/2} = \frac{1}{q\sqrt{S}} \tag{7.3.4}$$

This is true only when $m = 2$. As noted in Section 4.3, once we have $m \geq 3$, the point at which $\theta(\lambda) = 1/2$ depends, in a complicated manner, on all the correlations in the system. It can be seen from Fig. 7.2a that the location of $\lambda_{1/2}$ changes with m; for instance, when $m = 4$, $\lambda_{1/2} = 0.03$, and when $m = 5$, $\lambda_{1/2} = 0.023$.

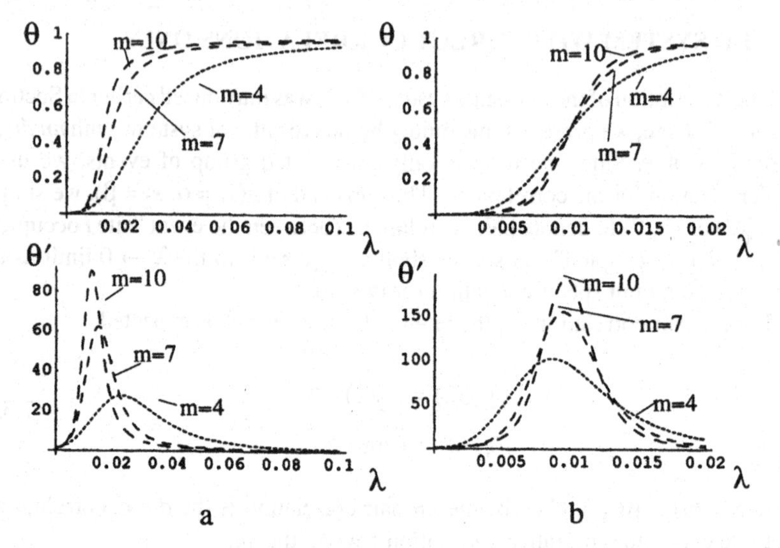

Figure 7.2. BI and the corresponding slopes for $m = 4, 7, 10$ (denoted next to each curve). (a) open and (b) closed systems. $q = 1$ and $S = 100$.

The behavior is quite different for the cyclic closed system. Here, as can be seen from Fig. 7.2b, the location $\lambda_{1/2}$ is common to any m and is

$$\lambda_{1/2} = \frac{1}{qS} \tag{7.3.5}$$

which, for $q = 1$ and $S = 100$, is $\lambda_{1/2} = 0.01$. Note, however, that the location of maximum slope, i.e., λ_{max} for which

$$\theta''(\lambda) = 0 \tag{7.3.6}$$

is only in the neighborhood of $\lambda_{1/2} = 0.01$ and, as shown in Fig. 7.2b, λ_{max} changes with m. As m increases, the slope at $\lambda_{1/2}$ increases, and for $m \sim 50$ it reaches a limiting value, as can be seen in Fig. 7.3.

The different behavior of the open and closed system is due to the difference in the pattern of correlations in the two systems; hence the average correlation differs for the two systems with the same m.

Figure 7.4 shows $\bar{g}(C)$ for cyclic systems with $m = 3, 4, 5, 6$ ($q = 1$ and $S = 100$, and $\lambda_0 = 1$, hence $k = 1$). It is seen that in the $C \to 0$ limit [where $\bar{g}(C)$ is determined by the pair correlation only], the larger is m the smaller is $\bar{g}(C \to 0)$. In Table 7.1 we list the pair correlations for a few m. Thus, for $m = 3$ we have only one pair correlation S. For $m = 4$ we have four nn pair correlations, each contributing S, and

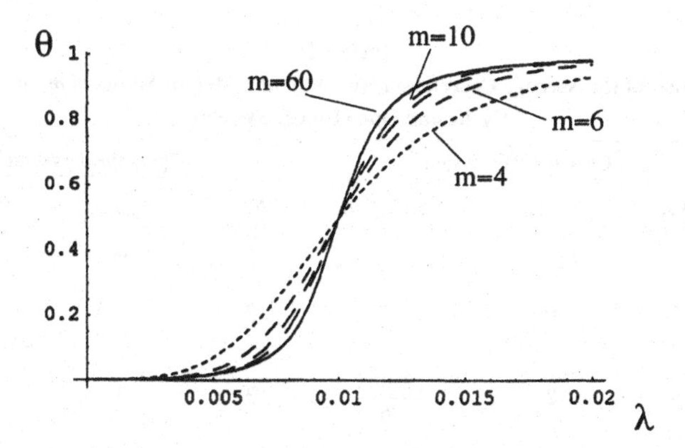

Figure 7.3. The BIs as in Fig. 7.2b, but for $m = 4, 6, 8, 10, 60$. It is seen that the curves converge to a limiting curve for $m \gtrsim 50$.

two non-*nn* pair correlations, which in this approximation contribute unity to $\overline{g}(C \to 0)$. In general, for the cyclic system of m sites there are m *nn* pair correlations. Altogether there are $m(m-1)/2$ pairs in the system. Hence the number of non-*nn* pair correlations is $m(m-1)/2 - m = m(m-3)/2$.

For the open linear system the pattern of pair correlations is different. We have again $m(m-1)/2$ total number of pairs, but now only $(m-1)$ of these are *nn* pairs, contributing S each, and the remaining $(m-1)(m-2)/2$ are non-*nn* pairs, contributing unity to the average correlation $\overline{g}(C \to 0)$; see Table 7.1.

In the $C \to \infty$ limit, all the sites are bound; the average correlation $\overline{g}(C \to \infty)$ is determined by the mth-order correlation function, which is S^m for the cyclic and S^{m-1} for the open linear system. This is true within the pairwise additive approximation for direct interaction, and neglecting long-range correlations.

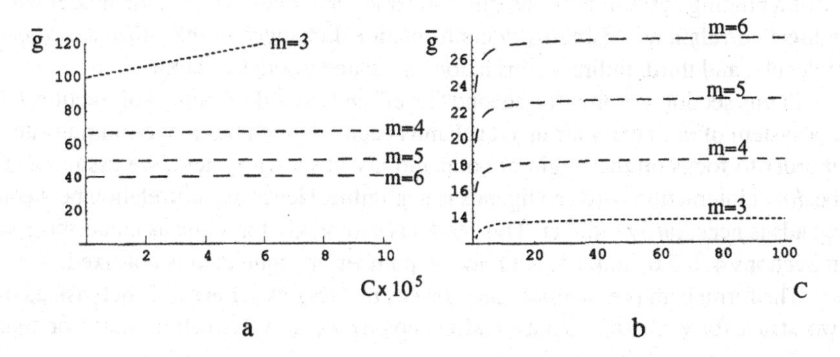

Figure 7.4. The average correlation $\overline{g}(c)$ for a cyclic system with $m = 3, 4, 5, 6$ ($q = 1$ and $S = 100$): (a) in the $C \to 0$ limit, (b) in the $C \to \infty$ limit.

Table 7.1
Limiting Values of the Average Correlation $\bar{g}(C \to 0)$ for Different Values of m, for the Closed Cyclic and Open Linear Systems

	Closed cyclic system[a]			Open linear system[a]		
m	N_{nn}	$N_{non\text{-}nn}$	$g(C \to 0)$	N_{nn}	$N_{non\text{-}nn}$	$\bar{g}(C \to 0)$
2	—	—	—	1	0	S
3	1	0	S	2	1	$\frac{1+2S}{3}$
4	4	2	$\frac{2+4S}{6}$	3	3	$\frac{3+3S}{6}$
5	5	5	$\frac{5+5S}{10}$	4	6	$\frac{6+4S}{10}$
6	6	9	$\frac{9+6S}{15}$	5	10	$\frac{10+5S}{15}$
7	7	14	$\frac{14+7S}{21}$	6	15	$\frac{15+6S}{21}$
m	m	$\frac{m(m-3)}{2}$	$\frac{m(m-3)+2mS}{m(m-1)}$	$(m-1)$	$\frac{(m-1)(m-2)}{2}$	$\frac{(m-2)+2S}{m}$
large m	m	$m^2/2$	1	m	$m^2/2$	1

[a] N_{nn} is the number of nn pairs $N_{non\text{-}nn}$ is the number of non-nn pairs.

7.4. A SYSTEM OF *m* LINEARLY ARRANGED SUBUNITS

In previous chapters we developed in several places three "levels" of complexity of a binding system: first, the direct-interaction model; second, the model with indirect correlations arising from conformational changes in the *entire* adsorbent molecule; and third, indirect correlations mediated through subunits.

In this section we omit the second "level" and extend the model of Section 7.1 to a system of m linearly arranged subunits each of which has a one-binding site.* In order to focus on the origin of cooperativity in such a system, we assume that the *direct* interaction between ligands is negligible. Hence any correlation between ligands is necessarily indirect. This model is essentially the same as that discussed in Sections 4.7, 5.6, and 6.3, but now the number of subunits m is not fixed.

The formalism is essentially the same as in Sections 7.1 and 7.2, but instead of two states for each site (empty and occupied) we now have four states or four

*One can always obtain the intermediate model by letting $\eta \to 0$. See, for example, Section 4.7.

degrees of freedom ($f = 4$), namely, empty H, empty L, occupied H, and occupied L. The 4×4 matrix is

$$\begin{pmatrix} Q_H Q_{HH} & Q_{HL}\sqrt{Q_H Q_L} & \sqrt{\lambda q_H}\, Q_H Q_{HH} & Q_{HL}\sqrt{\lambda q_L Q_H Q_L} \\ Q_{HL}\sqrt{Q_H Q_L} & Q_L Q_{LL} & Q_{HL}\sqrt{\lambda q_H Q_H Q_L} & \sqrt{\lambda q_L}\, Q_L Q_{LL} \\ \sqrt{\lambda q_H}\, Q_H Q_{HH} & Q_{HL}\sqrt{\lambda q_H Q_H Q_L} & \lambda q_H Q_H Q_{HH} & \lambda Q_{HL}\sqrt{q_H q_L Q_H Q_L} \\ Q_{HL}\sqrt{\lambda q_L Q_H Q_L} & \sqrt{\lambda q_L}\, Q_L Q_{LL} & \lambda Q_{HL}\sqrt{q_H q_L Q_H Q_L} & \lambda q_L Q_L Q_{LL} \end{pmatrix} \tag{7.4.1}$$

The corresponding vector $\langle v|$ is now given by

$$\langle v| = [Q_H^{1/2}, Q_L^{1/2}, (\lambda q_H Q_H)^{1/2}, (\lambda q_L Q_L)^{1/2}] \tag{7.4.2}$$

and we define the four unit vectors by [see Eq. (7.1.11)]

$$\langle e_1| = \langle H0| = (1, 0, 0, 0), \qquad \langle e_2| = \langle L0| = (0, 1, 0, 0)$$

$$\langle e_3| = \langle H1| = (0, 0, 1, 0), \qquad \langle e_4| = \langle L1| = (0, 0, 0, 1) \tag{7.4.3}$$

Again, the configuration of the entire system is denoted by the m-dimensional vector

$$\mathbf{s} = (s_1, s_2, s_3, \ldots, s_m)$$

where each s_i can be one of the states $H0$, $L0$, $H1$, or $L1$. The energy of each configuration $\mathbf{s}$ now differs from Eq. (7.1.3). Here, we do not include ligand–ligand interactions, so for the open linear case we have

$$E(\mathbf{s}) = \sum_{i=1}^{m} \delta_{s_i,H} E_H + \delta_{s_i,L} E_L + \delta_{i,L1} U_L + \delta_{s_i,H1} U_H$$

$$+ \sum_{i=1}^{m-1} \delta_{s_i,H}\delta_{s_{i+1},H} E_{HH} + \delta_{s_i,L}\delta_{s_{i+1},L} E_{LL} + (\delta_{s_i,L}\delta_{s_{i+1},H} + \delta_{s_i L}\delta_{s_{i+1},H}) E_{LH} \tag{7.4.4}$$

where

$$\delta_{s_i,H} = \delta_{s_i,H0} + \delta_{s_i,H1} \text{ and } \delta_{s_i,L} = \delta_{s_i,L0} + \delta_{s_i,L1} \tag{7.4.5}$$

Although Eq. (7.4.4) looks complicated, it is quite simple. We scan through all the subunits: each subunit in state H contributes E_H and, if occupied, also U_H; each pair of successive H, either empty or occupied, contributes E_{HH}, etc.

Having written the energy, we can express the GPF of the system exactly as in Section 7.1. Thus, for the open and closed systems we have

$$\xi_O = \langle v|\mathbf{M}^{m-1}|v\rangle \tag{7.4.6}$$

and

$$\xi_C = \mathrm{Tr}(\mathbf{M}^m) \tag{7.4.7}$$

For simplicity, we also introduce the notation

$$K = Q_H/Q_L, \qquad h = q_H/q_L, \qquad K' = \frac{Q_{HH}}{Q_{LL}}, \qquad \eta = \frac{Q_{HL}^2}{Q_{HH}Q_{LL}} \tag{7.4.8}$$

with $Q_{LH} = Q_{HL}$, as before. We also set $\overline{K} = \overline{KK'}$ whenever the two constants appear together. The quantity $\overline{K}^m$ may be interpreted as the equilibrium constant for the conversion of m subunits arranged in a cycle from L to H, i.e.,

$$m\,L \rightleftharpoons m\,H$$

$$\overline{K}^m = \frac{[HH\cdots H]}{[LL\cdots L]} = \left(\frac{Q_H Q_{HH}}{Q_L Q_{LL}}\right)^m \tag{7.4.9}$$

This is true in the case of the closed system for which the number of subunits and the number of subunit–subunit "bonds" are equal.

In terms of the constants h, $\overline{K}$, and η, the matrix M is written as

$$\begin{pmatrix} \overline{K} & \sqrt{\eta}\sqrt{\overline{K}} & \sqrt{h}\sqrt{x}\,\overline{K} & \sqrt{x}\sqrt{\eta}\sqrt{\overline{K}} \\ \sqrt{\eta}\sqrt{\overline{K}} & 1 & \sqrt{h}\sqrt{x}\sqrt{\eta}\sqrt{\overline{K}} & \sqrt{x} \\ \sqrt{h}\sqrt{x}\,\overline{K} & \sqrt{h}\sqrt{x}\sqrt{\eta}\sqrt{\overline{K}} & h\,x\,\overline{K} & \sqrt{h}\,x\sqrt{\eta}\sqrt{\ldots} \\ \sqrt{x}\sqrt{\eta}\sqrt{\overline{K}} & \sqrt{x} & \sqrt{h}\,x\sqrt{\eta}\sqrt{\overline{K}} & x \end{pmatrix} \tag{7.4.10}$$

where we have extracted the factor $Q_L Q_{LL}$ and also changed variables so that $x = \lambda/q_L$. By solving the secular equation

$$|\mathbf{M} - d\mathbf{I}| = 0 \tag{7.4.11}$$

where $\mathbf{I}$ is the 4×4 unit matrix, we obtain the four eigenvalues

$$d_1 = d_2 = 0$$

$$d_3 = \frac{1}{2}(1 + x + \overline{K} + h\overline{K}x - \sqrt{})$$

$$d_4 = \frac{1}{2}(1 + x + \overline{K} + h\overline{K}x + \sqrt{}) \tag{7.4.12}$$

where

$$\sqrt{} = \sqrt{4(\eta - 1)\bar{K}(1 + x)(1 + hx) + (1 + \bar{K} + x + h\bar{K}x)^2} \tag{7.4.13}$$

Note that the largest eigenvalue is d_4.*

The corresponding eigenvectors are, before normalization,

$$\left.\begin{aligned} a_1' &= (0, -\sqrt{x}, 0, 1) \\ a_2' &= (\sqrt{hx}, 0, 1, 0) \\ a_3' &= \left(\frac{A - \sqrt{}}{B\sqrt{x}}, \frac{1}{\sqrt{x}}, \frac{\sqrt{h}(A - \sqrt{})}{B}, 1\right) \\ a_4' &= \left(\frac{A + \sqrt{}}{B\sqrt{x}}, \frac{1}{\sqrt{x}}, \frac{\sqrt{h}(A + \sqrt{})}{B}, 1\right) \end{aligned}\right\} \tag{7.4.14}$$

where $\sqrt{}$ is given in Eq. (7.4.13) while $A = \bar{K} - h\bar{K}x - x - 1$ and $B = 2(1 + hx)\sqrt{\bar{K}\eta}$. These should be normalized, i.e.,

$$a_i = a_i'/\sqrt{a_i' \cdot a_i'} \tag{7.4.15}$$

before using the numerical calculations below.

If we expand the identity matrix in terms of its eigenvectors [see Eq. (7.2.9)], then the GPFs can be expressed in the form

$$\xi_O = \sum_{i=1}^{4} \langle v|a_i\rangle^2 d_i^{m-1} \quad \text{and} \quad \xi_C = \sum_{i=1}^{4} d_i^m \tag{7.4.16}$$

where d_i is the ith eigenvalue of the matrix **M**. In the limit of a very long chain ($m \to \infty$) for which d_{mx} is the largest eigenvalue, we have

$$\xi_O \to \langle v|a_{mx}\rangle^2 d_{mx}^{m-1} \quad \text{and} \quad \xi_C \to d_{mx}^m \tag{7.4.17}$$

Thus, all one needs to compute the thermodynamic properties of the system, including the BI, is d_{mx}. The factor $\langle v|a_{mx}\rangle^2$ is independent of m and does not affect the thermodynamics of the system in this limit.

For any finite m, it is more convenient to work with the closed cyclic systems. We have seen that a linear open system of four *identical* subunits has two different intrinsic binding constants ($k_1^{(1)}, = k_1^{(4)}, k_1^{(2)} = k_1^{(3)}$). Thus, in general, the sites are not identical, neither in the weak nor in the strict sense. (Recall, however, that in the

*The reason for having only two nonzero eigenvalues is that there are only two states, L and H, that, in this model, determine the energy of the system.

model of Section 7.1 the sites were identical in the weak sense.) In contrast, for the case of a closed cyclic system of m identical subunits the sites are all weakly identical, i.e., all sites have the same (first) intrinsic binding constant given by

$$k_1 = \frac{Q(1)}{mQ(0)}\lambda_0 = \frac{\lambda_0}{m}\lim_{\lambda\to 0}\frac{\partial \ln \xi}{\partial \lambda} \tag{7.4.18}$$

These are, of course, different for different m. For instance, when $m = 3$ and $m = 4$ we have

$$k_1(m=3) = \frac{1 + h\bar{K}^3 + \eta(2\bar{K} + h\bar{K} + 2h\bar{K}^2)}{1 + \bar{K}^3 + \eta(3\bar{K} + 3\bar{K}^2)}\lambda_0 \text{ and } k_1(m=4) = \frac{1 + h\bar{K}^4 + \eta(\cdots)}{1 + \bar{K} + \eta(\cdots)}\lambda_0 \tag{7.4.19}$$

and, in general,

$$k_1(m) = \frac{1 + h\bar{K}^m + \eta(\cdots)}{1 + \bar{K}^m + \eta(\cdots)}\lambda_0 \tag{7.4.20}$$

where $(\cdots)$ is a complicated expression involving $\bar{K}$, h, and η.

Clearly, when $\eta = 1$ the sites become independent and we have, for any m,

$$k_1 = \frac{1 + h\bar{K}}{1 + \bar{K}}\lambda_0 \tag{7.4.21}$$

On the other hand, when $\eta = 0$ all the subunits act concertedly, i.e., the whole system changes between L and H states; the corresponding intrinsic binding constants become

$$k_1(m, \eta = 0) = \frac{1 + h\bar{K}^m}{1 + \bar{K}^m} \tag{7.4.22}$$

The calculation of the correlation functions is essentially the same as in Section 7.2, except for the additional complexity due to the existence of more states. Thus, the singlet distribution, i.e., the probability of finding any specific single site occupied, is*

$$P(\textit{site } 1 \textit{ occupied}) = P(s_1 = L1 \textit{ or } s_1 = H1)$$
$$= \sum_{\mathbf{s}} P(s_1, \ldots, s_m)(\delta_{s_1,L1} + \delta_{s_1,H1}) \tag{7.4.23}$$

*In probability notation, this would be written as $P[(s_1 = L1) \cup (s_1 = H1]$.

Following similar steps, as in Section 7.2, we obtain

$$P(s_1 = L1 \textbf{ or } s_1 = H1) = \frac{1}{\xi}\left[\sum_{i=1}^{4} (\langle L1|a_i\rangle^2 d_i^m + \langle H1|a_i\rangle^2 d_i^m)\right] \tag{7.4.24}$$

Note that the sum over i is over all four states of a single subunit. The unit vectors $\langle L1|$ and $\langle H1|$ are constants. When $m \to \infty$, Eq. (7.4.24) reduce to

$$P(1) = P(s_1 = L1 \textbf{ or } s_1 = H1) = \langle L1|a_4\rangle^2 + \langle H1|a_4\rangle^2 = \langle e_4|a_4\rangle^2 + \langle e_3|a_4\rangle^2 \tag{7.4.25}$$

This represents the sum of the probabilities of finding any *specific* site, say $i = 1$, occupied *and* in either an L or H conformational state, respectively. Again, we note that $P(1)$ is simply the BI per site. Note that the expression for $P(1)$ in general (any m) involves both the eigenvalues and eigenvectors, but in the $m \to \infty$ limit only the eigenvector $|a_4\rangle$ corresponding to the largest eigenvalue is needed.

By employing a similar procedure as in Section 7.2, we can compute all the distribution functions and the corresponding correlation functions of this system. As an example we derive here the expression for the 1, l pair distribution, and the corresponding pair correlation. The procedure is the same as in Section 7.2, with the additional complexity that we now have *two* occupied states rather than one as in Section 7.2. [On the other hand, expressions (7.2.24) and (7.2.25) are more general in the sense that they apply to *any* state α and β. Here, we are interested in *specific* states, i.e., states such as "site i is occupied," for calculating cooperativities.] The probability of finding site 1 occupied *and* site l occupied is*

$$\begin{aligned} P_{1,l}(2) &= P[(s_1 = L1 \textbf{ or } s_1 = H1) \textbf{ and } (s_l = L1 \textbf{ or } s_l = H1)] \\ &= \sum_{\mathbf{s}} P(s_1, \ldots, s_m)[\delta_{s_1,L1} + \delta_{s_1,H1}][\delta_{s_l,L1} + \delta_{s_l,H1}] \\ &= \frac{1}{\xi}\sum_{\alpha,\beta} \langle\alpha|\mathbf{M}^{l-1}|\beta\rangle\langle\beta|\mathbf{M}^{m-l+1}|\alpha\rangle \\ &= \sum_{\alpha,\beta}\frac{1}{\xi}\left[\sum_i \langle\alpha|a_i\rangle\langle\beta|a_i\rangle d_i^{l-1}\right]\left[\sum_j \langle\beta|a_j\rangle\langle\alpha|a_j\rangle d_j^{m-l+1}\right] \\ &\xrightarrow{m\to\infty} \sum_{\alpha,\beta}\sum_i \langle\alpha|a_i\rangle\langle\beta|a_i\rangle \left(\frac{d_i}{d_{mx}}\right)^{l-1} \langle\alpha|a_{mx}\rangle\langle\beta|a_{mx}\rangle \end{aligned} \tag{7.4.26}$$

*In probability notation this would be written as $P[((s_1 = L1) \cup (s_1 = H1)) \cap ((s_l = L1) \cup (s_l = H1))]$.

This is the same as Eq. (7.2.26), except for the additional summation over α, β which runs over the *occupied* states, i.e., $L1$ and $H1$. The sum over i is, in general, over all f degrees of freedom, but since in this particular model $d_1 = d_2 = 0$ the sum over i reduces to $i = 3, 4$.

The 1, l pair correlation in the $m \to \infty$ limit is

$$g_{1,l}(2) = \frac{P_{1,l}(2)}{[P(1)]^2}$$

$$= 1 + \frac{\Sigma_{\alpha\beta} \langle\alpha|a_3\rangle\langle\beta|a_3\rangle\langle\alpha|a_4\rangle\langle\beta|a_4\rangle}{\Sigma_{\alpha\beta} \langle\alpha|a_4\rangle^2\langle\beta|a_4\rangle^2}\left(\frac{d_3}{d_4}\right)^{l-1} \tag{7.4.27}$$

which is the generalization of Eq. (7.2.27) ($d_4 = d_{mx}$, $|a_4\rangle = |a_{mx}\rangle$). Note that in calculating the correlation functions, as required in the binding isotherm, we must take the $\lambda \to 0$ limit of $g_{1,l}(2)$.

The general form of $g_{1,l}(2)$ is quite complicated. We shall present some numerical results of $g_{1,l}(2)$ as a function of parameters h, $\overline{K}$, and η. Here, it is instructive to present the formal form of the pair correlations (in the limits $m \to \infty$ and $\lambda \to 0$) for the case $\overline{K} = 1$. The first few pair correlations are

$$\left.\begin{aligned} g_{nn} = g_{1,2}(2) &= 1 - \frac{(\sqrt{\eta} - 1)}{(\sqrt{\eta} + 1)}\left(\frac{h-1}{h+1}\right)^2 \\ g_{1,3}(2) &= 1 + \frac{(\sqrt{\eta} - 1)^2}{(\sqrt{\eta} + 1)^2}\left(\frac{h-1}{h+1}\right)^2 \\ g_{1,4}(2) &= 1 - \frac{(\sqrt{\eta} - 1)^3}{(\sqrt{\eta} + 1)^3}\left(\frac{h-1}{h+1}\right)^2 \end{aligned}\right\} \tag{7.4.28}$$

and, in general,

$$g_{1,l}(2) = 1 + (-1)^{l-1}\frac{(\sqrt{\eta} - 1)^{l-1}\,(h-1)^2}{(\sqrt{\eta} + 1)^{l-1}\,(h+1)^2} \tag{7.4.29}$$

It is easy to see from Eq. (7.4.29) that "either $h = 1$ or $\eta = 1$" is a necessary and sufficient condition for no correlations, of any order. The general necessary and sufficient condition for no correlation is "$\overline{K} = 0$, or $h = 1$, or $\eta = 1$." In the above example we have written the correlations for the case $\overline{K} = 1$. The case $\overline{K} = 0$ may be checked directly from the definition of the correlation function expressed in terms of the canonical PF.

We have already seen in Section 5.7 that the extent of the pair correlation depends on the number of subunit–subunit boundaries across which information must be transformed. If we have, say, four *different* subunits, then the 1,4 pair correlation will have the form (for $h \neq 1$ and $\overline{K} \neq 0$)

$$g_{1,4}(2) = 1 + \gamma \prod_{i=1}^{3} (1 - \sqrt{\eta_{i,i+1}}) \tag{7.4.30}$$

Thus, we have a factor of $(1 - \sqrt{\eta_{i,i+1}})$ for each boundary. If one of the $\eta_{i,i+1}$ is unity, then the whole line of communication between sites 1 and 4 will become short and $g_{1,4}(2) = 1$. When the subunits are identical, we have the factor $(1 - \sqrt{\eta})^3$ and, in general, for any l we have the factor $(1 - \sqrt{\eta})^{l-1}$, as in Eq. (7.4.29). The most interesting aspect of the dependence on η is that it can also change the sign of the cooperativity. We recall that for a system of identical subunits, $\overline{K}$ and h determine only the *size* of the correlation but η determines also its sign. When $\eta < 1$, then $(1 - \sqrt{\eta})^{l-1} > 0$ and the correlation is always positive, i.e., $g_{1,l}(2) > 1$. On the other hand, for $\eta > 1$ the factor $(1 - \sqrt{\eta})^{l-1}$ changes sign, according to whether l is even or odd. The reason, as we saw in Section 4.7, is that for each crossing of the boundary the extent of conformational change induced by the ligand changes sign, hence also the sign of correlation* is changed. We shall present below a numerical demonstration of this effect.

The dependence of the higher-order correlations on h, $\overline{K}$, and η is more complicated. One example has already been examined in Section 5.5. We shall not examine this aspect here.

We turn now to the finite open and closed chain and compare the pair correlations obtained in the different systems. First, we note that in the $m \to \infty$ limit all the sites become identical in the weak sense, i.e., there is only one intrinsic binding constant, but different pair (and higher-order) correlations as shown in Eq. (7.4.28). It should be noted, however, that owing to the translational invariance of the infinite system there is only *one nn* pair correlation, only *one* second *nn* pair correlation, etc. In other words, it does not matter where in the chain we choose the pair of *nn* neighbors, or the second *nn* neighbors, etc. This translational invariance is lost in the finite open system.

Figure 7.5 shows how the *nn* pair correlation $g_{12}(2)$ changes with m. Starting with $m = 6$ we see that $g_{12}(2)$ almost reaches its "limiting" value at $m \cong 10$. The exact limiting value for the parameters $h = 0.01$, $\overline{K} = 4$, and $\eta = 0.1$ is 1.108.

In Fig. 7.6 we show diagrammatically the ($m \to \infty$ limit) pair correlations $g_{1,l}$ for $l = 2, 3, 4, 5, 6$, starting with the value $g_{1,2}(2) = 1.108$ for the nearest neighbors. We see how the correlation quickly decays to unity at $l \sim 6$. The particular rate of decay for the parameters chosen here is

$$g_{1,l}(2) \approx 1 + 1.73 \exp(-1.39l) \tag{7.4.31}$$

*As usual, changing the "sign" of the correlation means changing from smaller to larger than unity.

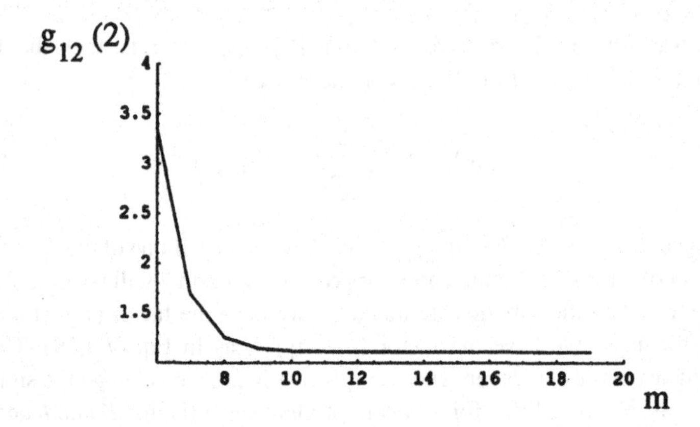

Figure 7.5. The dependence of $g_{12}(2)$ on m for a system of m subunits with parameters $h = 0.01$, $\bar{K} = 4$, and $\eta = 0.1$.

In the diagram of Fig. 7.7 we show all possible *pair* correlations for the finite open linear case with $m = 6$. The parameters are still the same: $h = 0.01$, $\bar{K} = 4$, and $\eta = 0.1$. Note that there are three *different nn* pair correlations: $g_{1,2} = g_{5,6}$, $g_{2,3} = g_{4,5}$, and $g_{3,4}$. There are two different second-*nn* pair correlations, $g_{1,3}$ and $g_{2,4}$ (only the different ones are indicated in the diagram). There are two third-*nn* pair correlations, $g_{1,4}$ and $g_{2,5}$. There is one fourth-*nn* pair correlation $g_{1,5}$ and one sixth-*nn* pair correlation $g_{1,6}$. It should also be noted that, as in the infinite chain ($m \to \infty$) case, the correlations decay with distance l.

A similar diagram for the cyclic case of $m = 6$ is shown in Fig. 7.8. Here, owing to the symmetry of the system, there is only one *nn* pair correlation, one second-*nn* pair correlation, etc. Note, however, that the sites are still identical only in the weak sense (i.e., there is one intrinsic binding constant k_1). The values of the different

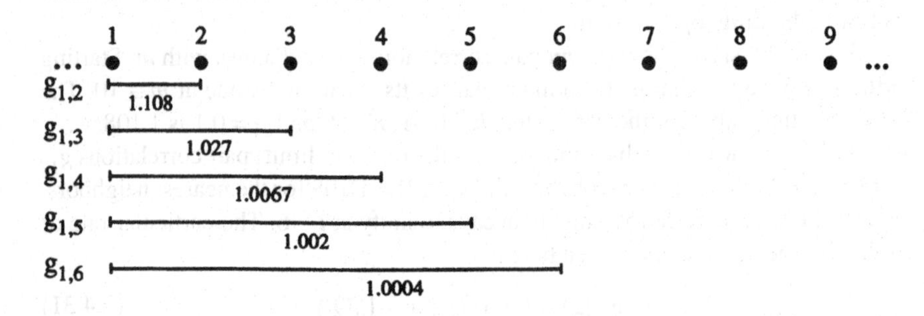

Figure 7.6. A section of an infinite linear system of subunits. We selected an arbitrary unit, denoted 1, and determined the pair correlations $g_{1,l}(2)$ for $l = 2, 3, 4, 5, 6$. The parameters are the same as in Fig. 7.5.

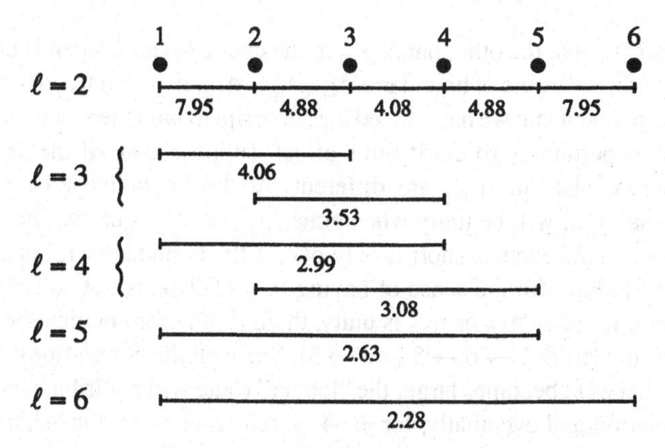

Figure 7.7. All correlation functions for an open linear system with $m = 6$. The parameters are as in Fig. 7.5. For $g_{12}(2)$, all correlations are shown; for $g_{1,l}(2)$, $l > 2$, only representative correlations are shown.

correlations in the cyclic case differ from the corresponding values in the open linear case (compare with Fig. 7.6). The reason is twofold: First, although the parameters h, $\overline{K}$, and η used in the calculations are the same, the response of the system to binding is different since the equilibrium constant for conformational changes is different, mainly because there are six subunit–subunit interactions in the cyclic case but only five in the open linear case. (Of course, for larger m these differences will become smaller and smaller.) Second, and perhaps a more interesting effect, is that there is only one path of transmitting information in the open linear case but two in the cyclic case. For instance, the pair correlation $g_{1,2}$ in the open linear case depends on the subunit–subunit boundary between 1 and 2, more specifically on

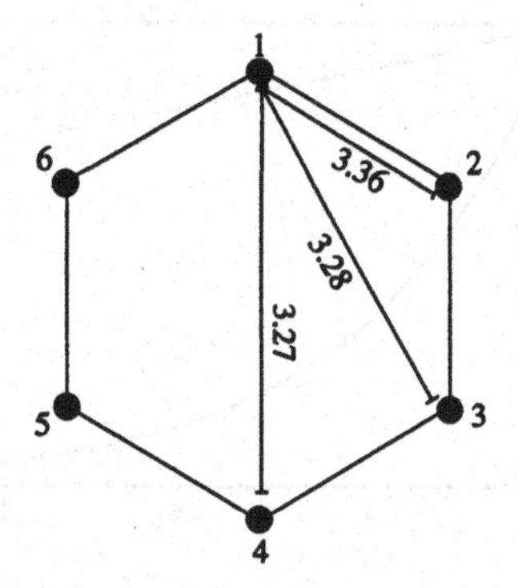

Figure 7.8. The three different pair correlations $g_{1,2}(2)$, $g_{1,3}(2)$, and $g_{1,4}(2)$ for the cyclic system with $m = 6$. The parameters are the same as in Fig. 7.5.

the parameter η_{12}. On the other hand, $g_{1,2}$ in the cyclic system depends both on η_{12} and on the "indirect" route of boundaries $\eta_{1,6}$, $\eta_{6,5}$, $\eta_{5,4}$, $\eta_{4,3}$, and $\eta_{3,2}$. (In this model all η values are equal, but we have added the subscript to indicate to which boundary the relevant η pertains.) To clarify this effect, suppose that all the subunits are different, hence also all $\eta_{i,i+1}$ are different. In the open linear case, the pair correlation, say $g_{1,3}$, will be unity when either $\eta_{1,2}$ or $\eta_{2,3}$ is unity. The communication between 1,3 becomes short if only one of the boundaries 1,2 or 2,3 cannot transmit information (in the sense of having $\eta = 1$). This is not so for $g_{1,3}$ in the cyclic system: if either $\eta_{1,2}$ or $\eta_{2,3}$ is unity, there is still communication along the alternative route (from $1 \rightarrow 6 \rightarrow 5 \rightarrow 4 \rightarrow 3$). Since all the correlations decay with l, it is clear that as m becomes larger the "longer" route will contribute less and less to the correlation and eventually, for $m \rightarrow \infty$, only one route, the "shorter" route, will be effective.

Finally, we examine the "distance" dependence of the pair correlation function. As we shall see below the distance dependence of $g_{1,l}(2)$ is not determined by the *real* distance between the ligands (as is, indeed, the case for the direct correlation discussed in Section 7.3), but only on the *number* l of subunit–subunit boundaries.

We recall that the parameters $\overline{K}$ and h can change the magnitude of the correlation but not its sign (i.e., whether $g > 1$ or $g < 1$; note, however, that for a system of *different* subunits the values of h for the different sites can determine the sign of the correlation). When $\eta < 1$, all the pair correlations $g_{1,l}(2)$ are larger than unity and they decay to unity as l increases. Figure 7.9 shows the dependence of $g_{1,l}(2)$ on l for $\eta = 0.1$, $\eta = \sqrt{0.1}$, and $\eta = 0$. Clearly, as η becomes smaller the range of the pair correlation increases—eventually, for $\eta = 0$, there is a *total* transmission

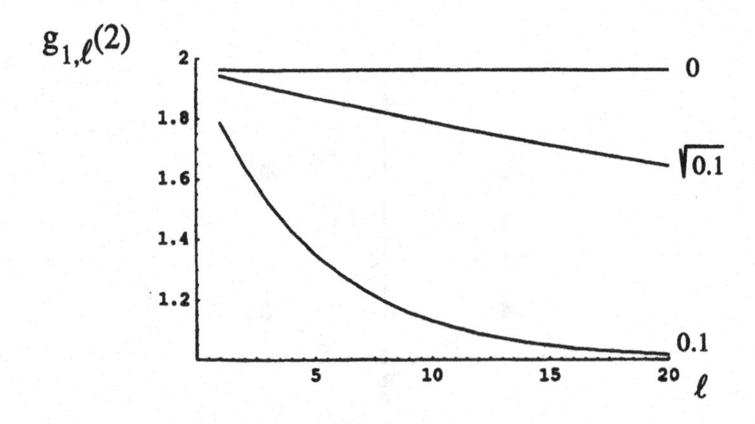

Figure 7.9. The dependence of the pair correlation function $g_{1,l}(2)$ on l for three different values of η. (In these curves, $\overline{K} = 1$ and $h = 0.01$.) Note the monotonic behavior when $\eta < 1$.

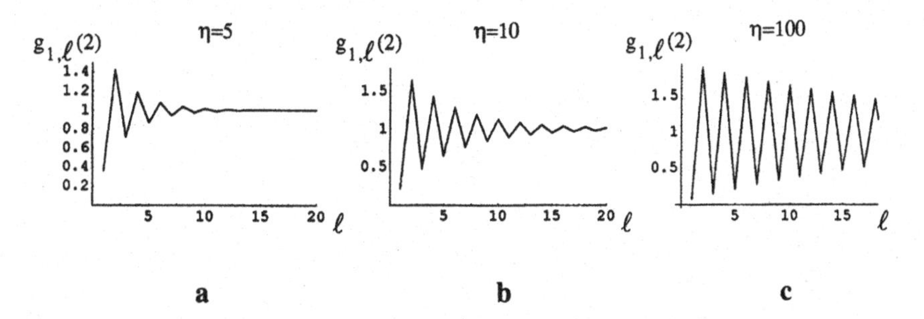

Figure 7.10. The same as Fig. 7.9, but for (a) $\eta = 5$, (b) $\eta = 10$, and (c) $\eta = 100$. Note the oscillatory behavior when $\eta > 1$. Compare with Fig. 7.9.

of information, i.e., all subunits act concertedly, namely, binding on site $i = 1$ affects equally the probability of binding on any other $i \neq 1$.

Perhaps the most interesting long-range behavior of the pair correlation $g_{1,l}(2)$ is when $\eta > 1$. For even l, i.e., an odd number of boundaries to cross, the value of $g_{1,l}(2)$ is smaller than unity, namely, *negative* cooperativity. For odd values of l, i.e., an even number of boundaries, the cooperativity is *positive*. Figure 7.10 shows how $g_{1,l}(2)$ oscillates between values above and below unity. These oscillations decay rapidly when η is small (but still $\eta > 1$); however, when η is very large we observe a very long-range correlation, and ultimately for $\eta \to \infty$ the range becomes infinite.

The interpretation of this behavior is already contained in Section 4.7, where we examined the relation between indirect correlations and structural changes induced in the subunit. We have seen that for $\eta < 1$, the structural change induced by the ligand is transmitted with the *same* sign, but with diminishing force each time we cross a boundary. For instance, if $d_L^{(1)}$ is the conformational change induced in the first subunit (on which the ligand binds), then the conformational change induced in every other subunit in the system will have the same sign as $d_L^{(1)}$, but a diminishing value as $(1 - \sqrt{\eta})^{l-1}$. In the case $\eta > 1$, it was seen in Section 4.7 that each time we cross a boundary the induced conformational change is in a different direction. Hence, also the cooperativity changes its sign each time we cross a boundary between two subunits.

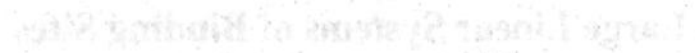

8

Regulatory Enzymes

8.1. INTRODUCTION AND HISTORICAL PERSPECTIVE

The history of regulatory enzymes has its roots in the history of enzymes in general. The theoretical framework of how enzymes work was founded by Emil Fischer in 1894. Fischer discovered that some enzymes can distinguish between two closely related substrates, such as two stereoisomeric sugars. On this basis he formulated the "lock and key" model which hypothesizes that enzymes have specific sites that can accommodate ligands which have a *complementary structure*. The lock and key model is based on the existence of a *geometrical fit* between the substrate (the key) and the binding site (the lock); see Fig. 8.1a. This model has undergone several modifications and generalizations. The first modification follows from the recognition that a *geometrical fit* does not necessarily imply the strongest affinity between the enzyme and the ligand. The more important quantity is the binding free energy. To achieve the largest (in absolute magnitude) binding free energy, one does not need to have a geometrical fit. A complementary pattern of functional groups (such as charged, hydrogen-bonding, and hydrophobic groups) can produce a strong affinity between the ligand and the site even without a *geometrical complementarity* (Fig. 8.1b). In fact, one can show that solvent effects can produce strong affinity even when there is neither a geometrical nor a complementary pattern fit (Fig. 8.1c). In this case, the possibility of the formation of hydrogen-bonded bridges by solvent molecules is the main driving force for selecting the binding site (see also Chapter 9).

The second generalization, developed mainly by Koshland, is based on the recognition that enzymes (like any protein) have a multitude of conformations at equilibrium. Since the ligand is likely to interact differently with the various conformations, one can expect a shift in the distribution of conformations induced by the binding process. This is the *induced fit* model. It states that the best fit (by either geometrical or by a complementary pattern) does not necessarily exist *before*

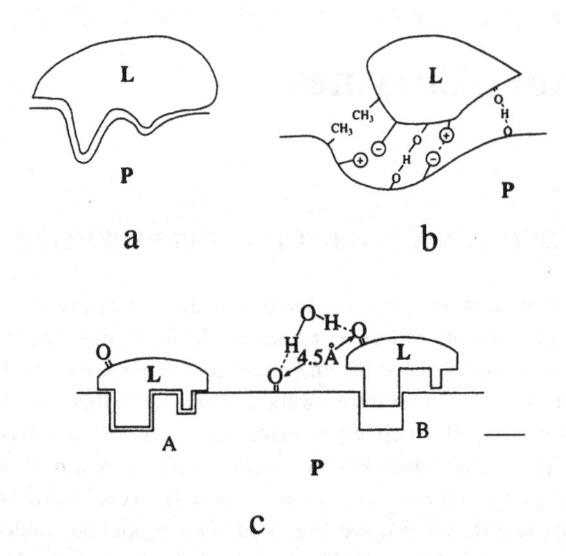

Figure 8.1. Lock and key model: (a) geometrical fit, (b) complementary pattern of functional groups, (c) site preference due to the solvent effect. The ligand **L** may better fit site *A*, but it binds preferentially to site *B* due to the solvent effect.

the binding, but can be produced *after* the binding has modified the conformational distribution. Figure 8.2 depicts a schematic process of the induced fit model.

The essential idea underlying the theory of regulatory enzymes, as developed in two classical papers by Monod, Changeux, and Jacob (MCJ) (1963) and by Monod, Wyman, and Changeux (MWC) (1965), is the same as the *induced fit* model. However, the "purpose" of the induced fit is different in the two theories. In the original theory, the induced fit is presumed to maximize the binding affinity between the ligand and the site on which it binds. In the theory of regulatory enzymes, the induced conformational change is presumed to change the affinity of the site to the substrate as well as the affinity of a *different* (allosteric) site to a different (regulatory) ligand.*

It was pointed out in Section 6.8 that the term "allosteric" as coined by MCJ and MWC has been used with three different meanings. In Chapter 6 we discussed the allosteric effect in hemoglobin (Hb). There, the two allosteric sites were identical; this has been referred to as the *homotropic* effect. When the two sites bind different ligands, the *heterotropic* effect, the induced fit by one ligand can either enhance or diminish the binding affinity of the second ligand (see the example in Section 4.5).

*The possibility that the induced fit model might also be used to explain the working mechanism of regulatory enzymes was already mentioned by Koshland himself (1962).

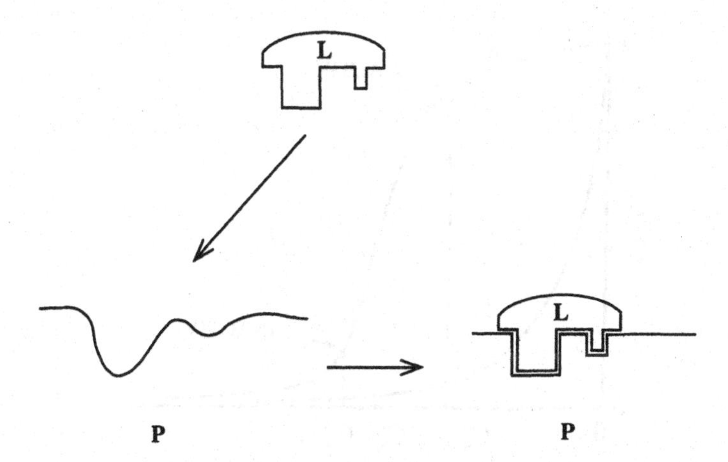

Figure 8.2. Induced fit model. The ligand **L** approaching the site on **P** will cause a change in **P** so that the fit between **L** and **P** improves.

Today, it is believed that many enzymes are regulated by the allosteric mechanism. Figure 8.3 shows a schematic illustration of the change in the activity of an enzyme with respect to a substrate **A** upon the addition of a different ligand **R**. The term "activity" of the enzyme is essentially a kinetic property, e.g., the rate of production of the product of the enzymatic reaction. We shall see in Section 8.2 that one can grasp the essence of the regulatory mechanism by studying an equilibrium system. For the illustration in Fig. 8.3 we assume that activity is simply the rate of the reaction (either the rate of disappearance of the substrate **A**, or the rate of appearance of the product **p**). Curve a in Fig. 8.3 shows that the activity of the enzyme drops sharply near $C_R \approx 0$. The second curve, b, shows that initially the activity of the enzyme changes slowly upon addition of **R**, but at some concentration around C_R^* the activity drops sharply, until at some higher concentration $C_R^* + \delta$ the enzyme becomes essentially inactive.

In the following sections we shall develop a few molecular models that exhibit similar behavior regarding the drop in "activity" of the enzyme upon the addition of a ligand **R** that differs from substrate **A**. The terms "activity" or "inactivity" of the enzymes should not be understood as sharply defined. Rather, we assume that the enzyme exists in at least two conformations denoted by H and L. (In the biochemical literature these are denoted by R, for relaxed, and T, for tensed.) One conformation is more active than the other. Hence, at any equilibrium state the "activity" of the enzyme is an average of the "activities" of the two conformations. Any ligand (or, in general, any change in the environment of the enzyme, see Chapter 9) that binds with different affinities

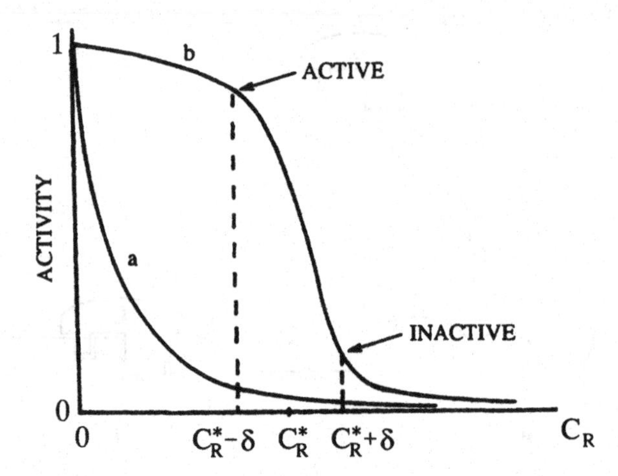

Figure 8.3. Schematic illustration of the change in the activity of an enzyme upon the addition of an effector **R**. (a) The sharp drop in activity occurs near $C_R \approx 0$. (b) The activity is initially unchanged upon addition of **R**, it drops sharply near a finite concentration, between $C_R^* - \delta$ and $C_R^* + \delta$..

to the two conformations will change their relative amounts and hence affect also the average "activity" of the enzyme.†

8.2. THE CONNECTION BETWEEN THE KINETIC EQUATION AND THE BINDING ISOTHERM

The basic kinetic scheme of an enzymatic reaction is

$$\mathbf{E}+\mathbf{A} \underset{k_1''}{\overset{k_1'}{\rightleftharpoons}} \mathbf{EA} \xrightarrow{k_2'} \mathbf{E}+\mathbf{p} \tag{8.2.1}$$

where **E**, **A**, and **p** are the enzyme, the substrate and the product, respectively; **EA** is the intermediary complex which, in our language, may be referred to as an adsorbent molecule (**E**), having a bound ligand (**A**) at the (active) site denoted by *A*. The various rate constants in Eq. (8.2.1) are primed to distinguish them from the intrinsic binding constants.

†Nowadays, it is a well-established fact that the effector **R** is not *isosteric* but *allosteric* to the substrate **A**. Originally, this conclusion was based on the finding that the enzyme can be desensitized, i.e., one can add substances that affect the regulatory site but not the activity of the enzyme with respect to the substrate.

In the steady-state approximation, the rate of formation of **EA** is equal to the rate of its decomposition, hence we have the equality

$$k_1' C_{\mathbf{E}} C_{\mathbf{A}} = k_1'' C_{\mathbf{EA}} + k_2' C_{\mathbf{EA}} \tag{8.2.2}$$

If C_T denotes the total enzyme concentration, then the fractional saturation at site A is

$$\theta_{\mathbf{A}} = \frac{C_{\mathbf{EA}}}{C_T} = \frac{C_{\mathbf{EA}}}{C_{\mathbf{E}} + C_{\mathbf{EA}}} \tag{8.2.3}$$

If we set $k_A = k_1'/(k_1'' + k_2')$, we can rewrite Eq. (8.2.3), using Eq. (8.2.2) in the form

$$\theta_{\mathbf{A}} = \frac{k_{\mathbf{A}} C_{\mathbf{A}}}{1 + k_{\mathbf{A}} C_{\mathbf{A}}} = \frac{C_{\mathbf{A}}}{K_m + C_{\mathbf{A}}} \tag{8.2.4}$$

where the first form on the rhs of Eq. (8.2.4) is the familiar Langmuir BI, with $k_{\mathbf{A}}$ the binding constant of **A** to site A, while the second form is the more conventional form, where $K_m = k_{\mathbf{A}}^{-1}$ is the Michaelis-Menten constant.

The rate of formation of the product **p** is

$$\nu = \frac{dC_{\mathbf{p}}}{dt} = k_2' C_{\mathbf{EA}} = k_2' C_T \theta_{\mathbf{A}} \tag{8.2.5}$$

Since $k_2' C_T$ is independent of $C_{\mathbf{A}}$, Eq. (8.2.5) shows that the rate of formation of the product is simply proportional to $\theta_{\mathbf{A}}$, i.e., the fractional saturation at the active site A. Since $\theta_{\mathbf{A}} \le 1$, or $C_{\mathbf{EA}} \le C_T$, the quantity $k_2' C_T$ may be referred to as the maximal rate and is denoted by $\nu_{\max}$.

Equation (8.2.5) establishes the connection between the *rate* of the enzymatic reaction within the steady-state approximation, and the equilibrium binding isotherm.

Suppose now that the enzyme has another site, denoted by R, which is at a different location from the active site. The site R binds a different ligand, denoted by **R**, which will be referred to as an effector or a regulator. Clearly, the presence of **R** may affect all the rate constants in Eq. (8.2.1), hence $k_{\mathbf{A}}$; however, the general form of Eq. (8.2.5) will remain unchanged. Therefore, Eq. (8.2.5) is rewritten in the form

$$\nu = \nu_{\max} \frac{k_{\mathbf{A}}(C_{\mathbf{R}}) C_{\mathbf{A}}}{1 + k_{\mathbf{A}}(C_{\mathbf{R}}) C_{\mathbf{A}}} \tag{8.2.6}$$

where we emphasize the dependence of $k_{\mathbf{A}}$ on the concentration of the regulator, $C_{\mathbf{R}}$.

As an extreme example, suppose that **R** binds to the active site A itself. Hence, any complex formed by **E** and **R** removes some of the enzymes from our consideration in Eq. (8.2.1). The total enzyme concentration is now

$$C_T = C_{\mathbf{E}} + C_{\mathbf{EA}} + C_{\mathbf{ER}} \tag{8.2.7}$$

so Eq. (8.2.3) is replaced by

$$\theta_{\mathbf{A}} = \frac{C_{\mathbf{EA}}}{C_T} = \frac{C_{\mathbf{EA}}}{C_{\mathbf{E}} + C_{\mathbf{EA}} + C_{\mathbf{ER}}} \tag{8.2.8}$$

The form of Eq. (8.2.5) is retained with the modified C_T and $\theta_{\mathbf{A}}$ of Eqs. (8.2.7) and (8.2.8). If we substitute

$$k_{\mathbf{A}} = \frac{C_{\mathbf{EA}}}{C_{\mathbf{E}}C_{\mathbf{A}}} \quad \text{and} \quad k_{\mathbf{R}} = \frac{C_{\mathbf{ER}}}{C_{\mathbf{E}}C_{\mathbf{R}}} \tag{8.2.9}$$

in Eq. (8.2.8) then the rate of formation of the product is

$$v = k_2' C_T \frac{k_{\mathbf{A}}C_{\mathbf{A}}}{1 + k_{\mathbf{A}}C_{\mathbf{A}} + k_{\mathbf{R}}C_{\mathbf{R}}} = k_2' C_T \frac{k_{\mathbf{A}}(C_{\mathbf{R}})C_{\mathbf{A}}}{1 + k_{\mathbf{A}}(C_{\mathbf{R}})C_{\mathbf{A}}} \tag{8.2.10}$$

where the explicit form of the effective **R**-dependent constant $k_{\mathbf{A}}(C_{\mathbf{R}})$ is

$$k_{\mathbf{A}}(C_{\mathbf{R}}) = \frac{k_{\mathbf{A}}}{1 + k_{\mathbf{R}}C_{\mathbf{R}}} \tag{8.2.11}$$

Clearly, when $C_{\mathbf{R}} \to 0$, $k_{\mathbf{A}}(C_{\mathbf{R}}) \to k_{\mathbf{A}}$ and we recover Eq. (8.2.4). In the subsequent sections we shall examine a few other examples where the effective constant depends on the concentration of the effector, or the regulator **R**. In this section we have made a distinction between the notations for the sites, say A or R, and for the corresponding names of the ligand, **A** or **R**. In the following sections we shall maintain this distinction, except when the letter A or R appears as a subscript.

8.3. THE REGULATORY CURVE AND THE CORRESPONDING UTILITY FUNCTION

In all the examples discussed in this and subsequent sections, we shall construct a BI which depends on the two concentrations C_A and C_R, i.e., we shall examine the function $\theta_A(C_A, C_R)$. Note that this is always the BI of the substrate **A** (as indicated by the subscript A in θ_A). This function depends on both C_A and C_R. We shall always require that when $C_R \to 0$, θ_A reduces to the BI of **A**, in the absence of the effector **R**.

When **R** is present, we expect that the entire BI of A will change. The extent of this change will depend on C_R. We shall examine, in particular, the rate of change of θ_A as a function of C_R, along a line of fixed C_A. The resulting curve will be referred to as the *regulatory curve*. The choice of the fixed value of C_A, although arbitrary, is made as follows.

We first choose an arbitrary value of θ_A near the saturation value $\theta_A = 1$, say $\theta_A = 0.8$. We may loosely say that whenever $\theta_A \geq 0.8$, the enzyme is "fully active." Similarly, when θ_A falls below a certain value, say $\theta_A = 0.2$, we shall say that the enzyme is "fully inactive." Starting with the BI of A in the absence of **R**, i.e., $\theta_A(C_A, C_R = 0)$, and choosing the value $\theta_A = 0.8$ determines the concentration C_A^*, which is the solution of the equation

$$\theta_A^* \equiv \theta_A(C_A^*, C_R = 0) = 0.8 \tag{8.3.1}$$

We next keep C_A^* fixed and follow the variation in θ_A as a function of C_R. This defines a new function

$$Rg(C_R) \equiv \theta_A(C_A^*, C_R) \tag{8.3.2}$$

which we call the *regulatory curve*.

Figure 8.4 demonstrates the effect of adding an effector **R** on the BI of **A**. When $Rg(C_R) > Rg(0) = \theta_A^*$, we say that **R** is an *activator*. In this case, addition of **R** causes θ_A to increase, hence the rate of production, Eq. (8.2.5), increases. When $Rg(C_R) < Rg(0)$, we say that **R** is an *inhibitor*, i.e., θ_A increases, hence the rate of production of **p** decreases. Both these effects are demonstrated in Fig. 8.4.

The special case of interest in regulatory enzymes is when the product **p** itself is the effector, i.e., **p** itself binds to the regulatory, or the allosteric, site R. In this case, C_R is identified with C_p and we have the differential equation

$$\nu = \frac{\partial C_p}{\partial t} = \nu_{max}\theta_A(C_A, C_p) \tag{8.3.3}$$

The qualitative behavior of this equation is simple. First, suppose that **p** is an *activator*. If, after the steady state has been established, C_p fluctuates upward, then

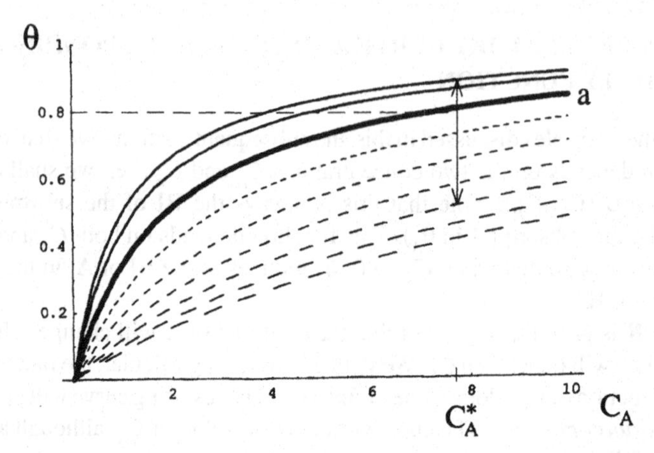

Figure 8.4. Schematic illustration of the effect of adding an effector. Curve a (bold line) is the BI of the enzyme in the absence of an effector. We choose the arbitrary value $\theta_A = 0.8$, corresponding to the concentration C_A^*. We follow the variation in θ_A along the vertical line at C_A^*. The addition of an activator **R** shifts the BI upward (thin lines). The addition of an inhibitor shifts the BIs downward (dashed lines).

θ_A will increase and the rate of production of **p** will increase, which again causes an increase in θ_A, followed by an increase in the rate, and so on. The opposite behavior will occur when C_p fluctuates downward.

The more interesting case is when **p** is an *inhibitor*. In this case, a different behavior is expected. An upward fluctuation in C_p will cause a *drop* in the fractional occupancy of the site A, i.e., θ_A will decrease, hence the rate of production of **p** will decrease. Similarly, a downward fluctuation in C_p will increase θ_A, hence increasing the rate of production. Thus, in both cases any fluctuation in C_p will be restored by this feedback mechanism. The net effect is that the level of concentration of **p** will be kept nearly fixed. The actual magnitude of the variation in C_p will depend on how effective is the effector in changing θ_A. We shall examine a few examples of the regulatory curve $Rg(C_R)$ in the following sections. The sharper the transition between an active enzyme (say, $\theta_A \gtrsim 0.8$) to an inactive enzyme (say, $\theta_A \lesssim 0.2$), the more precise will be the regulation of the production of **p**, i.e., the narrower the range of variations in C_p.

This qualitative description of the effect of an inhibitor on the BI of A, hence the rate of production of **p**, leads us to define the utility function for the regulatory enzyme. The idea is essentially the same as that introduced in connection with hemoglobin (Section 6.8 and Appendix K).

In the case of hemoglobin, we started with a given pressure difference $P_2 - P_1$, (at the loading and unloading terminals). The utility function was defined as the efficiency of transporting oxygen between these two limiting pressures, i.e., $\theta(P_2) - \theta(P_1)$. One could also ask the inverse question: Given two limits, loaded θ_2 and unloaded θ_1, of the adsorbent molecule, what is the minimal pressure difference under which the transporting system can operate?

Similarly, in the regulatory enzyme one may define the utility function in two equivalent ways. One can first fix the limits of activity of the enzyme, say $\theta_{A,2} = 0.8$ and $\theta_{A,1} \approx 0.2$, for the active and inactive enzyme, and ask for the efficiency of regulation. More precisely, given the difference $\theta_{A,2} - \theta_{A,1}$ what is the minimal variation allowed for the concentration of the end product? Alternatively, suppose we are *required* to keep the concentration of product **p** (which is the same as the effector **R**) within given limits, say $C_R^* - \delta \leq C_R \leq C_R^* + \delta$ (i.e., for fixed values of C_R^* and δ). The question is: what will be the maximal limits of variation in the regulatory curve $Rg(C_R)$ for a fixed interval about C_R^*? In the next few sections we shall examine the relation between the efficiency of maintaining a sharp concentration of a product and the form of the regulatory curve.

8.4. THE COMPETITIVE REGULATION

The simplest case of regulation is when **A** and **R** compete for the same site A (Fig. 8.5). This case was discussed earlier in Section 2.5. The GPF of the system is

$$\xi = Q(0) + Q(A)\lambda_A + Q(R)\lambda_R \tag{8.4.1}$$

and the binding isotherm for **A** is

$$\theta_A = \lambda_A \frac{\partial \ln \xi}{\partial \lambda_A} = \frac{k_A C_A}{1 + k_A C_A + k_R C_R} \tag{8.4.2}$$

where $k_A = q_A \lambda_{0A}$ and $k_R = q_R \lambda_{0R}$.

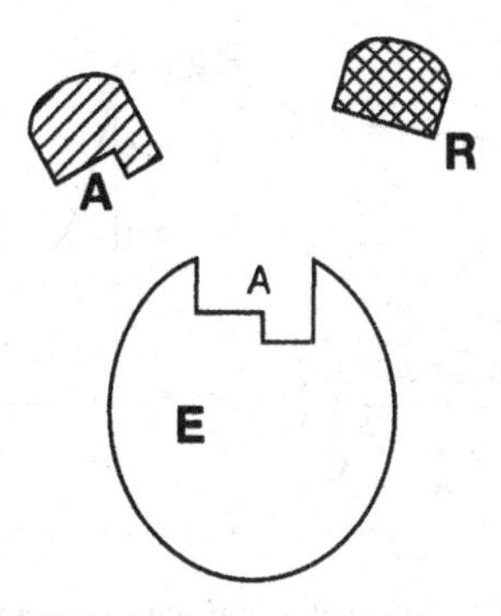

Figure 8.5. Competitive binding of two ligands **A** and **R** on the same site A. The two ligands bind with different affinities to the site A.

We define C_A^* by the equation

$$\theta_A^* = \frac{k_A C_A^*}{1 + k_A C^*} = 0.8 \tag{8.4.3}$$

Solving Eq. (8.4.3) for $C_A^* = 4/k_A$ and substituting in Eq. (8.4.2), we obtain the regulator curve

$$Rg(C_R) = \theta_A(C_A^*, C_R) = \frac{4}{5 + k_R C_R} \tag{8.4.4}$$

Clearly, the regulator curve in this case is a hyperbola, which can be transformed into a Langmuir-type curve by

$$0.8 - Rg(C_R) = \frac{0.8 k_R C_R}{5 + k_R C_R} \tag{8.4.5}$$

Since $k_R \geq 0$, the regulatory function is a monotonically decreasing function of C_R, i.e., the effector acts as an inhibitor. Figure 8.6 shows a series of BI, θ_A, for different values of C_R, and the corresponding regulatory curve. Note the sharp initial drop of $Rg(C_R)$ at $C_R \approx 0$.

8.5. A SIMPLE ALLOSTERIC REGULATION

We consider here the case where the absorbent molecule (the enzyme) has one active site A for binding the substrate **A**, and one regulator site R for the effector **R** (Figure 8.7). We assume that the two sites A and R are far apart, so that direct

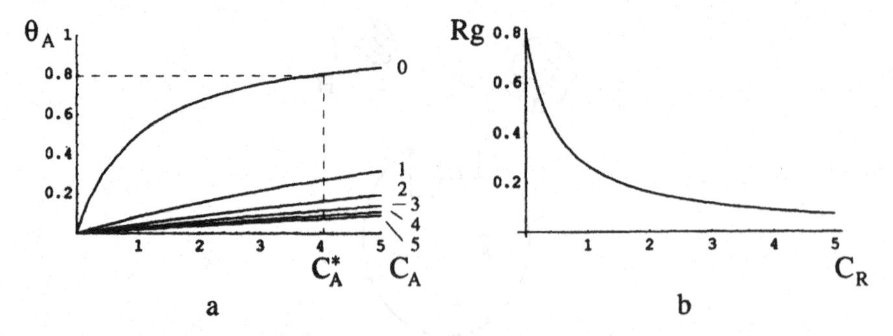

Figure 8.6. (a) Demonstration of the effect of a competitive effector **R** on the BI of **A**. The value of C_A^* (corresponding to $\theta_A^* = 0.8$ on the BI of A) is $C_A = 4$. (b) The regulatory curve $Rg(C_R)$ is the function $\theta_A(C_A^*, C_R)$, i.e., the change in θ_A along the vertical line at C_A^*.

Figure 8.7. Schematic illustration of an enzyme with a catalytic site for binding **A** and a regulatory (allosteric) site for binding **R**.

interaction between the ligands **A** and **R** may be neglected. To obtain indirect correlation, we assume the two-conformational model, as in Section 4.5. The GPF of the system is

$$\xi = Q(0,0) + Q(A,0)\lambda_A + Q(0,R)\lambda_R + Q(A,R)\lambda_A\lambda_R \tag{8.5.1}$$

where

$$\begin{aligned} Q(0,0) &= Q_L + Q_H, \quad Q(A,0) = Q_L q_{LA} + Q_H q_{HA} \\ Q(0,R) &= Q_L q_{LR} + Q_H q_{HR}, \quad Q(A,R) = Q_L q_{LA} q_{LR} + Q_H q_{HA} q_{HR} \end{aligned} \tag{8.5.2}$$

Note that we did not include any direct correlation in $Q(A, R)$. Defining

$$k_A C_A = \frac{Q(A,0)}{Q(0,0)}\lambda_A, \quad k_A C_R = \frac{Q(0,R)}{Q(0,0)}\lambda_R, \quad y_{AR} = \frac{Q(A,R)Q(0,0)}{Q(A,0)Q(0,R)} \tag{8.5.3}$$

we can express the BI for the substrate A as

$$\theta_A(C_A, C_R) = \frac{k_A C_A + k_A k_R y_{AR} C_A C_R}{1 + k_A C_A + k_R C_R + k_A k_R y_{AR} C_A C_R} \tag{8.5.4}$$

The regulatory curve is derived by again solving the equation

$$\theta_A^* = \theta_A(C_A^*, 0) = 0.8 \tag{8.5.5}$$

to obtain C_A^*. The regulatory curve is then

$$Rg(C_R) = \frac{4(1 + k_R y_{AR} C_R)}{5 + 4k_R C_R + 4k_R y_{AR} C_R} \tag{8.5.6}$$

where

$$y_{AR} = 1 + \frac{K(h_A - 1)(h_R - 1)}{(1 + Kh_A)(1 + Kh_R)} \qquad (8.5.7)$$

Here K, h_A, and h_R are defined as usual [see Eq. (8.5.9) below]. We see that if $(h_A - 1)$ and $(h_R - 1)$ have the same sign, $y_{AR} > 1$, and we have positive cooperativity and $Rg(C_R) \geq 0.8$ for any C_R [$Rg(0) = 0.8$]. In this case **R** is an activator for the site A, and $Rg(C_R)$ increases monotonically with C_R.

When $(h_A - 1)$ and $(h_R - 1)$ have different signs, $y_{AR} < 1$, and we have negative cooperativity between **A** and **R**. In this case **R** is an inhibitor to the site A. We have defined the fully active enzyme whenever $\theta_A \gtrsim 0.8$. Similarly, we may choose a lower bound on θ_A, say 0.2, to define the inactive enzyme. In order to have an efficient regulation we require that as C_R increases, Rg falls below ~0.2. The limiting values of Rg are

$$Rg(C_R = 0) = 0.8, \quad Rg(C_R = \infty) = \frac{y_{AR}}{1 + y_{AR}} \qquad (8.5.8)$$

Thus, in order to fall below the lower bound of 0.2, y_{AR} must be at most 0.25. We demonstrate in Fig. 8.8 the behavior of a system with y_{AR} above and below this

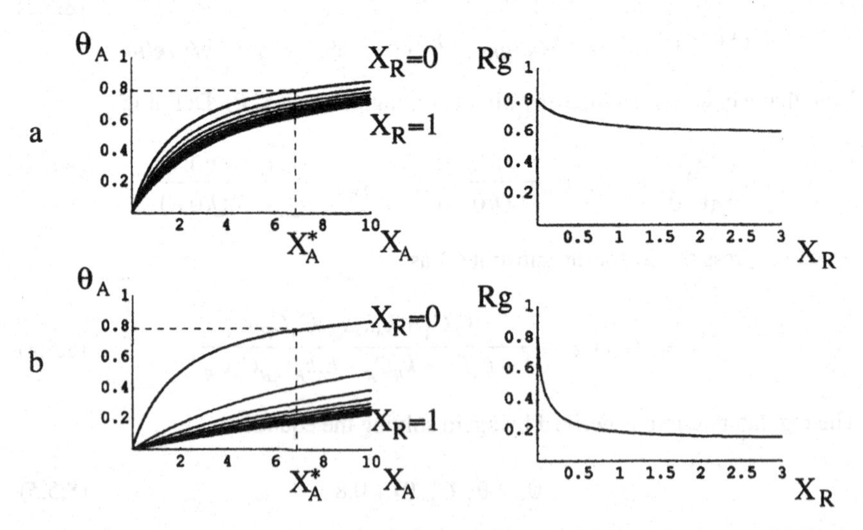

Figure 8.8. BIs and the regulatory curve for the model of Section 8.5. The parameters are (a) $K = 1$, $h_A = 0.1$, $h_R = 10$ (small negative cooperativity); (b) $K = 1$, $h_A = 0.01$, $h_R = 100$ (large negative cooperativity). Both of the regulatory curves are hyperbolic. They differ in the limiting value of Rg when $C_R \to \infty$ [Eq. (8.5.8)]: $Rg(C_R \to \infty) = 0.6$ and $Rg(C_R \to \infty) = 0.13$, respectively. In both (a) and (b) the curves are drawn for $X_R = 0, 0.1, \ldots, 1.0$.

value. The curves are drawn as a function of the variables $X_A = q_{LA}\lambda_A = k_{LA}C_A$ and $X_R = q_{LR}\lambda_R = k_{LR}C_R$. Both $\theta(C_A, C_R)$ and $Rg(C_R)$ are expressed in terms of the molecular variables

$$h_A = \frac{q_{HA}}{q_{LA}}, \qquad h_R = \frac{q_{HR}}{q_{LR}}, \qquad K = \frac{Q_H}{Q_L} \tag{8.5.9}$$

Figure 8.8 shows the corresponding regulatory curves for (a) $K = 1$, $h_A = 0.1$, $h_R = 10$ and (b) $K = 1$, $h_A = 0.01$, $h_R = 100$. In both cases $y_{AR} < 1$, i.e., negative cooperativity between **A** and **R** (in other words **A** favors the L form while **R** favors the H form). In both cases the regulatory curves (8.5.6) are hyperbolic. (This is true also when $y_{AR} > 1$, i.e., when there is positive cooperativity.) In case (a) the negative cooperativity is relatively small, $y_{AR} = 0.33$; the value of Rg drops from 0.8 to about 0.6. This can hardly be considered as an effective regulation. On the other hand, in the second case (b) there is a strong negative cooperativity, $y_{AR} = 0.04$, and the limiting value of Rg is about 0.13, in which case the value of $\theta_A(X_A^*, X_R)$ changes from a fully active enzyme, $\theta_A = 0.8$, to a very inactive enzyme at $\theta_A \lesssim 0.2$. In both cases the regulation is achieved near $X_R \sim 0$, not at any arbitrarily chosen value of X_R (or C_R). All the binding isotherms θ_A are drawn for ten values of $X_R = 0, 0.1, 0.2, \ldots, 1$. In both cases the main drop in θ_A is at $X_R \sim 0$, but the rate at which θ_A drops is different in the two cases. Thus, in spite of the fact that the BI is cooperative (either positive or negative), the regulatory curve is essentially an inverted Langmuir type, i.e., a hyperbolic curve. In order to obtain an inverted S-shaped regulator curve, we need at least two regulatory sites. This is examined in the next section.

8.6. ONE ACTIVE AND TWO REGULATORY SITES

We extend the model of the previous section. Instead of one regulatory site we now have two sites, say R_1 and R_2, which are identical and bind the effector **R** (Fig. 8.9). We still have one active site A, hence in the absence of the effector the binding isotherm $\theta_A(C_A, C_R = 0)$ is a simple Langmuir curve.

The GPF for this system is

$$\begin{aligned}\xi = {} & Q(0,0,0) + Q(A,0,0)\lambda_A + 2Q(0,R,0)\lambda_R + Q(0,R,R)\lambda_R^2 \\ & + 2Q(A,R,0)\lambda_A\lambda_R + Q(A,R,R)\lambda_A\lambda_R^2 \end{aligned} \tag{8.6.1}$$

The canonical PFs, Q, are expressed in terms of the molecular quantities in the usual manner. For instance, $Q(A, R, 0)$ is given by

$$Q(A, R, 0) = Q_L q_{LA} q_{LR} + Q_H q_{HA} q_{HR} \tag{8.6.2}$$

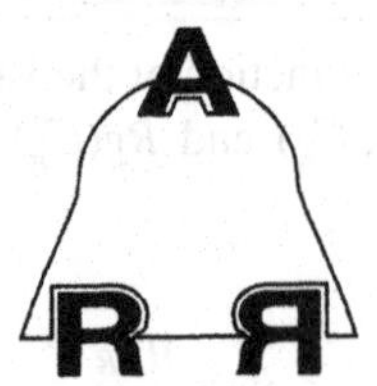

Figure 8.9. An enzyme with one catalytic site for A and two regulatory (allosteric) sites for R.

The intrinsic binding constants and the correlation functions are defined by

$$\left.\begin{aligned} k_A C_A &= \frac{Q(A,0,0)}{Q(0,0,0)}\lambda_A, \quad k_R C_R = \frac{Q(0,R,0)}{Q(0,0,0)}\lambda_R \\ y_{AR} &= \frac{Q(A,R,0)Q(0,0,0)}{Q(A,0,0)Q(0,R,0)}, \quad y_{RR} = \frac{Q(0,R,R)Q(0,0,0)}{[Q(0,R,0)]^2} \\ y_{ARR} &= \frac{Q(A,R,R)Q(0,0,0)^2}{Q(A,0,0)Q(0,R,0)^2} \end{aligned}\right\} \tag{8.6.3}$$

Here, we have three different correlations. In terms of these correlations the BI of A is

$$\theta_A = \frac{k_A C_A + 2k_A k_R y_{AR} C_A C_R + k_A k_R^2 y_{ARR} C_A C_R^2}{1 + k_A C_A + 2k_R C_R + k_R^2 y_{RR} C_R^2 + 2k_A k_R y_{AR} C_A C_R + k_A k_R^2 y_{ARR} C_A C_R^2} \tag{8.6.4}$$

The corresponding regulatory curve is

$$Rg(C_R) = \frac{4 + 8k_R y_{AR} C_R + 4k_R^2 y_{ARR} C_R^2}{5 + 2k_R C_R + k_R^2 y_{RR} C_R^2 + 8k_R y_{AR} C_R + 4k_R^2 y_{ARR} C_R^2} \tag{8.6.5}$$

the limiting values of which are

$$Rg(C_R = 0) = 0.8 \text{ and } Rg(C_R = \infty) = \frac{4y_{ARR}}{y_{RR} + 4y_{ARR}} \tag{8.6.6}$$

The limiting slope and the limiting curvature at $C_R = 0$ are

$$\left.\frac{\partial Rg}{\partial C_R}\right|_{C_R=0} = 0.32\, k_R(y_{AR} - 1) \tag{8.6.7}$$

and

$$\left.\frac{\partial^2 Rg}{\partial C_R^2}\right|_{C_R=0} = 1.024\, k_R^2[0.3125\,(y_{ARR} - y_{RR}) - y_{AR}^2 + 0.75\, y_{AR} + 0.25] \tag{8.6.8}$$

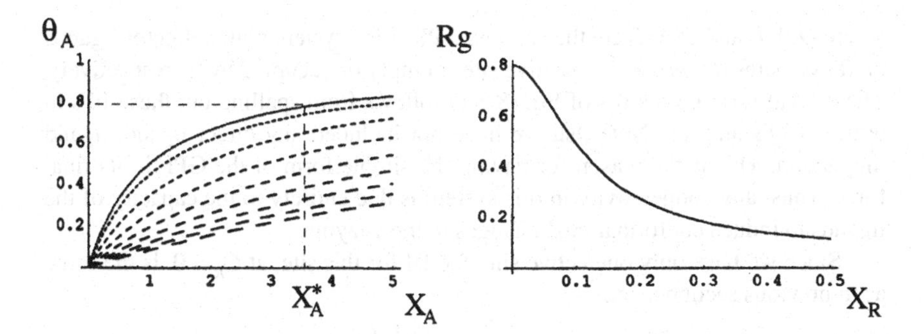

Figure 8.10. The BIs and the regulatory curve for the model discussed in Section 8.6. Note the inverted S-shaped regulatory curve.

Since the two regulatory sites are identical y_{RR} must be larger than unity (see Section 4.5). Therefore, in order to have an inverted S-shaped regulatory curve, the curve must start with a (negative) slope close to zero, hence $y_{AR} \lesssim 1$, and a negative curvature at $C_R \to 0$, i.e., $y_{ARR} << y_{RR}$. Therefore, if we choose a binding system with $y_{AR} \lesssim 1$ and such that $16y_{ARR} - y_{RR} \lesssim 0$, we can guarantee that in the $C_R \to \infty$ limit, Rg will fall below $\theta_A(C_A^*, C_R = \infty) \lesssim 0.2$. Of course, in this model we cannot choose y_{AR}, y_{RR}, and y_{ARR} independently. The independent variables are h_A, h_R, and K. For a choice of $K = 10^{-6}$, $h_A = 0.01$, and $h_R = 2 \times 10^4$, we obtain an initial slope of about zero with negative cooperativity between **A** and **R** ($y_{AR} = 0.98$) and a very strong positive cooperativity $y_{RR} \approx 385$ much larger than $16y_{ARR}$ ($y_{ARR} = 4.8$). The corresponding BI and the regulatory curve are shown in Fig. 8.10. Note how the BI drops initially (at $X_R \sim 0$) slowly, then more rapidly, then again slowly. This is a result of the inverted S-shaped regulatory curve. To obtain a sharper inverted S-shaped regulatory curve, we shall need the cooperation of more regulatory sites. This will be demonstrated in Sections 8.7 and 8.8.

8.7. ONE ACTIVE AND *m* REGULATORY SITES

The generalization of the model in Section 8.6 to three or more regulatory sites is quite straightforward. We still assume a two-conformation enzyme (L, H) with one active site and m identical (in the strict sense) regulator sites.

The GPF for such a system is

$$\xi = \sum_{i=0}^{m} Q(A, i)\lambda_A \lambda_R^i + \sum_{i=0}^{m} Q(0, i)\lambda_R^i$$

$$= Q_L(1 + q_{LA}\lambda_A)(1 + q_{LR}\lambda_R)^m + Q_H(1 + q_{HA}\lambda_A)(1 + q_{HR}\lambda_R)^m \quad (8.7.1)$$

where $Q(0, i)$ and $Q(A, i)$ are the canonical PFs for a system with i effector ligands on (any) i sites ($i \leq m$), and the active site is empty or occupied by **A**, respectively. The second form on the rhs of Eq. (8.7.1) follows from spelling out these PFs in terms of Q's and q's. Note that we have not included any direct ligand–ligand interaction. This is the reason for having the simple form of the GPF in its final form. Thus, any cooperativity in this system is due entirely to the capacity of the ligands to induce conformational changes in the enzyme.

Since we have only one active site, the BI for this site, at $C_R = 0$, is the same as in previous sections, i.e.,

$$\theta_A(C_A, C_R = 0) = \frac{k_A C_A}{1 + k_A C_A} \tag{8.7.2}$$

The regulatory curves are again obtained by solving for C_A^*,

$$\theta_A(C_A^*, C_R = 0) = 0.8 \tag{8.7.3}$$

$$C_A^* = \frac{4}{k_A} = \frac{4(Q_H + Q_L)}{Q_H q_{HA} + Q_L q_{LA}} \tag{8.7.4}$$

and substituting C_A^* in $\theta_A(C_A, C_R)$, to obtain the regulatory curve

$$Rg(C_R) = \theta_A(C_A^*, C_R) \tag{8.7.5}$$

For computational purposes we use again the transformation of variables, as in Section 8.5, to obtain the functions

$$\theta_A(X_A, X_R) = \frac{X_A[(1 + X_R)^m + h_A K(1 + h_R X_R)^m}{(1 + X_R)^m + X_A(1 + X_R)^m + K(1 + h_R X_R)^m + h_A K X_A(1 + h_R X_R)^m} \tag{8.7.6}$$

and

$$Rg(X_R) = \frac{0.8(1 + K)[(1 + X_R)^m + h_A K(1 + h_R X_R)^m]}{(1 + X_R)^m + h_A K^2(1 + h_R X_R)^m + 0.2K[(1 + X_R)^m (4 + h_A) + (1 + h_R X_R)^m (4h_A + 1)]} \tag{8.7.7}$$

Figure 8.11 shows the binding isotherms for the same parameters as in Section 8.6, namely,

$$K = 10^{-6}, \qquad h_A = 0.01, \qquad h_R = 2 \times 10^4 \tag{8.7.8}$$

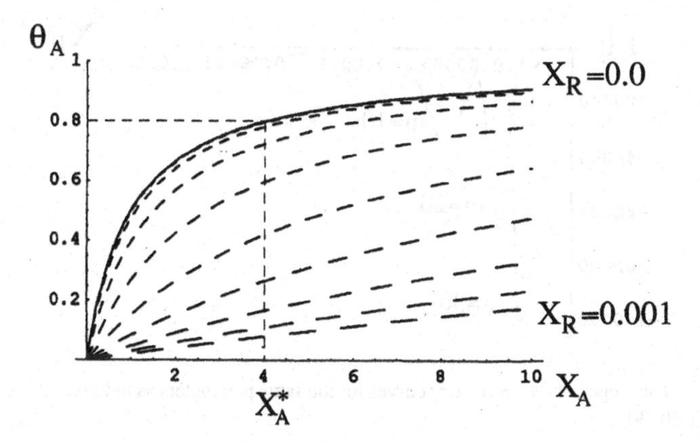

Figure 8.11. BIs for the model in Section 8.7, with parameters given in Eq. (8.7.8) and $m = 6$. The values of X_R are between 0 to 0.001, at intervals of 0.0002.

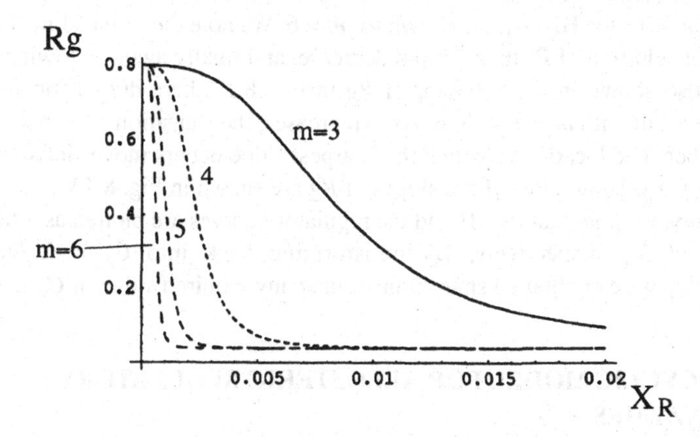

Figure 8.12. Regulatory curves for the same parameters as in Fig. 8.11 but for $m = 3, 4, 5, 6$. The larger m, the sharper the transition between the active and inactive enzyme. Note the shift of the transition point to lower values of X_R.

As before, these parameters produced an inverted S-shaped regulatory curve, while $y_{AR} = 0.98$ (for the first **R**) and $y_{RR} = 385$ (for the first pair of **R**s). However now, due to the presence of m regulatory sites, we have higher-order correlations. For instance, with $m = 4$, we find

$$g_{AR} = 0.98, \quad g_{ARR} = 4.8, \quad g_{ARRR} = 7.5 \times 10^4, \quad g_{ARRRR} = 1.4 \times 10^9$$
$$g_{RR} = 3.85, \quad g_{RRR} = 7.5 \times 10^6, \quad g_{RRRR} = 1.48 \times 10^{11} \tag{8.7.9}$$

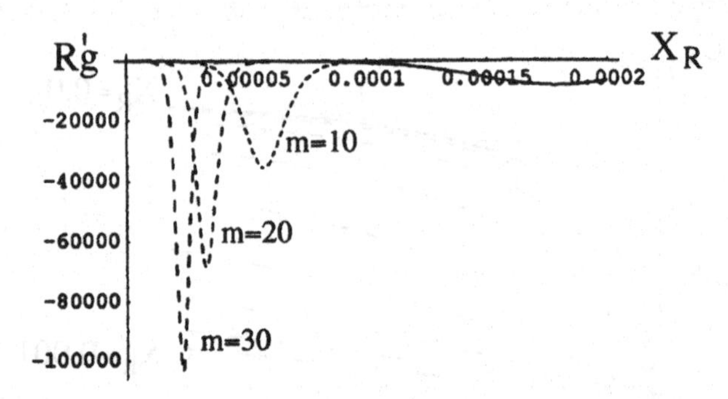

Figure 8.13. The slopes of the regulatory curves for the same parameters as in Figs. 8.11 and 8.12, but with m = 10, 20, 30.

We see that the correlations grow rapidly with m, leading to very sharp inverted S-shaped regulatory curves.

In Fig. 8.11 the BIs, θ_A, are drawn for $m = 6$. We note the initial slow decrease in θ_A upon addition of **R**, then a rapid decrease, and finally again a slowing down. This is also shown in the behavior of Rg in Fig. 8.12, here drawn for the same parameters but with $m = 3, 4, 5, 6$. As m increases, the transition becomes sharper and sharper. The location at which the sharpest slope occurs moves leftward, i.e., decreasing X_R. Some plots of the slopes of Rg are shown in Fig. 8.13.

Finally, we note that the BI and the regulatory curves are plotted as a function of X_A and X_R, respectively. By transforming back into $C_A = X_A/k_{LA}$ and $C_R = X_R/k_{LR}$ we can obtain a sharp transition at any required value of C_R.

8.8. A CYCLIC MODEL FOR ALLOSTERIC REGULATORY ENZYMES

In this section we extend the models of Chapter 7, using the formalism of the matrix method. This allows us to study the dependence of regulatory curves on the number of subunits m. The model itself is a simple generalization of the model treated in Section 7.4: instead of one site per subunit we now have two sites, one for binding the substrate **A** and another, allosteric site, for binding the effector **R** (Fig. 8.14).* The formalism presented in previous sections is the same, only the number of degrees of freedom increases; we now have $f = 8$, i.e., each unit has eight possible states. Thus, to characterize the state of a unit in the chain we need to know its conformational state L or H, whether it is empty or occupied at the active site A, and whether it is empty or occupied at the regulatory site R. The corresponding unit

*More details on this model will be found in Bohbot and Ben-Naim (1995).

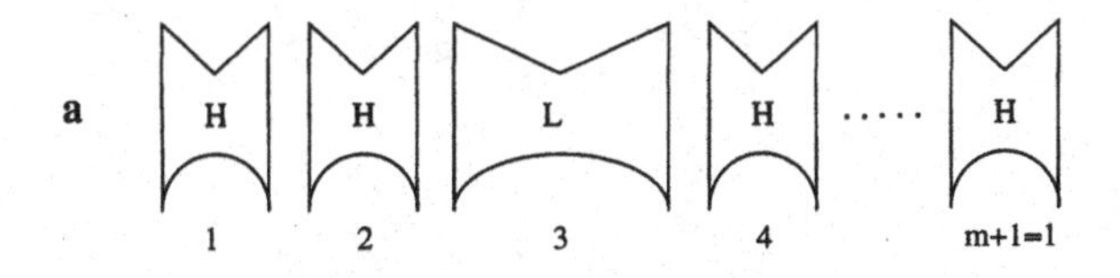

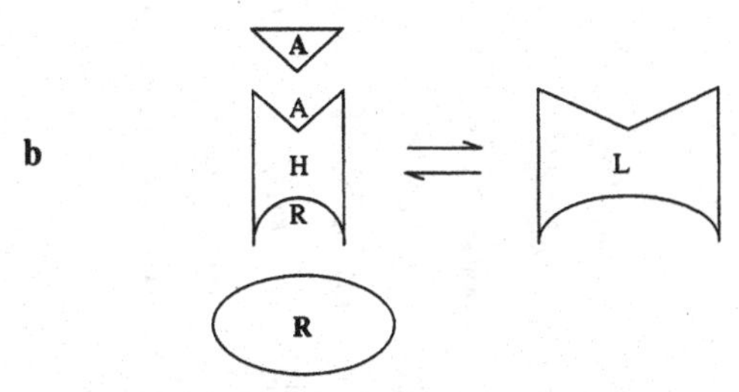

Figure 8.14. (a) A schematic multisubunit enzyme. Each subunit contains one catalytic (A) and one regulatory (R) site. The subunits are arranged cyclically, i.e., the $m+1$ subunit is the same as the first. (b) A single subunit in the two conformational states L and H.

vectors are [see Eq. (7.4.3)]

$$\left.\begin{array}{c}\langle e_1| = \langle L,0,0| = (1,0,0,0,0,0,0,0), \quad \langle e_2| = \langle H,0,0| \\ \langle e_3| = \langle L,\mathbf{A},0|, \quad \langle e_4| = \langle L,0,\mathbf{R}|, \quad \langle e_5| = \langle H,\mathbf{A},0|, \\ \langle e_6| = \langle H,0,\mathbf{R}|, \quad \langle e_7| = \langle L,\mathbf{A},\mathbf{R}|, \quad \langle e_8| = \langle H,\mathbf{A},\mathbf{R}| \end{array}\right\} \tag{8.8.1}$$

where $\langle e_i|$ is the ith unit vector in an eight-dimensional vector space; only $\langle e_1|$ is shown explicitly. Thus, a typical element of the matrix $\mathbf{M}$ has one factor $Q_{\alpha\beta}$ ($\alpha, \beta = L, H$) due to subunit–subunit interaction, a factor $(Q_\alpha Q_\beta)^{1/2}$ corresponding to the "energy levels" of the subunits in conformational states α, β, and a factor of the type $(\lambda_i\lambda_j\lambda_k\lambda_l q_{\alpha i} q_{\alpha j} q_{\beta k} q_{\beta l})^{\delta/2}$ where $\delta = \delta_{i,A} + \delta_{j,R} + \delta_{k,A} + \delta_{l,R}$. The general element of the matrix is

$$\langle \alpha, i, j|\mathbf{M}|\beta, k, l\rangle = (\lambda_i\lambda_j\lambda_k\lambda_l q_{\alpha i} q_{\alpha j} q_{\beta k} q_{\beta l})^{\delta/2} (Q_\alpha Q_\beta)^{1/2} Q_{\alpha\beta} \tag{8.8.2}$$

For example, a pair of sites in states $\langle L, \mathbf{A}, 0|$ and $\langle H, 0, \mathbf{R}|$ will contribute the matrix element

$$\langle L, \mathbf{A}, 0|\mathbf{M}|H, 0, \mathbf{R}\rangle = (\lambda_A\lambda_R q_{LA} q_{HR})^{1/2} (Q_L Q_H)^{1/2} Q_{LH} \tag{8.8.3}$$

As in Section 7.4, we transform into variables K, h_A, h_R, and η and introduce $X_A = q_{LA}\lambda_A = k_{LA}C_A$ and $X_R = q_{LR}\lambda_R = k_{LR}C_R$ to rewrite the matrix $\mathbf{M}$ in the form

$$\begin{bmatrix}
1 & (K\eta)^{1/2} & X_A^{1/2} & X_R^{1/2} & (K\eta h_A)^{1/2}X_A^{1/2} & (K\eta h_R)^{1/2}X_R^{1/2} & X_A^{1/2}X_R^{1/2} & (K\eta h_A h_R)^{1/2}X_A^{1/2}X_R^{1/2} \\
(K\eta)^{1/2} & K & (K\eta)^{1/2}X_A^{1/2} & (K\eta)^{1/2}X_R^{1/2} & Kh_A^{1/2}X_A^{1/2} & Kh_R^{1/2}X_R^{1/2} & (K\eta)^{1/2}X_A^{1/2}X_R^{1/2} & Kh_A^{1/2}h_R^{1/2}X_AX_R^{1/2} \\
X_A^{1/2} & (K\eta)^{1/2}X_A^{1/2} & X_A & X_A^{1/2}X_R^{1/2} & (K\eta h_A)^{1/2}X_A & (K\eta h_R)^{1/2}X_A^{1/2}X_R^{1/2} & X_AX_R^{1/2} & (K\eta h_A h_R)^{1/2}X_AX_R^{1/2} \\
X_R^{1/2} & (K\eta)^{1/2}X_R^{1/2} & X_A^{1/2}X_R^{1/2} & X_R & (K\eta h_A)^{1/2}X_A^{1/2}X_R^{1/2} & (K\eta h_R)^{1/2}X_R & X_A^{1/2}X_R & (K\eta h_A h_R)^{1/2}X_AX_R^{1/2} \\
(K\eta h_A)^{1/2}X_A^{1/2} & Kh_R^{1/2}X_R^{1/2} & (K\eta h_A)^{1/2}X_A & (K\eta h_A)^{1/2}X_A^{1/2}X_R^{1/2} & Kh_AX_A & Kh_A^{1/2}h_R^{1/2}X_A^{1/2}X_R^{1/2} & (K\eta h_A)^{1/2}X_AX_R^{1/2} & Kh_Ah_R^{1/2}X_AX_R^{1/2} \\
(K\eta h_R)^{1/2}X_A^{1/2} & Kh_R^{1/2}X_R^{1/2} & (K\eta h_R)^{1/2}X_A^{1/2}X_R^{1/2} & (K\eta h_R)^{1/2}X_R & Kh_A^{1/2}h_R^{1/2}X_A^{1/2}X_R^{1/2} & Kh_AX_R & (K\eta h_R)^{1/2}X_A^{1/2}X_R & Kh_A^{1/2}h_RX_A^{1/2}X_R \\
X_A^{1/2}X_R^{1/2} & (K\eta)^{1/2}X_A^{1/2}X_R^{1/2} & X_AX_R^{1/2} & X_A^{1/2}X_R & (K\eta h_A)^{1/2}X_AX_R^{1/2} & (K\eta h_R)^{1/2}X_A^{1/2}X_R & X_AX_R & (K\eta h_A h_R)^{1/2}X_AX_R \\
(K\eta h_A h_R)^{1/2}X_A^{1/2}X_R^{1/2} & Kh_A^{1/2}h_R^{1/2}X_AX_R^{1/2} & (K\eta h_A h_R)^{1/2}X_AX_R^{1/2} & (K\eta h_A h_R)^{1/2}X_A^{1/2}X_R & Kh_Ah_R^{1/2}X_AX_R^{1/2} & Kh_A^{1/2}h_RX_A^{1/2}X_R & (K\eta h_A h_R)^{1/2}X_AX_R & Kh_Ah_RX_AX_R
\end{bmatrix}$$

where K denotes KK' (8.8.4)

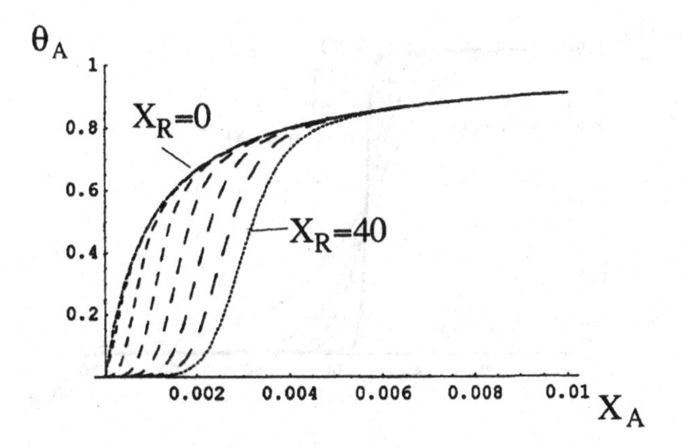

Figure 8.15. BIs for the model of Section 8.8, with parameters given in Eq. (8.8.10), and $m = 10$. The curves from left to right correspond to increasing values of $X_R = 0, 5, 10, 15, \ldots, 40$. Note the transition from the noncooperative BI for $X_R = 0$ to a highly (positive and homotropic) cooperative curve for $X_R = 40$.

The secular, or characteristic, equation is now

$$|\mathbf{M} - d\mathbf{I}| = 0 \tag{8.8.5}$$

where **I** is the 8×8 unit matrix.

Equation (8.8.5) can be written in the simplified form

$$d^6(d^2 - F_1 d - F_2) = 0 \tag{8.8.6}$$

This equation has six roots which are zeros,* while the other two roots are given by

$$d_{\pm} = \frac{1}{2}\left(F_1 \pm \sqrt{F_1^2 + 4F_2}\right) \tag{8.8.7}$$

where

$$\begin{aligned} F_1 &= 1 + X_A + X_R + X_A X_R + K(1 + h_A X_A)(1 + h_R X_R) \\ F_2 &= K(\eta - 1)(1 + X_A)(1 + X_R)(1 + h_A X_A)(1 + h_R X_R) \end{aligned} \tag{8.8.8}$$

Thus, for any finite m we can compute the GPF of the system by

$$\xi(m) = \mathrm{Tr}(\mathbf{M}^m) = d_+^m + d_-^m \tag{8.8.9}$$

For very large systems ($m \to \infty$), it is sufficient to consider only the largest root d_+.

In Fig. 8.15 we present some numerical results for this model with the parameters

$$K = 10, \qquad h_A = 10^3, \qquad h_R = 10^{-3}, \qquad \eta = 0.1 \tag{8.8.10}$$

*See footnote on p. 245.

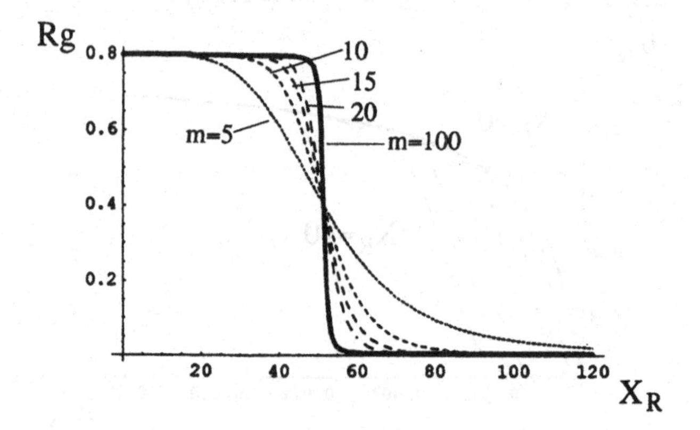

Figure 8.16. Regulatory curves for the same model as in Fig. 8.15 but with $m = 5, 10, 15, 20$. Note that the larger is m, the sharper is the transition from an active to an inactive enzyme. The full line corresponds to $m = 100$.

In contrast to the examples in Chapter 7, where in all cases the BI, θ_A, was for only *one* active site, hence there were no **A–A** cooperativities. Here, we have m active sites and m regulatory sites. Therefore, for the particular choice of parameters in (8.8.10), we expect positive cooperativity between **A** and **A** as well as between **R** and **R**, but negative cooperativity between **A** and **R**.

Figure 8.15 shows how the BI, θ_A, changes from noncooperative for $X_R = 0$ to increasingly cooperative curves when X_R is increased. The corresponding regulatory curve, Fig. 8.16, becomes sharper as we increase m. Ultimately, for $m \to \infty$ we have a very sharply inverted S-shaped regulatory curve. Note, however, that even for $m \to \infty$ the regulatory curve is not a step function. This is because the cooperativity is finite for the particular choice of parameters in (8.8.10). Figure 8.17 shows the slopes of the regulatory curves for the same parameters as in Fig. 8.16.

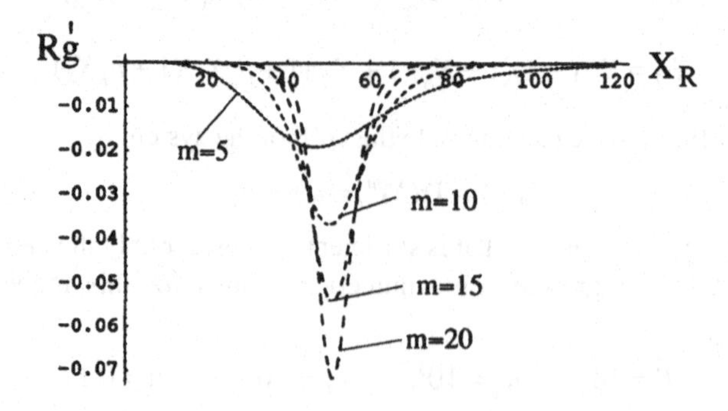

Figure 8.17. Slopes of the regulatory curves of Fig. 8.16 for $m = 5, 10, 15, 20$.

8.9. ASPARTATE TRANSCARBAMOYLASE (ATCase)

Aspartate transcarbamoylase (ATCase) from *Escherichia coli* is the most studied and best known regulatory enzyme. Yates and Pardee (1956) were the first to propose that the activity of ATCase is controlled by end product inhibition. This feedback inhibition was later studied in more detail by Gerhart and Pardee (1961, 1962, 1963). The three-dimensional structure of ATCase was determined by Lipscombe and his coworkers [Wiley *et al.* (1971), Wiley and Lipscomb (1968), Warren *et al.* (1973)].

Figure 8.18 shows a schematic spatial arrangement of the subunits in ATCase. It consists of two catalytic trimers $2C_3$, separated by three regulatory dimers $3R_2$. Altogether, the enzyme has six catalytic and six regulatory subunits: $(C_3)_2(R_2)_3$, or simply C_6R_6. The electronic micrographs, obtained by Richards and Williams (1972), viewed along the threefold symmetry axis of the molecule, show an inner solid equilateral triangle and a circumscribing larger triangle rotated by 60°. The structure, shown schematically in Fig. 8.18, is very reminiscent of the Star of David. It is also known [Kantrowitz and Lipscomb (1988, 1990)] that the conformational changes involve both expansion of the ATCase along the threefold symmetry axis as well as intramolecular rotations of the catalytic and regulatory subunits.

ATCase catalyzes the first step in the biosynthesis of cytidine triphosphate (CTP). The sequence of reactions leading from the reactants, aspartate and carbamoyl phosphate, to CTP is shown in Fig. 8.19.

Figure 8.20 shows the inhibitory effect of CTP, as well as the activation of ATCase by ATP. Based on measurements by Gerhart and Pardee (1962, 1963), the reaction rate is here plotted as a function of the concentration of aspartate. It is clear

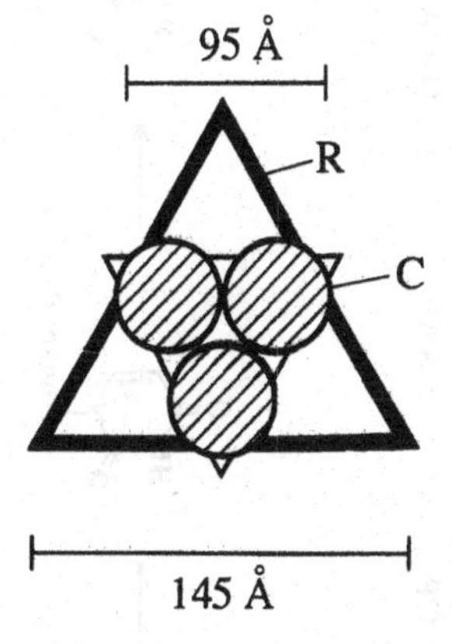

Figure 8.18. A schematic view of ATCase based on electron micrographs, viewed along the threefold symmetry axis. The outer equilateral triangle has an edge of 145 Å. The (almost) inscribed solid triangle with edge 95 Å is rotated by 60° relative to the large triangle.

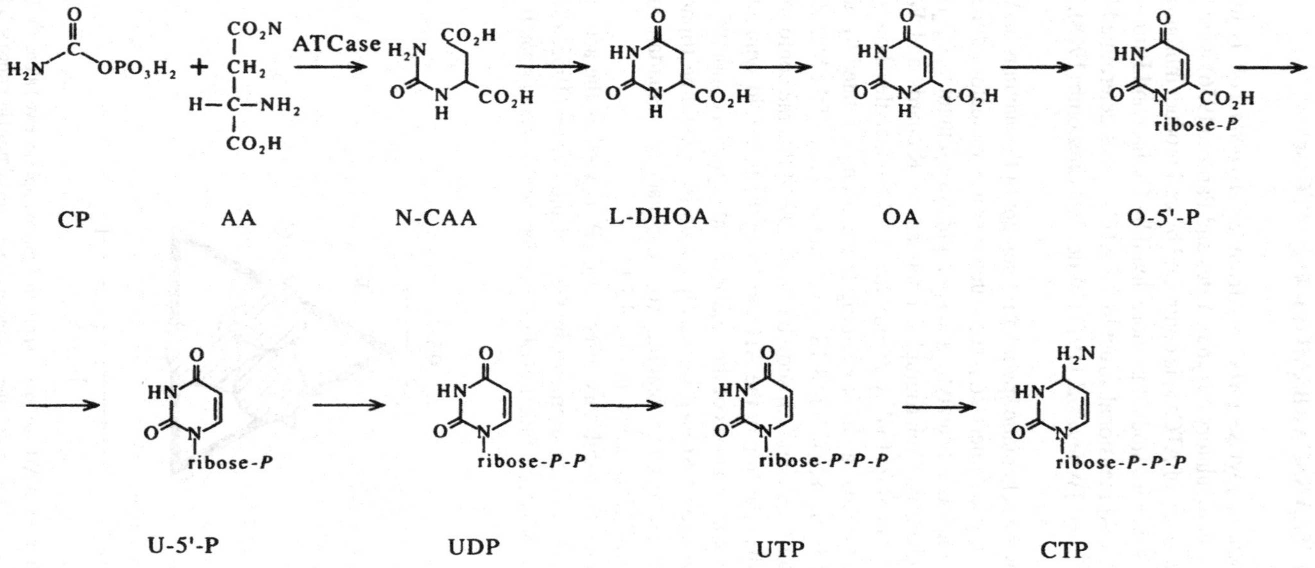

CP AA ATCase N-CAA L-DHOA OA O-5'-P U-5'-P UDP UTP CTP

Figure 8.19. Sequence reactions from aspartic acid (AA) and carbamoyl phosphate (CP) to the end product, cytidine triphosphate (CTP). The first reaction is catalyzed by ATCase. The intermediary compounds are N-carbamoyl aspartic acid (N-CAA), L-dihydroorotic acid (L-DHOA), orotic acid (OA), orotidine 5′-phosphate (O-5′-P), uridine 5′-phosphate (U-5′-P), uridine diphosphate (UDP), and uridine triphosphate (UTP).

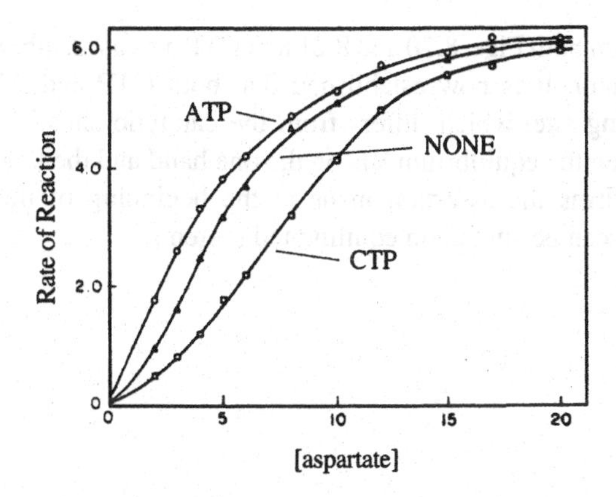

Figure 8.20. Schematic illustration of the effect of ATP and CTP on the rate of reaction of ATCase as a function of the concentration of aspartate (in mM). Redrawn with changes from Gerhart and Pardee (1962, 1963).

that for any given concentration of aspartate the reaction rate *decreases* upon addition of CTP, but increases upon addition of ATP.

Binding measurements of the bisubstrate analogue N-(phosphoacetyl)-L-aspartate (PALA) in the presence and absence of ATP were reported by Newell, *et al.*, (1989). Figure 8.21 shows the binding isotherm of ATCase, a plot of $\theta = [PALA]_{bound}/[ATCase]$ as a function of $[PALA]_{free}$ (at 23 °C and in a buffer solution).

It is evident that addition of CTP lowers the BI while addition of ATP elevates the BI of the effector-free ATCase.

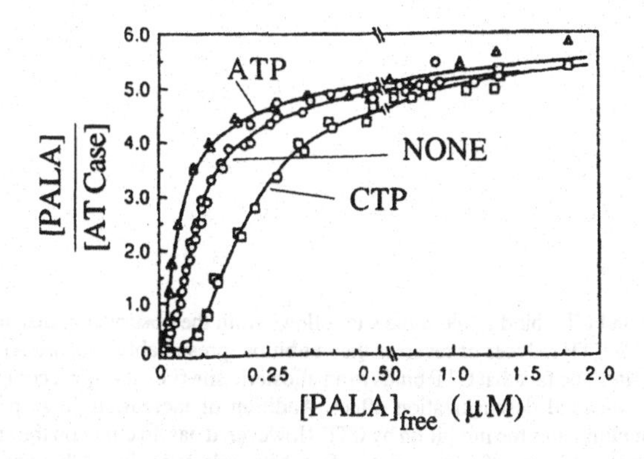

Figure 8.21. Binding isotherms of PALA to ATCase in the presence of ATP and CTP. Redrawn with changes from Newell *et al.* (1989).

It is clear from both Figs. 8.20 and 8.21 that CTP acts as an inhibitor while ATP acts as an activator. It is now established that both CTP and ATP bind to the regulatory binding site, which differs from the catalytic sites.* The remarkable similarity between the equilibrium BIs on the one hand and the kinetic data on the other hand confirms the assertion made at the beginning of this chapter, that allosteric effects can be studied in equilibrated systems.

*The fact that ATP and CTP bind to the same site follows from the observation that adding ATP to the inhibited enzyme by CTP reduces or reverses the inhibition, presumably because ATP competes with CTP for the same site. The fact that CTP binds to an allosteric site (i.e., it is not a competitive inhibitor) follows from the so-called desensitization effect. Addition of mercurials [e.g., p-mercuribenzoate (PMB)] reduces or eliminates the inhibition by CTP. However, it has no effect on the enzymatic activity of ATCase, presumably because the mercurials affect the regulatory subunits but not the catalytic site. As for the mechanism of cooperativity (both positive and negative), it is known that CTP does induce changes in the quaternary structure of the enzyme.

9

Solvent Effects on Cooperativity

9.1. INTRODUCTION

All binding processes in real-life systems occur in some solvent. The solvent is, in general, a mixture of many components, including water electrolytes and nonelectrolytes.* At present, it is impossible to account for all possible solvent effects, even when the solvent is pure water. Yet, the solvent, whether a single or multi-component, cannot be ignored. Any serious molecular theory of cooperativity must deal with solvent effects. We shall see in this chapter that this is not an easy task even when the solvent is inert, such as argon, or a simple hydrocarbon liquid.†

In all the theoretical developments in the previous chapters we have assumed that the systems operate in vacuum (except for the case of alkylated succinic acid, Section 4.8). This assumption has enormously simplified the theory. Strictly speaking, all we have learned so far about cooperativity applies only to vacuum systems. One might justifiably wonder whether we have not wasted our time and effort in studying systems that do not exist in reality. In fact, we shall soon see that the introduction of the solvent does change the theory of cooperativity. But the changes are such that the formal structure of the results obtained for the vacuum system is preserved. Formally, if $g^g(1, 1)$ is the pair correlation function discussed

*Colombo *et al.* (1992) examined the water effect on the cooperativity of hemoglobin. They found that about 60 water molecules are involved in the transition between the oxy and the deoxy conformations of hemoglobin. Unfortunately, this paper does not discuss the solvent effect on the free energies involved in the cooperativity of hemoglobin.

†DiCera's treatment in his book (1996) begins by defining the system of a macromolecule with "*N* water molecules (the solvent)." However, nowhere in the book does he mention any specific effect of water as a solvent. On page 47 the author switches, without giving any reason, to an "inert" solvent. Reading through the book reveals that not even an inert-solvent effect is discussed. Since in most of DiCera's book the phenomenological approach is used, and since the binding constants in this approach contain implicitly *any* solvent effect: inert, water, or physiological fluids, the reader might wonder why the author has chosen water, in the first place, and switched from water to an inert solvent in the second place.

in the previous sections, i.e., the vacuum case, then we have found that

$$g^g(1, 1) = f(K, h, \eta) \tag{9.1.1}$$

i.e., $g^g(1, 1)$ is a function of the parameters K, h, and η. When a solvent is present, the corresponding correlation $g^l(1, 1)$ will have the general form

$$g^l(1, 1) = y_s f(K^*, h^*, \eta^*) \tag{9.1.2}$$

where y_s is a new source of cooperativity. This is the same contribution to $g(1, 1)$ that is found in the theory of liquids.[†] It is a correlation transmitted through the solvent molecules. When the solvent is absent, $y_s = 1$. The second modification is in the parameters K, h, and η in Eq. (9.1.1). These will now include the solvation Gibbs energies of all the molecules involved in the binding process. The general procedure for modifying K, h, η into K^*, h^*, η^* is discussed in Section 9.2. Once we have done that, we can use the *same* functional form f in Eq. (9.1.2) as in Eq. (9.1.1). Thus, although solvation effects will significantly change the correlation functions, the formal dependence on the fundamental parameters K, h, and η is unchanged. In this sense, the theoretical results developed in the previous chapters are preserved. We shall demonstrate this solvation effect on the vacuum theory for a particular model in Section 9.4. All the above comments about the pair correlation can be extended to any higher-order correlation.

The most general approach to a theory of binding is to start with the grand partition function of a multicomponent system, and then take the low-density limit to obtain the partition function of a dilute system in a solvent. This approach was carried out by T. L. Hill (1985). Since we are interested only in the dilute limit, i.e., when all the adsorbent molecules are independent, we can use a shortcut to obtain the required modification from the vacuum to the solvent system. It will be seen in Sections 9.2 and 9.3 that this is possible because each of the parameters K, h, and η is an equilibrium constant of a specific "reaction." By transferring the entire reaction from the vacuum into the solvent, one can immediately obtain the required modifications of these parameters. The actual implementation of the formal result is difficult. It requires knowledge of the distribution of functional groups on the surface of the ligands and on the adsorbent molecules. We shall devote Sections 9.5 and 9.6 to examining some specific ingredients of the solvation to the ligand–ligand correlations.

9.2. SOLVATION EFFECT ON THE EQUILIBRIUM CONSTANTS

The three fundamental parameters K, h and η that determine the indirect cooperativity are essentially equilibrium constants, corresponding to three well-defined processes. In this section we explore the modification that we are required to make in these parameters when the same processes are carried out in a solvent.

[†]See, for example, Chapters 2 and 8 in Ben-Naim (1992).

We begin with a general process leading from an initial (i) to a final (f) state

$$i \rightarrow f \tag{9.2.1}$$

where i and f represent all the reactants and all the products of the process, respectively.

In an ideal gas (g) phase, the free-energy change associated with this process is denoted by $\Delta G^g(i \rightarrow f)$. This free-energy change consists, in general, of three parts: one depending on the internal energy states of all the particles involved in the process, the second depending on the translational and rotational degrees of freedom, and the third depending on the concentrations of all the species involved. Since we are interested in the *change* in the free energy associated with the same process when carried out in a solvent, and since the second and third parts are presumed to be unchanged when the process is carried out in a solvent,* we can assume, for simplicity, that the process in Eq. (9.2.1) is carried out while all the particles involved in the process, in both the initial and final states, are at some fixed positions, i.e., all the particles are devoid of translational degrees of freedom. This assumption eliminates the second and third contributions to ΔG^g, yet will have no effect on the modification introduced by transferring the same process from the ideal gas phase (g) into the liquid phase (l). We denote by $\Delta G^l(i \rightarrow f)$ the free-energy change of the same process, (9.2.1), carried out in the liquid phase l. To obtain the connection between ΔG^g and ΔG^l, we follow the cyclic process indicated in Fig. 9.1, for which the free energy is zero,

$$\Delta G^l(i \rightarrow f) - \Delta G_f^* - \Delta G^g(i \rightarrow f) + \Delta G_i^* = 0 \tag{9.2.2}$$

Let us define the solvent effect on the reaction (9.2.1) by

$$\delta G(i \rightarrow f) = \Delta G^l(i \rightarrow f) - \Delta G^g(i \rightarrow f) = \Delta G_f^* - \Delta G_i^* \tag{9.2.3}$$

with ΔG_α^* the solvation Gibbs energy of the molecule α, defined below. We see that the solvent effect on the Gibbs energy change for any process $i \rightarrow f$ is determined by the difference in the *solvation* Gibbs energies of the products and reactants. This is a very general result. We shall now proceed with some specific examples of Eq. (9.2.3) that are of interest in the binding systems.

(a) The simplest process is the conversion $H \rightleftharpoons L$ between the two conformations of the empty adsorbent molecule. The process is written symbolically as

$$(L; 0, 0) \xrightarrow{I} (H; 0, 0) \tag{9.2.4}$$

*In a classical system, the translational and rotational degrees of freedom are unchanged by the presence of a solvent. The third contribution is unchanged because of the requirement that the *same* process be carried out in the two phases; this includes the specification of all the concentrations of the species involved in the process.

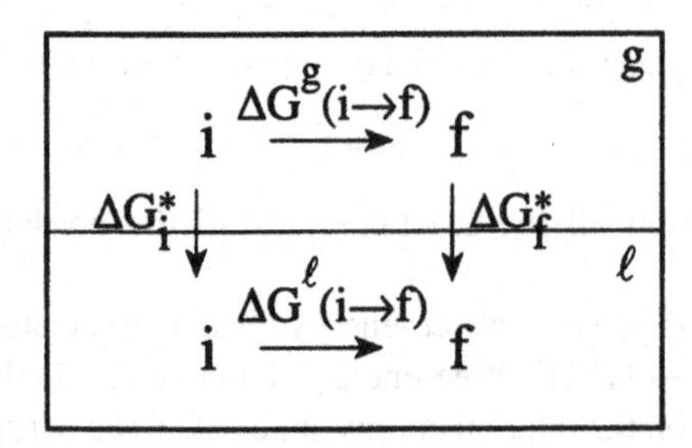

Figure 9.1. A cyclic process for which the total free-energy change is zero [Eq. (9.2.2)].

for which the equilibrium constant in an ideal gas phase is

$$K = K^g = \frac{Q_H}{Q_L} = \exp[-\beta(E_H - E_L)] \tag{9.2.5}$$

By applying the general cyclic process of Fig. 9.1 to this process, we obtain the modified equilibrium constant

$$K^* = K^l = K \exp[-\beta\delta G(I)] = K \exp[-\beta(\Delta G_H^* - \Delta G_L^*)] \tag{9.2.6}$$

where ΔG_α^* is the solvation Gibbs energy of the species $\alpha = L, H$ (see Section 9.4 for a precise definition). We see that the parameter K defined in terms of energy levels E_L and E_H is now modified by the solvent effect $\delta G(I)$, consisting of the difference in the solvation Gibbs energies of H and L. We recall that E_L and E_H, as defined in Chapter 2, were presumed to be strictly energy levels. This was done to stress the emergence of free-energy levels, when the same process ($H \rightleftharpoons L$) is carried out in a solvent. In general, E_L and E_H are themselves free energies (see Appendix B). In this case, the solvation Gibbs energies only shift from the original values E_L and E_H to the new free-energy levels $E_L + \Delta G_L^*$ and $E_H + \Delta G_H^*$, respectively.

(b) The second process is the binding of a ligand **L** to an empty adsorbent molecule, say in state L. The process is

$$(L; 0, 0) + \mathbf{L} \xrightarrow{II} (L; 1, 0) \tag{9.2.7}$$

for which we have defined

$$q_L = \exp(-\beta U_L) \tag{9.2.8}$$

where U_L is the binding energy of the ligand **L** to the L form of the empty system. The modification of q_L, when the same process is carried out in a solvent, is obtained by applying the cyclic process of Fig. 9.1 to process (9.2.7),

$$q_L^* = q_L \exp[-\beta\delta G(II)] = q_L \exp[\beta(\Delta G_L^*(1, 0) - \Delta G_L^*(0, 0) - \Delta G_{\mathbf{L}}^*)] \tag{9.2.9}$$

where $\Delta G_L^*(0, 0)$ and $\Delta G_L^*(1, 0)$ are the solvation Gibbs energies of the empty and singly occupied system in the L form while $\Delta G_{\mathbf{L}}^*$ is the solvation Gibbs energy of the ligand **L**. By writing a similar expression for q_H^*, we obtain the modified form of the parameter h, i.e.,

$$h = \frac{q_H}{q_L} \tag{9.2.10}$$

and

$$h^* = h \exp[-\beta(\Delta G_H^*(1, 0) + \Delta G_L^*(0, 0) - \Delta G_L^*(1, 0) - \Delta G_H^*(0, 0))] \tag{9.2.11}$$

We note again that U_L, the binding energy of **L** on the L form, was presumed to be an *energy* of binding. As noted above, this quantity is in general a binding free energy. However, in Eq. (9.2.9) we observe how an energy is modified into a free energy by the addition of the solvation Gibbs energies. Note also that while q_L (or q_L^*) corresponds to the process (9.2.7) of binding, the parameter h (or h^*) corresponds to a difference in such processes, namely,

$$(L; 0, 1) + (H; 0, 0) \xrightarrow{III} (H; 0, 1) \rightarrow (L; 0, 0) \tag{9.2.12}$$

We should also note that if the two binding sites are identical, then q_L as well as q_{H} are the same, whether the ligand binds to the *first* or to the *second* site. Therefore, we have only one parameter h defined in Eq. (9.2.10). When a solvent is present, q_L and q_H, and hence also h, might be modified differently for the first and for the second site. For the second site the process (9.2.7) is replaced by

$$(L; 1, 0) + \mathbf{L} \rightarrow (L; 1, 1) \tag{9.2.13}$$

for which q_L is the same as in Eq. (9.2.8). However, the solvent modification for the second site is

$$q_L^{**} = q_L \exp[-\beta(\Delta G_L^*(1, 1) - \Delta G_L^*(1, 0) - \Delta G_{\mathbf{L}}^*)] \tag{9.2.14}$$

Now compare Eqs. (9.2.14) and (9.2.9). A similar modification applies for q_H, hence the modified parameter h is now

$$h^{**} = \frac{q_H^{**}}{q_L^{**}} = h \exp[-\beta(\Delta G_H^*(1, 1) + \Delta G_L^*(1, 0) - \Delta G_L^*(1, 1) - \Delta G_H^*(1, 0))] \tag{9.2.15}$$

The corresponding process is

$$(L; 1, 1) + (H; 1, 0) \xrightarrow{IV} (H; 1, 1) + (L; 1, 0) \tag{9.2.16}$$

which should be compared with (9.2.12).

(c) The third parameter η, which also determines the cooperativity, is related to the process and written symbolically as

$$LL + HH \xrightarrow{V} 2LH \tag{9.2.17}$$

The corresponding equilibrium constant is defined by

$$\eta = \frac{Q_{LH}^2}{Q_{LL}Q_{HH}} = \exp[-\beta(2E_{LH} - E_{LL} - E_{HH})] \tag{9.2.18}$$

Note again that in Eq. (9.2.18), E_{LL}, E_{HH}, and E_{LH} are presumed to be *energy* parameters. The modification of η for the process (9.2.17) carried out in a solvent is

$$\eta^* = \eta \exp[-\beta(2\Delta G_{LH}^* - \Delta G_{LL}^* - \Delta G_{HH}^*)] \tag{9.2.19}$$

The specific processes discussed above are all special cases of the general process (9.2.1). In all of these cases we have seen the explicit modification of the equilibrium constant of the corresponding process. As indicated in Eq. (9.2.3), the general modification requires knowledge of the solvation Gibbs energies of all the components involved in the process. For macromolecules such as proteins or nucleic acid, none of these is known, however. Nevertheless, some specific solvation effects are examined in Sections 9.4 and 9.5.

We conclude this section by presenting the general statistical mechanical expression for the solvation Gibbs energy of any solute α,*

$$\Delta G_\alpha^* = -k_B T \ln \langle \exp(-\beta B_\alpha) \rangle_0 \tag{9.2.20}$$

where the symbol $\langle\ \rangle_0$ stands for an average over all the configurations of all the molecules in the system, except the *single* α-molecule for which the solvation Gibbs energy is calculated. In general, by "all the molecules" we also include any other α-molecules present in the system. However, since we always deal with independent adsorbing systems, we may neglect interactions between the α-molecules, in which case by "all the molecules" we simply mean all the solvent molecules. The solvent might have c components, with composition $\mathbf{N} = \{N_1, N_2, \ldots, N_c\}$, where N_i is the number of molecules of species i (excluding species α as part of the solvent, i.e., $i \neq \alpha$).

There are several ways of carrying out the average over all configurations of the solvent molecules. The most common one is a system at constant temperature T, pressure P, and solvent composition $\mathbf{N}$. The appropriate average is carried out in the so-called $T, P, \mathbf{N}$ ensemble, i.e., for any function of the solvent configuration

*For more details, see Ben-Naim (1987, 1992).

$\mathbf{X}_1, \ldots, \mathbf{X}_N$, we write

$$f = \langle f(\mathbf{X}_1, \ldots, \mathbf{X}_N) \rangle = \int dV \int d\mathbf{X}_1 \cdots d\mathbf{X}_N P_0(V, \mathbf{X}^N) f(\mathbf{X}^N) \qquad (9.2.21)$$

where $\mathbf{X}^N$ is a shorthand notation for a specific configuration of all solvent molecules $\mathbf{X}^N = \mathbf{X}_1, \ldots, \mathbf{X}_N$, and $\mathbf{X}_i$ denotes the configuration (normally the six coordinates of locations and orientation) of the single molecule i.

The probability density $P_0(V, \mathbf{X}^N)$ is

$$P_0(V, \mathbf{X}^N) = \frac{\exp(-\beta PV - \beta U_N)}{\int dV \int d\mathbf{X}_1 \cdots d\mathbf{X}_N \exp(-\beta PV - \beta U_N)} \qquad (9.2.22)$$

where U_N is the total interaction energy among all the solvent molecules. The subscript "0" is used to stress that this is the probability density of solvent configurations *before* introducing α into the system. This is also referred to as the probability density of the *pure* solvent.

The specific average (9.2.20) is over the function $\exp(-\beta B_\alpha)$, where B_α is the so-called binding energy of α to the solvent (not to be confused with binding energies, such as U_L and U_H). This is simply defined as the difference in the total interaction energy of a system before and after introducing α at some fixed configuration $\mathbf{X}_\alpha$. Thus,

$$B_\alpha = U_{N+1}(\mathbf{X}_\alpha, \mathbf{X}^N) - U_N(\mathbf{X}^N) \qquad (9.2.23)$$

which, for the pairwise additive interaction energy, reduces to

$$B_\alpha = \sum_i U(\mathbf{X}_\alpha, X_i) \qquad (9.2.24)$$

where $U(\mathbf{X}_\alpha, \mathbf{X}_i)$ is the interaction energy between α at $\mathbf{X}_\alpha$ and the ith solvent molecule at $\mathbf{X}_i$. The summation in Eq. (9.2.24) is over all solvent molecules.

In Section 9.4 we shall decompose the solvation Gibbs energy of a macromolecule α into various components or ingredients, which will allow us to examine and, in principle, estimate some specific contributions of the solvation to cooperativity.

9.3. SOLVENT EFFECT ON THE LIGAND–LIGAND PAIR CORRELATION

In Section 4.2 we defined the ($\lambda \to 0$ limit) pair correlation by

$$g(1, 1) = \frac{Q(1, 1)Q(0, 0)}{[Q(1, 0)]^2} \qquad (9.3.1)$$

where Q is the canonical PF of the adsorbent molecule having zero, one, or two bound ligands, as specified in the brackets. Here, we assume for simplicity that the two sites are identical. The pair correlation is related to the free-energy change for the process, which is written symbolically as

$$2(1, 0) \rightarrow (0, 0) + (1, 1) \tag{9.3.2}$$

In the previous section the general rule was found for converting the free-energy change of a process, carried out in vacuum, into a modified free-energy change for the same process carried out in a solvent. We shall now apply this rule for the process (9.3.2) to obtain the pair correlation for two ligands (hence the cooperativity) in a two-site system. We shall specifically discuss the two models introduced in Sections 4.3 and 4.5. These examples are sufficient to illustrate all the modifications that are brought about by the solvent on the cooperativity of any binding system.

The first model is that discussed in Section 4.3. This may be called the one-macrostate approximation. In this model the adsorbent molecule has only one state, and the binding process does not induce any conformational changes. Hence, the ligand–ligand pair correlation is due only to the *direct* ligand–ligand interaction

$$g(1, 1) = \exp[-\beta U(1, 1)] \tag{9.3.3}$$

where $U(1, 1)$ is the pair interaction energy between the two ligands as if they were in vacuum (at the same distance and relative orientation as on the binding sites). Here, as in Section 4.3, we assume that $U(1, 1)$ is strictly an *energy* of interaction. In general, the free energy of the process (9.3.2) can have an entropic contribution even when the process is carried out in vacuum, for instance, if the two ligands change their conformation when brought from infinite separation to the final configuration on the two sites. Here, as in the previous section, we neglect any entropic contribution to $U(1, 1)$ of the process (9.3.2). We do this only to stress the conversion from *energy* of interaction $U(1, 1)$ into free interaction energy $W(1, 1)$ when we perform the same reaction (9.3.2) in a solvent. By using the same arguments as in Section 9.2, we obtain (see Fig. 9.2)

$$W(1, 1) = U(1, 1) + \Delta G^*(1, 1) + \Delta G^*(0, 0) - 2\Delta G^*(1, 0) \tag{9.3.4}$$

or, equivalently,

$$\begin{aligned} g^l(1, 1) &= g^g(1, 1) \exp\{-\beta[\Delta G^*(1, 1) + \Delta G^*(0, 0) - 2\Delta G^*(1, 0)]\} \\ &= g^g(1, 1) \frac{\psi(1, 1)\psi(0, 0)}{\psi(1, 0)^2} \end{aligned} \tag{9.3.5}$$

where we set

$$\psi = \exp(-\beta\Delta G^*) \tag{9.3.6}$$

for any species indicated in the parentheses.

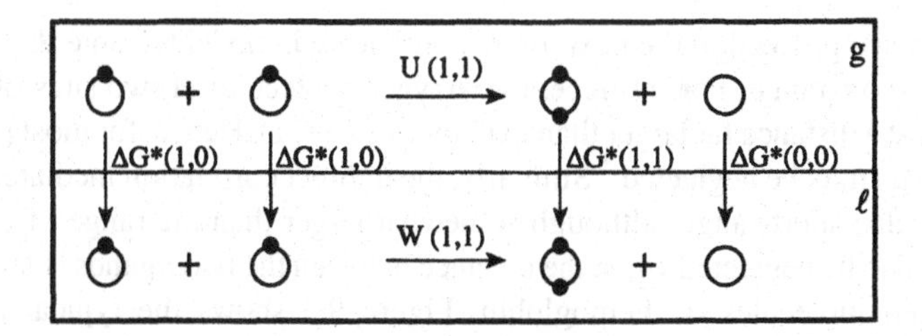

Figure 9.2. The process (9.3.2) carried out in an ideal gas (g) and in a liquid (l) phase. The relation between $W(1, 1)$ and $U(1, 1)$ in Eq. (9.3.2) is obtained by transferring all solutes from g to l.

As before, the modified pair correlation has one factor which is the same as the vacuum pair correlation, and a second contribution which has the same *form* as the solute–solute pair correlation in a solvent. It is not identical, however, to the solute–solute pair correlation in a solvent, due to the presence of the adsorbent molecule. Therefore, this factor has been referred to as a *conditional* pair correlation.* The exact statistical-mechanical expression for the conditional pair correlation is quite complicated. We shall not discuss this aspect here. The reader can understand the qualitative difference between the two factors on the rhs of Eq. (9.3.5) from the following considerations.

The *direct* interaction energy $U(1, 1)$ is defined as the difference in energy when the two ligands are brought from infinity to the final configuration in vacuum. The direct correlation defined by $S = \exp(-\beta U)$ is therefore the same whether or not the adsorbent molecule is present. In other words $U(1, 1)$, hence S, is unaffected by the presence of the adsorbent molecule. It does not change also in the presence of a solvent. However, in the presence of a solvent the pair correlation $g(1, 1)$ depends also on the solvation Gibbs energies of all the species involved in the process (9.3.2). Since each of the quantities $\psi(i, j)$ in Eq. (9.3.5) is an average over all the configurations of the solvent molecules [see Eq. (9.2.21)], and since this average depends on the distribution of the solvent configurations (9.2.22), we should expect these average quantities to be affected by the presence of the adsorbent molecule. This is why we have called the solvent contribution to g in Eq. (9.3.5) a *conditional* pair correlation, the condition being the presence of the adsorbent molecule.

We conclude that in this simple model there are two contributions to the ligand–ligand correlation, hence to the cooperativity: one due to *direct* ligand–ligand interaction $U(1, 1)$ and the other due to ligand–solvent interaction. The latter part of the indirect correlation is transmitted by *solvent molecules*. The extent of this correlation depends on the presence of the adsorbent molecule.

*See Chapter 8 in Ben-Naim (1992) and Ben-Naim (1972).

Before we proceed to the next, more complicated, model we note that $U(1, 1)$ is usually short-range. For instance, two oxygen molecules at two sites of hemoglobin are at a distance far larger than the range of $U(1, 1)$. Hence, for most practical cases $U(1, 1)$ may be neglected.* Similarly, the indirect correlation mediated by the solvent is also short-range [although somewhat larger than the range of $U(1, 1)$]. This may also be neglected when the distance between the two ligands is large, e.g., two oxygen molecules on hemoglobin. Figure 9.3 shows the typical distance dependence of $U(R)$ and $W(R)$.

In the model discussed below we shall encounter another solvent effect that in general could be significant even when the two ligands are far apart and for which the above-mentioned solvent effect [Eq. (9.3.5)] is negligible.

The model is that discussed in Section 4.5, for which the ligand–ligand pair correlation has the form

$$g(1, 1) = \frac{[Q_L(1, 1) + Q_H(1, 1)][Q_L(0, 0) + Q_H(0, 0)]}{[Q_L(1, 0) + Q_H(1, 0)]^2} \tag{9.3.7}$$

In vacuum we had the explicit expression

$$g^g(1, 1) = \frac{(Q_L q_L^2 S_L + Q_H q_H^2 S_H)(Q_L + Q_H)}{(Q_L q_L + Q_H q_H)^2}$$

$$= S\frac{(1 + Kh^2)(1 + K)}{(1 + Kh)^2} \tag{9.3.8}$$

The second form on the rhs of Eq. (9.3.8) is valid for the case $S_L = S_H = S$.

The expression for the pair correlation in a solvent can be obtained by using the same cyclic process as in Fig. 9.2. The free-energy change in vacuum is (presuming, for simplicity, $S_L = S_H = S$)

$$\Delta G^g = -k_B T \ln \frac{Q^g(1, 1)Q^g(0, 0)}{[Q^g(1, 0)]^2}$$

$$= -k_B T \ln \left[\frac{S(Q_L q_L^2 + Q_H q_H^2)(Q_L + Q_H)}{(Q_L q_L + Q_H q_H)^2} \right] \tag{9.3.9}$$

Note that ΔG^g is a *free energy* even in vacuum.

The corresponding free-energy change in the liquid phase is

$$\Delta G^l = \Delta G^g + [\Delta G^*(1, 1) + \Delta G^*(0, 0) - 2\Delta G^*(1, 0)] \tag{9.3.10}$$

*This is not true for two protons on, say, succinic acid.

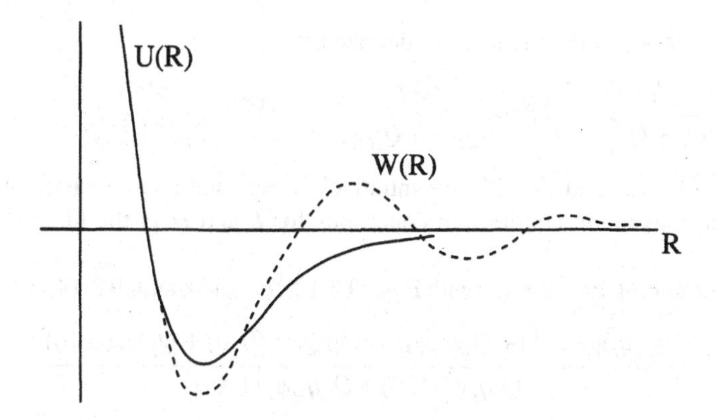

Figure 9.3. Schematic comparison of the direct pair potential $U(R)$ and the potential of average force $W(R)$, as a function of the ligand–ligand separation.

and the pair correlation in the liquid is

$$g^l(1, 1) = g^g(1, 1)\frac{\psi(1, 1)\psi(0, 0)}{[\psi(1, 0)]^2} \tag{9.3.11}$$

This is formally the same as Eq. (9.3.5). However, it should be stressed that here $g^g(1, 1)$ is related to the *free energy* ΔG^g, while in Eq. (9.3.5) it is related to the ligand–ligand interaction *energy* $U(1, 1)$. Also, the quantities ψ in Eq. (9.3.11) are now averages of ψ_L and ψ_H, as discussed below. Equation (9.3.11) has a simple interpretation. The ligand–ligand correlation g^l is factorized into two components: the correlation in vacuum (which may also be factorized into direct and indirect parts), and a solvent contribution part which contains the solvation Gibbs energies of the adsorbent molecules in the various occupancy states [the square brackets in Eq. (9.3.10)].

We now derive an alternative, but more useful, expression similar in form to Eq. (9.3.8), by the following considerations. As noted above, each of the quantities ψ in Eq. (9.3.11) is an average of ψ_L and ψ_H. The relations are

$$\begin{aligned}\psi(0, 0) = \exp[-\beta\Delta G^*(0, 0)] &= X_L^{(0)} \exp[-\beta\Delta G_L^*(0, 0)] + X_H^{(0)} \exp[-\beta\Delta G_H^*(0, 0)] \\ &= X_L^{(0)}\psi_L(0, 0) + X_H^{(0)}\psi_H(0, 0)\end{aligned} \tag{9.3.12}$$

and, similarly,

$$\begin{aligned}\psi(1, 0) &= X_L^{(1)}\psi_L(1, 0) + X_H^{(1)}\psi_H(1, 0) \\ \psi(1, 1) &= X_L^{(2)}\psi_H(1, 1) + X_H^{(2)}\psi_H(1, 1)\end{aligned} \tag{9.3.13}$$

where the various mole fractions are defined by

$$X_L^{(0)} = \frac{Q_L}{Q_L + Q_H}, \qquad X_L^{(1)} = \frac{Q_L q_L}{Q_L q_L + Q_H q_H}, \qquad X_L^{(2)} = \frac{Q_L q_L^2}{Q_L q_L^2 + Q_H q_H^2} \tag{9.3.14}$$

Relations (9.3.12) and (9.3.13) are intuitively clear* and can be easily obtained using thermodynamic equilibrium conditions for L and H in the three states of occupancy.

On combining Eq. (9.3.11) with Eqs. (9.3.12), (9.3.13), and (9.3.14), we obtain

$$g^l(1,1) = S\frac{[Q_L q_L^2 \psi_L(1,1) + Q_H q_H^2 \psi_H(1,1)][Q_L \psi_L(0,0) + Q_H \psi_H(0,0)]}{[Q_L q_L \psi_L(1,0) + Q_H q_H \psi_H(1,0)]^2} \tag{9.3.15}$$

This should be compared with the first form on the rhs of Eq. (9.3.8). The last form may also be written as

$$g^l(1,1) = S\frac{\psi_L(1,1)\psi_L(0,0)}{[\psi_L(1,0)]^2}\frac{[1 + Kh^2\psi_H(1,1)/\psi_L(1,1)][1 + K\psi_H(0,0)/\psi_L(0,0)]}{[1 + Kh\psi_H(1,0)/\psi_L(1,0)]^2}$$

$$= S\frac{\psi_L(1,1)\psi_L(0,0)}{[\psi_L(1,0)]^2}\frac{(1 + K^*h^*h^{**})(1 + K^*)}{(1 + K^*h^*)^2} \tag{9.3.16}$$

where in the last form of Eq. (9.3.16) we have used the modified forms of K and h. (Note that in the solvent, h^* and h^{**} correspond to the *first* and *second* h parameters, which were defined in Section 9.2.)

The pair correlation function is now seen to be in the form of Eq. (9.1.2). The ligand–ligand pair correlation has three contributions: one due to the *direct* ligand–ligand interaction (presumed to be the same for the L and H forms); a second due to correlation transmitted by the solvent, namely, the conditional pair correlation due to the solvent in the presence of the adsorbent molecule in the L form; and a third contribution due to conformational changes induced in the adsorbent molecule. Note that the latter term has the same formal structure as the corresponding expression in vacuum [see Eq. (9.3.8)]. The difference between this contribution and that in Eq. (9.3.8) arises from the presence of a solvent. The indirect correlation in Eq. (9.3.8) depends on the difference in the *energy* levels of L and H (through K), as well as on the difference in the binding energies to L and H (through h). In the presence of the solvent, both K^* and h^* depend on the corresponding *free* energies. Thus, when a ligand binds to the adsorbent molecule it might change its conformation. When $K \neq 0$, it is sufficient that $h \neq 1$ to obtain indirect correlation in Eq. (9.3.8). In a solvent, the change in conformation may cause a *change* of *solvation* of L and H. Therefore, even when $h = 1$ (i.e., the binding energies are equal, $U_L = U_H$) we can obtain solvent-induced correlation due to solvation effects.

*For the derivation, see Section 3.3 in Ben-Naim (1987).

Note, however, that this solvent effect is different from the solvent-mediated correlation denoted by $\psi_L(1, 1)\ \psi_L(0, 0)/\psi_L(1, 1)$ in Eq. (9.3.16). The latter is in general of relatively short range. For instance, in the case of two oxygen molecules bound to hemoglobin this contribution may be neglected. On the other hand, the solvent effects carried out through the third factor in Eq. (9.3.16) has the same range as the indirect correlation in vacuum. Thus, two oxygen molecules at two distant sites in hemoglobin could be strongly correlated if the solvation Gibbs energies of the L and H forms are very different and, of course, if the binding of oxygen induces conformational changes in hemoglobin. Thus, the first solvent-mediated effect is due to solvation Gibbs energies of *one* conformation [L in Eq. (9.3.6)] in the different occupancy states. The second solvent effect depends on the *difference* between the solvation Gibbs energies of the two forms L and H in the various occupancy states. The two effects are fundamentally different with respect to their range. They are also fundamentally different with respect to their additivity. This is similar to our discussion of nonadditivity in Chapter 5, but we shall not elaborate on this aspect here.

9.4. DECOMPOSITION OF THE SOLVATION GIBBS ENERGY OF MACROMOLECULES

Before we examine some specific solvation effects on cooperativity we must first consider various aspects of the solvation Gibbs energy of a macromolecule α. We present here one possible decomposition of ΔG^*_α which will be useful for our purposes. Consider a globular protein α which, for simplicity, is assumed to be compactly packed so that there are no solvent molecules within some spherical region to which we refer as the hard core of the protein. The interaction energy between α and the ith solvent molecule (the solvent is presumed to be water, w) is written as

$$U_{\alpha w}(\mathbf{X}_\alpha, \mathbf{X}_i) = U^H_{\alpha w}(\mathbf{X}_\alpha, \mathbf{X}_i) + U^S_{\alpha w}(\mathbf{X}_w, \mathbf{X}_i) + \sum_k U(k, \mathbf{X}_i) \qquad (9.4.1)$$

where U^H is the hard-core interaction, i.e., the infinite repulsion exerted on a solvent molecule whenever it penetrates the hard-core region of α. The second, U^S, may be referred to as the soft part of the solute–solvent interaction. We include here van der Waals interactions between α and a water molecule. The last sum over k includes all interactions that result from specific functional groups on the surface of α, such as charged or hydrogen-bonding groups. The split of the solute–solvent pair potential is, of course, not unique. There are infinite ways of dividing $U_{\alpha w}$ into a sum of different terms.* We shall choose a particular simplified form of Eq. (9.4.1) to illustrate the various possible solvent effects on cooperativity. It will be seen in

*For more details, see Chapter 8 in Ben-Naim (1992).

the next two sections that the best, or most convenient, way of expressing the pair potential $U_{\alpha w}$ depends on the specific system we wish to discuss. Here, it is assumed that α is essentially a hard sphere, i.e., we neglect the soft part of the interaction U^S (which, in principle, can always be added to the sum over the "functional groups"). The functional groups will be of two types: either hydrophobic ($H\phi O$) groups, such as methyl or ethyl, or hydrophilic ($H\phi I$) groups, such as carboxyl, hydroxyl, or carbonyl. A schematic illustration of the solute–solvent pair potential is shown in Fig. 9.4. With this description of the solute–solvent pair potential we proceed to write the solvation Gibbs energy of the solute α as

$$\Delta G^*_\alpha = \Delta G^{*H}_\alpha + \sum_i \Delta G^{*i/H}_\alpha + \sum_{j,k} \Delta G^{j,k/H}_\alpha + \cdots \tag{9.4.2}$$

This expanded form of ΔG^*_α may be derived exactly from the definition (9.2.20) and from the specific form of the pair potential (9.4.1).† We shall not derive this expression here. Instead, we present a qualitative description of the various terms on the rhs of Eq. (9.4.2) that must sum to the total solvation Gibbs energy ΔG^*_α.

The solvation process, for which ΔG^*_α is its free-energy change, is defined as the process of transferring a single α molecule (having a fixed conformational state) from a fixed point in vacuum (or an ideal gas phase) into the liquid phase at some fixed point. The same process is now carried out piecewise. We first cut off all the functional groups. Theoretically, we imagine that we can turn off all the interactions $U(k, \mathbf{X}_i)$ between the functional groups on the surface of α and all solvent molecules. We can now solvate all the parts of the molecule α in steps. First, we solvate the hard core. The resulting solvation Gibbs energy is ΔG^{*H}_α [the first terms on the rhs of Eq. (9.4.2)].

Next, we solvate all the functional groups [equivalently, we turn on the interactions $U(k, \mathbf{X}_i)$ that were turned off in the first step]. One way to do this is to solvate *all* the functional groups simultaneously. The resulting free-energy change would be $\Delta G^{*FG/H}_\alpha$ and the total solvation Gibbs energy would have been written as

$$\Delta G^*_\alpha = \Delta G^{*H}_\alpha + \Delta G^{*FG/H}_\alpha \tag{9.4.3}$$

Note that the first term is an average of the type (9.2.20) or (9.2.21), i.e., with a probability density of the *pure* solvent. The second quantity is a *conditional* average, i.e., we must use a conditional distribution instead of $P_0(V, \mathbf{X}^N)$. Since the hard core has already been transferred into the solvent, the distribution of solvent

†Note, however, that the terms in Eq. (9.4.2) do not correspond to the terms in the potential (9.4.1). Even when (9.4.1) is exact, the Gibbs energy of solvation of the functional groups is not additive. For more details, see Chapter 8 in Ben-Naim (1992).

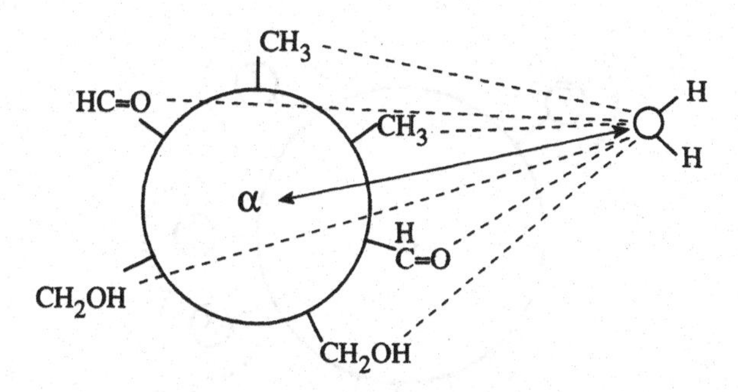

Figure 9.4. Schematic description of the solute–solvent pair potential. The double-arrowed line indicates the hard (repulsive) interaction between α and a water molecule. The dashed lines indicate the interaction between groups on the surface of α and a water molecule, the sum of which is the last term on the rhs of Eq. (9.4.1).

configurations now differs from $P_0(V, \mathbf{X}^N)$, hence we obtain a conditional average in $\Delta G_\alpha^{*FG/H}$, the condition being the presence of the hard core of α.

Although Eq. (9.4.3) is an exact expression for the solvation Gibbs energy (presuming the specific form of the pair potential function), it is not useful for studying the specific effect of different functional groups. To obtain a more detailed expansion of ΔG_α^*, we must solvate each group *separately*. In doing so we must be careful in the *order* of solvating the functional groups. For instance, in the example depicted in Fig. 9.5, if we solvate group 1 first and then group 7, the corresponding solvation Gibbs energies are $\Delta G_\alpha^{*1/H}$ and $\Delta G_\alpha^{*7/H,1}$, respectively. But since groups 1 and 7 are far apart the *condition* "1" in $\Delta G_\alpha^{*7/H,1}$ can be dropped. In this case the order of solvating these two groups is of no importance. We say that groups 1 and 7 are independently solvated and write

$$\Delta G_\alpha^{*1,7/H} = \Delta G_\alpha^{*1/H} + \Delta G_\alpha^{*7/H} \tag{9.4.4}$$

This is not the case for groups 2 and 3, for which the order of solvation is important, i.e., we obtain different conditional solvation Gibbs energies when we first solvate 2 and then 3, or first 3 and then 2. The corresponding solvation free energies are

$$\Delta G_\alpha^{*2,3/H} = \Delta G_\alpha^{*2/H} + \Delta G_\alpha^{*3/H,2}$$

$$= \Delta G_\alpha^{3/H} + \Delta G_\alpha^{*2/H,3} \tag{9.4.5}$$

In this case we say that groups 2 and 3 are correlated, i.e., turning on the interaction of, say, $U(2, \mathbf{X}_i)$ will affect the distribution of solvent molecules around group 3. Therefore, the condition "2" in $\Delta G_\alpha^{*3/H,2}$ cannot be overlooked. In the expansion of ΔG_α^* in Eq. (9.4.2), we first solvate the hard core. Next, we solvate all the

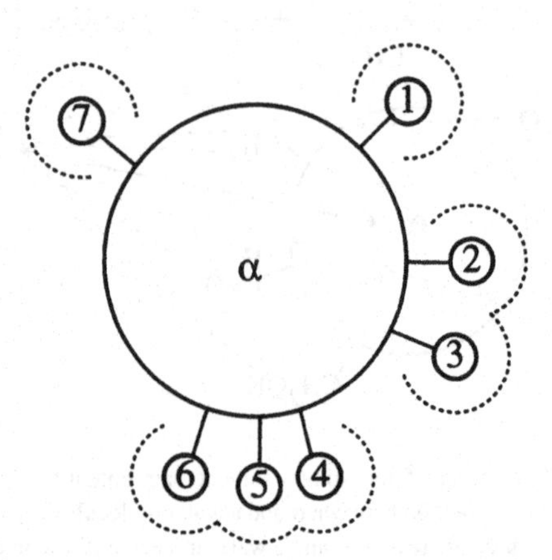

Figure 9.5. Schematic illustration of a distribution of functional groups on the surface of α. Groups 1 and 7 are independently solvated (there is no overlap between the solvation spheres, indicated by the dashed curves). Groups 2 and 3 are pair-correlated. Groups 4, 5, and 6 are triply correlated.

independently solvated groups—the corresponding Gibbs-energy change is the sum over i on the rhs of Eq. (9.4.2). Next, we solvate all the pairs of pair-correlated groups—the resulting terms are of the form $\Delta G^{*i,j/H}$, and so on for triply correlated and higher-order correlations.

Clearly, for any specific order of turning on the interaction $U(k, X_i)$, we shall obtain a different expansion on the rhs of Eq. (9.4.2). In the particular expansion written on the rhs of Eq. (9.4.2) we have classified all the functional groups on the surface of α (i.e., those FGs that are exposed to the solvent) into different classes. The first consists of all the FGs that are independently solvated. The second consists of all *pairs* of correlated FGs, and so on. We shall see in the next two sections that this particular form of expansion of ΔG^*_α is convenient for a qualitative analysis of the types of solvent effect we may expect on cooperativity.

It should be noted that the first term on the rhs of Eq. (9.4.2) is fundamentally different from all the other terms that constitute ΔG^*_α. This term depends only on the sizes of solute α and of the solvent molecules. We shall further elaborate in the next section on the meaning assigned to the size, or to the volume, of a molecule. Here, we stress the fact that once we have determined or assigned a size to α and to the solvent molecule, the value of ΔG^{*H}_α is determined. On the other hand, all the other terms on the rhs of Eq. (9.4.2) depend on the type of functional groups, their distribution on the surface of α, and on the specific interaction with a solvent molecule. We shall therefore discuss in the next section the volume effect which is common to any solvent (having roughly the same size as, say, water molecules). The solvation effects arising from the other terms on the rhs of Eq. (9.4.2) will be examined in Section 9.6.

Before proceeding to discuss the effect of the solvent on the ligand–ligand correlation, we present here a simple probabilistic interpretation of ΔG_α^{*H}. In the canonical ensemble† the solvation Helmholtz energy is

$$\exp(-\beta\Delta A_\alpha^*) = \int \cdots \int P_0(\mathbf{X}^N) \exp(-\beta B_\alpha) d\mathbf{X}^N \tag{9.4.6}$$

where $P_0(\mathbf{X}^N)$ is the distribution density of solvent configurations in the absence of α, B_α is the total binding energy (see Section 9.2) of α to all solvent molecules, and the integration is carried out over all possible configurations of the solvent molecules.

For the particular quantity ΔA_α^{*H} in Eq. (9.4.2), the function $\exp(-\beta B_\alpha)$ is simply a step function. It is zero whenever the centers of *all* solvent molecules are outside the repulsive region produced by α, and unity when at least one center of a solvent molecule penetrates into this region. We denote this excluded volume by V_α^{EX} and write

$$\exp(-\beta\Delta A_\alpha^{*H}) = \int \cdots \int P_0(\mathbf{X}^N) \exp(-\beta B_\alpha^H) d\mathbf{X}^N$$

$$= \int_{V-V_\alpha^{EX}} \cdots \int P_0(\mathbf{X}^N) d\mathbf{X}^N = Pr(V_\alpha^{EX}) \tag{9.4.7}$$

Thus, the original integral over all configurations of solvent molecules reduces to an integral of $P_0(\mathbf{X}^N)$ in the restricted region $V - V_\alpha^{EX}$. (Note that here we use the same symbol to denote a *region* and its volume.) Since $P_0(\mathbf{X}^N)$ is a probability density, the last integral on the rhs of Eq. (9.4.7) is the probability of the event that all centers of the solvent molecules be in $V - V_\alpha^{EX}$. This is the same as the probability of finding the region V_α^{EX} empty.

A particularly simple form of $Pr(V_\alpha^{EX})$ is obtained for the solvation Helmholtz energy of α in a solvent consisting of N hard-sphere solvent particles of diameter σ in a volume V. If the density N/V is very small, so that one can neglect solvent–solvent interactions, the probability density $P_0(\mathbf{R}^N)$ is simply V^{-N} and the integral on the rhs of Eq. (9.4.7) reduces to

$$Pr(V_\alpha^{EX}) = \frac{1}{V^N} \int_{(V-V_\alpha^E)} \cdots \int d\mathbf{R}^N = \left(\frac{V - V_\alpha^{EX}}{V} \right)^N$$

$$= \left(1 - \frac{V_\alpha^{EX}}{V} \right)^N \approx 1 - \frac{N V_\alpha^{EX}}{V} = 1 - \rho V_\alpha^{EX} \tag{9.4.8}$$

†In Section 9.2 we have defined the Gibbs energy of solvation ΔG_α^* in the T, P, N ensemble. In the T, V, N (canonical) ensemble the appropriate quantity is ΔA_α^*, the Helmholtz energy of solvation. It can be shown that the two are equal for macroscopic systems, provided the volume V in the T, V, N ensemble is equal to the *average* volume of a system in the T, P, N ensemble.

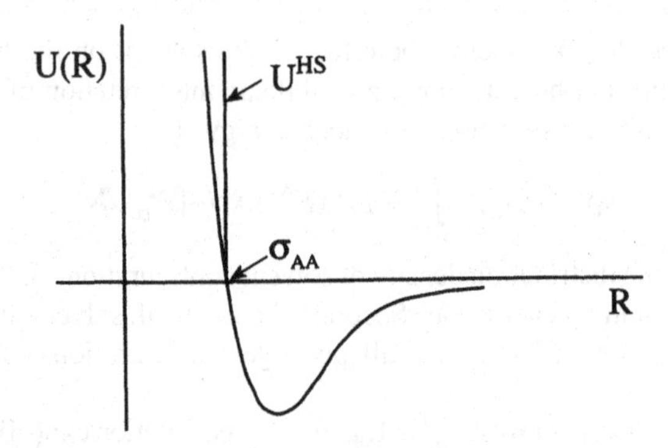

Figure 9.6. A schematic form of the pair potential $U(R)$ for two real spherical molecules. The hard-sphere potential $U^{HS}(R)$ corresponding to these molecules is the bold vertical line at $R = \sigma_{AA}$.

The last term on the rhs of this equation is intuitively clear. The probability of finding a *specific* solvent molecule in any region $d\mathbf{R} = dxdydz$ is simply $d\mathbf{R}/V$. In particular, the probability of finding a specific solvent molecule in V_α^{EX} is V_α^{EX}/V. The probability of finding it outside V_α^{EX} is $(1 - V_\alpha^{EX}/V)$. If all solvent molecules are independent, the probability of finding *all* solvent molecules outside V_α^{EX} is $(1 - V_\alpha^{EX}/V)^N$. For $V_\alpha^{EX}/V << 1$, this is approximately equal to $1 - \rho V_\alpha^{EX}$. Thus, in Eq. (9.4.8), we have found a relation between the probability of finding the region V_α^{EX} empty, and the actual volume of the region, also denoted by V_α^{EX}. This is, of course, true only in the low-density limit.

9.5. EFFECT OF SIZE ON THE COOPERATIVITY

We discuss in this section a relatively simple solvent effect that depends only on the "size" or "volume" of the particles involved. It will be seen below that since this type of effect depends only on the sizes of the particles and not on any specific interactions between the solutes and the solvent molecules, it may be referred to as the nonspecific solvent effect.*

First, we need to elaborate on the concept of the "radius" or "diameter" of the molecules involved in the binding process. Real molecules do not have well-defined boundaries as do geometrical objects such as spheres or cubes. Nevertheless, one can assign to each molecule an *effective* radius. This assignment depends on the form of the intermolecular potential function between any pair of real particles.

*An inert solvent, such as argon or methane, would behave as a nonspecific solvent. In this section we consider only the volume of the solvent molecules.

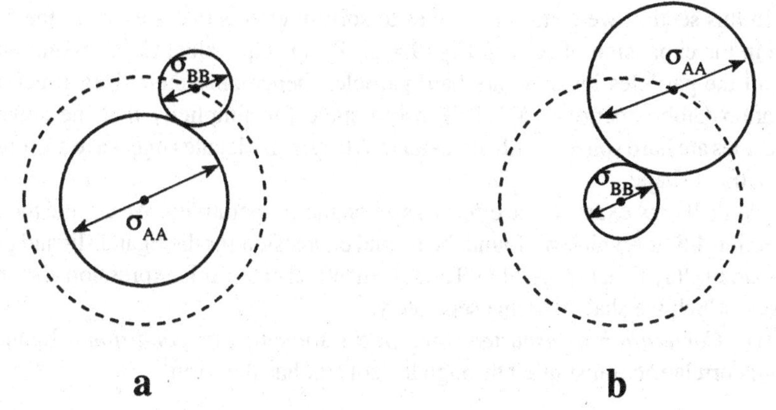

Figure 9.7. (a) The excluded volume (dashed curve) of A with respect to B. (b) The excluded volume of B with respect to A.

Consider first two simple spherical atoms, say argon atoms. The pair interaction potential has the general form depicted in Fig. 9.6. Note that for $R < \sigma_{AA}$ the potential function becomes very steep, i.e., it is strongly repulsive. This means that a large amount of energy must be used to push the two atoms to a distance shorter than σ_{AA}. Thus, although it is possible for the two particles to be found at a distance $R < \sigma_{AA}$, the probability of finding such an event is negligibly small (for systems at normal temperatures and pressures). Therefore, we regard σ_{AA} as an *effective* diameter of the argon atoms and $\pi\sigma_{AA}^3/6$ as its *effective* volume. The idealization of hypothetical particles having a precise volume are hard-sphere particles. These are defined through their pair potential

$$U^{HS}(R) = \begin{cases} 0 \text{ for } R > \sigma_{AA} \\ \infty \text{ for } R \leq \sigma_{AA} \end{cases} \tag{9.5.1}$$

Two hard spheres cannot be pushed to a distance $R < \sigma_{AA}$, since this would require infinite energy. Clearly, this is only approximately true for any real particles.

Once we have assigned a diameter to any spherical particle, we can define the *excluded volume* of a particle A with respect to particle B. This is simply the spherical region of radius $(\sigma_{AA} + \sigma_{BB})/2$ around the center of A. Clearly, this region (Fig. 9.7a) is (effectively) excluded for the center of any B particle. Similarly, Fig. 9.7b shows the excluded volume of B with respect to any A particle.

It should be noted that the excluded volume is a property of a *pair* of particles. If a solute α is in a mixed solvent of type A and B, then there is a different excluded volume of α with respect to A and with respect to B.*

*For more details, see Chapter 5 in Ben-Naim (1992).

In this section we restrict ourselves to solvent effects that are due to the first term in the expansion of ΔG^*_α in Eq. (9.4.2). This is equivalent to the assumption that all the particles involved are hard particles, hence only their sizes affect the solvation Gibbs energies. We shall also assume for simplicity that the solvent molecules are hard spheres with diameter σ. All other molecules may have any other geometrical shape.

We shall now examine the effect of size on the cooperativity. We use the model of Section 4.5, for which we found the formal expression for the ligand–ligand pair correlation $g^l(1, 1)$ in Eq. (9.3.16). The solvent effect enters this expression via three factors, which we shall examine separately.

(i) *Correlation transmitted through the solvent*: The *conditional* ligand–ligand correlation transmitted through the solvent has the form†

$$y_L(1, 1) = \frac{\psi_L(1, 1)\psi_L(0, 0)}{[\psi_L(1, 0)]^2}$$

$$= \exp\{-\beta[\Delta G_L^{*H}(1, 1) + \Delta G_L^{*H}(0, 0) - 2\Delta G_L^{*H}(1, 0)]\} \qquad (9.5.2)$$

Note that the rhs of Eq. (9.5.2) contains only the solvation Gibbs energies of the hard part of the L form. In view of relation (9.4.7) we may rewrite $y_L(1, 1)$ as

$$y_L(1, 1) = \frac{Pr[V_L^{EX}(1, 1)]Pr[V_L^{EX}(0, 0)]}{Pr[V_L^{EX}(1, 0)]^2} \qquad (9.5.3)$$

The various probabilities on the rhs of Eq. (9.5.3) depend on the shape of the adsorbent molecule in each of the occupancy states. It is very difficult to compute these probabilities for arbitrary shapes. It is, however, intuitively clear that the whole term $y_L(1, 1)$ becomes nearly unity when the two ligands are small compared with the size of the adsorbent molecule, and when the separation between the sites is large compared with the diameter of the solvent molecules. The whole term $y_L(1, 1)$ will be unity when the ligands are buried within the adsorbent molecule, in which case there is no excluded volume change in the reaction

$$2\,(1, 0) \rightarrow (1, 1) + (0, 0) \qquad (9.5.4)$$

In the low-density limit region $\rho \rightarrow 0$ the quantity $y_L(1, 1)$ reduces to [see Eq. (9.4.8)]

$$y_L(1, 1) \approx 1 - \rho[V_L^{EX}(1, 1) + V_L^{EX}(0, 0) - 2V_L^{EX}(1, 0)] \qquad (9.5.5)$$

†Note that subscript L refers to *state* L, and superscript H refers to the *hard* part of the interaction.

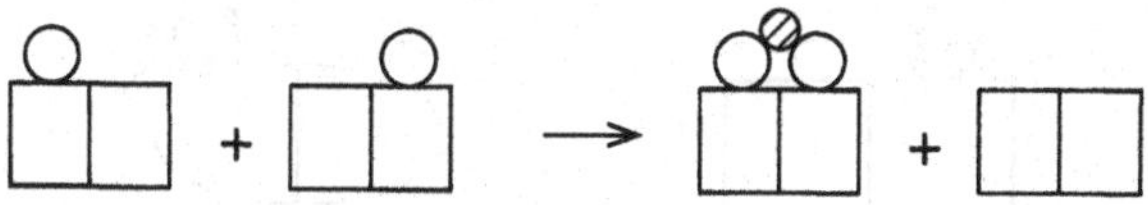

(1,0) + (0,1) → (1,1) + (0,0)

Figure 9.8. A schematic illustration of reaction (9.5.4). The two ligands occupying the two sites [in the state (1,1)] are correlated when a solvent particle (hatched circle) interacts simultaneously with the two ligands.

Clearly, the excluded volume change in reaction (9.5.4) will be noted only when a solvent molecule can interact simultaneously with the two ligands occupying the two sites† (Fig. 9.8). This is exactly the same condition for the solvent-induced correlation in the theory of liquids.‡

(ii) *Solvent effect on K*: We recall that K changes into K^* in the presence of a solvent [Eq. (9.3.16)]. For the particular case of hard particles, the solvent effect is

$$\frac{K^*}{K} = \exp[-\beta(\Delta G_H^{*H} - \Delta G_L^{*H})]$$

$$= \frac{Pr[V_H^{EX}(0, 0)]}{Pr[V_L^{EX}(0, 0)]} \tag{9.5.6}$$

This is perhaps the most important excluded volume effect on the cooperativity. It depends on the probability ratio of finding the excluded volumes of the L and H forms empty. The larger the difference between the shape and size of the two forms, the larger will be this ratio. In contrast to the previously discussed solvent effect (as well as that discussed below), this effect does not depend on the relative sizes of the ligands and the adsorbent molecules, nor on the ligand–ligand separation. It could be large even when the ligands are well buried in the interior of the adsorbent molecule. The low-density limit $\rho \to 0$ of Eq. (9.5.6) is

$$\frac{K^*}{K} = 1 - \rho[V_H^{EX}(0, 0) - V_L^{EX}(0, 0)] \tag{9.5.7}$$

which depends on the difference in the actual excluded *volumes* of the H and L forms (Fig. 9.9).

†This is true only in the low-density limit (9.5.5). At higher solvent densities, correlation between the ligands may occur at a somewhat larger range of distances.
‡For more details, see Chapter 5 in Ben-Naim (1992).

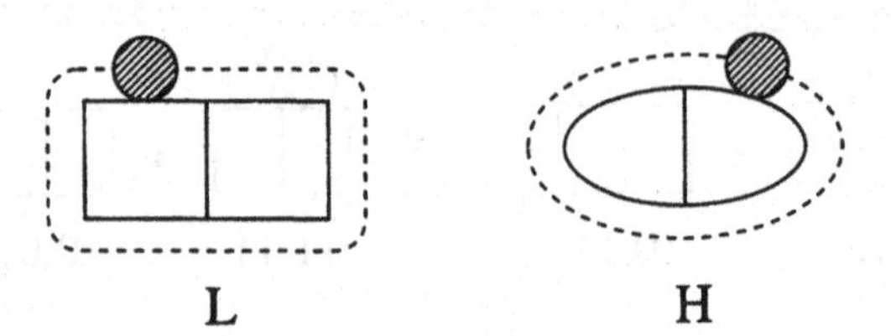

Figure 9.9. The excluded volume (dashed lines) of the L and H forms with respect to a solvent molecule (hatched circle).

(iii) *Solvent effect on h*: In Section 9.2 we have seen two possible modifications of h into h^* and h^{**}, depending on whether we add the first or second ligand, respectively. The excluded volume effect, say for the first ligand, is

$$\frac{h^*}{h} = \exp\{-\beta[\Delta G_H^{*H}(1,0) + \Delta G_L^{*H}(0,0) - \Delta G_L^{*H}(1,0) - \Delta G_H^{*H}(0,0)]\} \quad (9.5.8)$$

Note that subscripts L and H refer to the two forms of the adsorbent molecule, while superscript H refers to the hard part of the interaction. Here, again, we do not expect a large solvent effect when the size of the ligand is small compared with the adsorbent molecule. There will be no effect when the ligand is buried in the interior of the adsorbent molecule. The low-density limit ($\rho \to 0$) is now

$$\frac{h^*}{h} = 1 - \rho[V_H^{EX}(1,0) + V_L^{EX}(0,0) - V_L^{EX}(1,0) - V_H^{EX}(0,0)] \quad (9.5.9)$$

Thus, the change in the excluded volume in reaction (9.2.12) or (9.2.16) is not expected to be large, unless the binding of a ligand changes significantly the shape and size of one form (L or H) relative to the other.

In conclusion, whenever the size of the ligand is very small compared with the size of the adsorbent molecule (e.g., oxygen and hemoglobin), we do not expect a significant solvent effect unless the excluded volumes of the L and H forms differ significantly. The actual estimate of the solvent effect in this case requires calculation of the difference in the Gibbs energies of solvation of the hard-core part of the H and L forms of the adsorbent molecule. This effect is independent of either the size or the separation between the ligand molecules.

9.6. SOME SPECIFIC SOLVENT EFFECTS

Having dealt with the excluded volume effect arising from the first term, ΔG_α^{*H}, on the rhs of Eq. (9.4.2), we now examine a few other solvent effects associated with the remaining terms on the rhs of this equation. These are referred to as *specific* solvent effects, since they arise from the specific functional groups (FG) distributed on the surface of α, as well as on the specific interactions (other than repulsive) between these FGs and the solvent molecules.

Clearly, since these effects depend on the type and distribution of the FGs, there is no general method of dealing with an arbitrary binding system. We shall therefore treat one, relatively simple example. The extension to any other specific system should become clear by generalization of the procedure performed in this example.

In the following model example, we assume that each species involved in the binding process has a spherical shape and that the FGs on its surface are distributed in such a way that each pair of FGs on the surface (i.e., exposed to the solvent) is independently solvated. In other words, the conditional solvation Gibbs energy of the ith FG (given the hard core H) is independent of the presence or absence of any other FGs. Formally, this is equivalent to taking only the first sum over i in the expansion on the rhs of Eq. (9.4.2).

As in the previous section, we shall discuss each of the three types of solvent effects separately.

(i) *Correlation transmitted through the solvent*: Here, the important quantity is the change in the solvation Gibbs energy of the L form in the process

$$2(1, 0) \rightarrow (0, 0) + (1, 1) \tag{9.6.1}$$

As in the case of the volume effect, when the two ligands are far apart the change in the solvation Gibbs energy in reaction (9.6.1) will be negligibly small. It will be zero when the ligands are buried in the interior of the adsorbent molecule. When the ligands on the sites are close enough, there might be (conditional) correlations transmitted through the solvent. In the most general case one must write the conditional solvation Gibbs energy of each ligand before and after the process (9.6.1). It can be shown that the largest contribution to this correlation occurs when a solvent molecule interacts simultaneously with both ligands. We shall demonstrate this effect for a very simple solvent—a single water molecule. For this case, the solvation Gibbs energy of any species α is

$$\Delta G_\alpha^* = -k_B T \ln \int P_0 \exp(-\beta B_\alpha) d\mathbf{X}_w \tag{9.6.2}$$

where $P_0 d\mathbf{X}_w$ is the probability of finding the water molecule in any specific configuration between $\mathbf{X}_w$ and $\mathbf{X}_w + d\mathbf{X}_w$. In this particular case

$$P_0 = \frac{1}{\int d\mathbf{X}_w} \tag{9.6.3}$$

and

$$B_\alpha = U(\mathbf{X}_\alpha, \mathbf{X}_w) = U_{\alpha w}^H + \sum_k U(k, \mathbf{X}_w) \tag{9.6.4}$$

The solvation Gibbs energy can be rewritten as

$$\Delta G^*_\alpha = \Delta G^{*H}_\alpha + \sum_k \Delta G^{*k/H}_\alpha \tag{9.6.5}$$

where

$$\Delta G^{*H}_\alpha = -k_B T \ln \int P_0 \exp(-\beta U^H_{\alpha w}) d\mathbf{X}_w \tag{9.6.6}$$

and

$$\Delta G^{*k/H}_\alpha = -k_B T \ln \int P(\mathbf{X}_w/H) \exp[-\beta U(k, \mathbf{X}_w)] d\mathbf{X}_w \tag{9.6.7}$$

The quantity ΔG^{*H}_α is the solvation Gibbs energy of the hard part of the interaction and has been dealt with in the previous section. The second expression is the *conditional* solvation Gibbs energy of the kth FG given that the hard part of the interaction has already been solvated. The conditional probability density is

$$P(\mathbf{X}_w/H) = \frac{\exp(-\beta U^H)}{\int \exp(-\beta U^H) d\mathbf{X}_w} \tag{9.6.8}$$

where U^H is the hard part of the solute–solvent pair interaction. Note that since we have assumed that all the FGs are independently solvated, the sum over k in Eq. (9.6.4) is over the same groups as the sum over k in Eq. (9.6.5).

For this particular solvent we can now combine the solvation Gibbs energies for the process (9.6.1) to obtain

$$y_L = \frac{\psi_L(1, 1)\psi_L(0, 0)}{\psi_L(1, 0)^2} = y^H_L y^{FG/H}_L \tag{9.6.9}$$

where y^H_L is the hard part of y_L that we have discussed in the previous section; $y^{FG/H}_L$ is the conditional correlation due to all the FGs and has the general form

$$y^{FG/H}_L = \frac{\psi_L(FG/(1, 1))\psi_L(FG/(0, 0))}{\psi_L(FG/(0, 1))^2} \tag{9.6.10}$$

where each of the factors on the rhs depends on the conditional solvation Gibbs energies of the species indicated as a condition [i.e., (1, 1), (0, 0), and (0, 1)]. This is a complicated product of many terms, each of which depends on one FG. Nevertheless, we expect that there will be many cancellations of factors in this expression. In fact, it is easy to see that a FG, the solvation of which does not change in reaction (9.6.1), will not contribute to this correlation. The only contribution to $y^{FG/H}_L$ arises from those FGs whose solvation has changed during the reaction (9.6.1). Figure 9.10 shows one such reaction, where the ligand has two FGs denoted

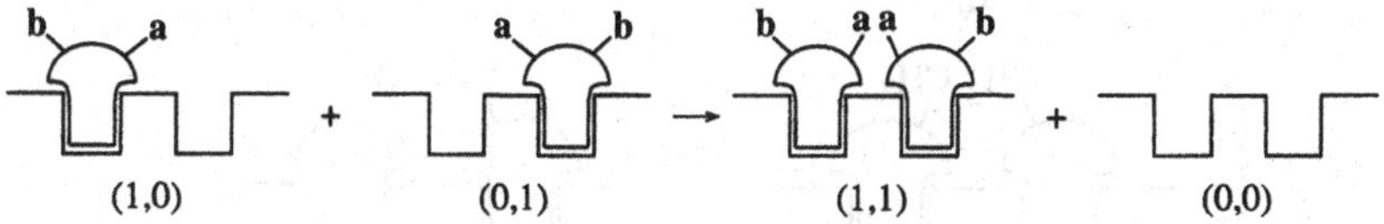

Figure 9.10. Reaction (9.6.1) for which the solvation of groups of type b is not changed in the process. The solvation of type a is changed in this process.

by **a** and **b**. When the process (9.6.1) occurs, the two groups **b** (one on each ligand) are well separated in both the initial and final state of the process. Therefore, these groups will have the same factor in the numerator and denominator of Eq. (9.6.10). On the other hand, the two groups **a** that are initially separated in the initial state come close to each other in the final state. The contribution of these two groups to the correlation (9.6.10) is

$$y_L(\mathbf{a},\mathbf{a}) = \frac{\int P(\mathbf{X}_w/(1,1))\exp[-\beta U(\mathbf{a},w) - \beta U(\mathbf{a},w)]d\mathbf{X}w}{\left\{\int P(\mathbf{X}_w/(1,0))\exp[-\beta U(\mathbf{a},w)]d\mathbf{X}_w\right\}^2} \tag{9.6.11}$$

Note here that the numerator is an average over a product of two functions $\exp[-\beta U(\mathbf{a}, w)]$ and $\exp[-\beta U(\mathbf{a}, w)]$, each of which belongs to a different ligand. When the two ligands are far apart, this average will be factorized into a product of two factors. If we further ignore the slight difference in the conditional densities $P(\mathbf{X}_w/(1, 1))$ and $P(\mathbf{X}_w/(1, 0))$, we find that the whole term becomes unity. However, when these two FGs are close enough so that a solvent molecule can interact simultaneously with them, the correlation $y_L(\mathbf{a}, \mathbf{a})$ will not be unity. Figure 9.11 shows two possibilities for the FGs. In one the groups are *HφO*, and in the other they are *HφI*. It has been estimated that the correlation in the second case, referred to as the *HφI* interaction, is stronger than in the former case.*

In a real solvent there could be many possibilities for solvent molecules to "bridge" the two functional groups. For the *HφI* interaction, the most favorable situation would be when the distance between the oxygens on the *HφI* groups is about 4.5 Å and the orientations of these groups are such that a water molecule can form a hydrogen-bonded bridge connecting the two groups. It has been estimated that the contribution to the solvation Gibbs energy for the process (9.6.1) is on the order of −3 kcal/mol per pair of such FGs.* The corresponding value of the correlation function $y_L(1, 1)$ would be on the order of about $\exp(3.0/0.6) = \exp(5) \sim 150$. This value is for one pair of FGs at the most favorable distance and orientation. In the most general case, one should account for the formation of pairs,

*For more details, see Chapters 7 and 8 in Ben-Naim (1992).

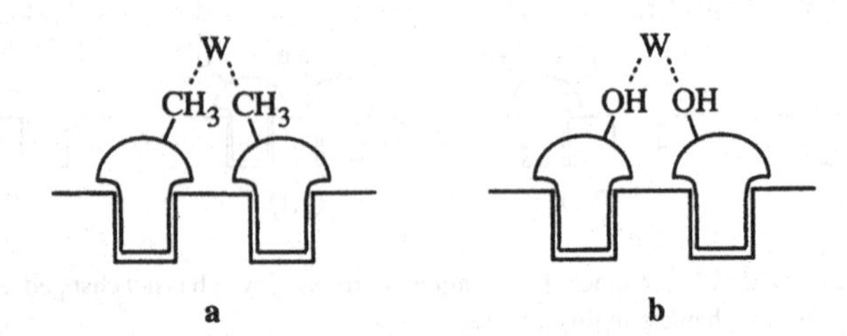

Figure 9.11. Two possible correlations transmitted by a single water molecule (w). In **a** the correlation is between two hydrophobic ($H\phi O$) groups, and in **b** between two hydrophilic ($H\phi I$) groups.

triplets, and higher-order correlations between FGs on one ligand and FGs on the second ligand.

(ii) *Solvent effect on K*: In our simplified model where all FGs are independent, both in the L and H forms of the adsorbent molecules, the (conditional) solvation Gibbs energies of each FG will not change in the reaction $L \rightleftharpoons H$. In this case, K^* will equal K. However, in the more general case, the change in conformation could either form or eliminate correlations between FGs on the surface of the adsorbent molecule. In Fig. 9.12 we show that groups **a** and **b** as well as groups **c** and **d**, which are far apart, hence uncorrelated in L, become correlated in H. Furthermore, groups **e**, **f**, and **g** which are uncorrelated in L become triply correlated in H. Clearly, the conversion $L \rightleftharpoons H$ can produce or eliminate higher-order correlations among the FGs. The specific solvent effect must be calculated for each specific example in which the distribution of FGs exposed to the solvent is known.

(iii) *Solvent effect on h*: Here, the relevant reaction is either (9.2.12) (for the first ligand) or (9.2.16) (for the second ligand). Again, the estimate of the precise

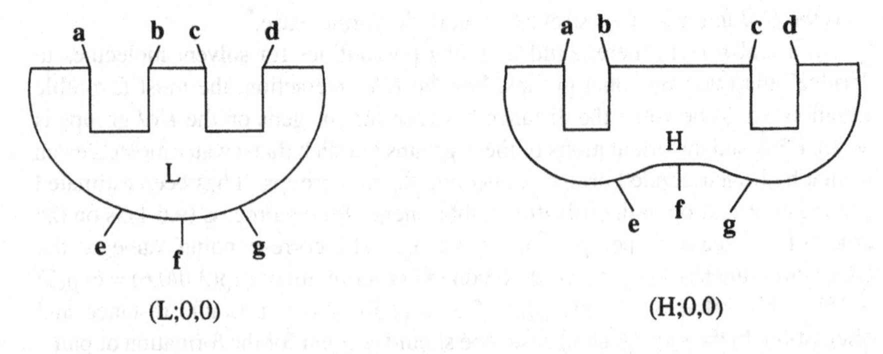

Figure 9.12. Redistribution of functional groups on the surface of the adsorbent molecule in the conversion $L \rightarrow H$. Groups **a** and **b**, groups **c** and **d**, and groups **e**, **f**, and **g** are far apart, hence uncorrelated in L but become correlated in H. Groups **b** and **c** which are correlated in L become uncorrelated in H.

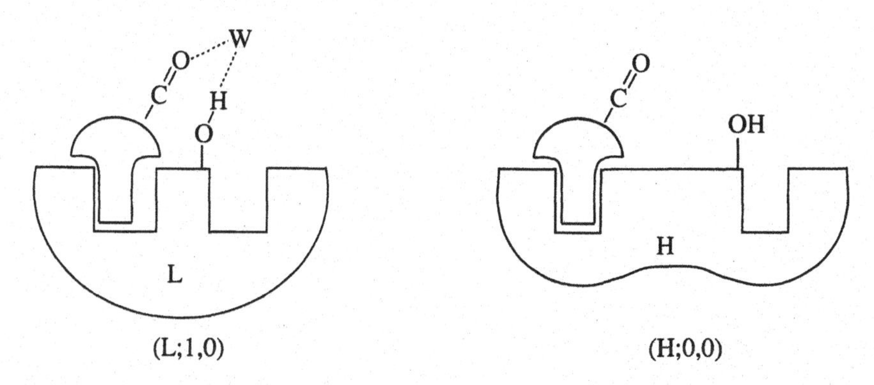

Figure 9.13. A ligand fits better the site on *H*, so that the *H*-form is stabilized in the absence of water. In the presence of water, the possibility of formation of a HB bridge will stabilize the *L*-form.

solvent effect on h depends on knowledge of the distribution of all FGs in the various species in the process (9.2.12) or (9.2.16). There is no way of giving a "general" result that will be appropriate for any system. In Fig. 9.13, we show one example where the ligand fits better the sites on H than L, so that $h = q_H/q_L > 1$. This means that in the absence of a solvent the ligand will shift the equilibrium between the two conformations, in favor of H. In the presence of water, the possibility of forming a hydrogen-bonded bridge in L, but not in H, will make the difference in the solvation Gibbs energy $\Delta G_H^*(1,0) - \Delta G_L^*(1,0)$ positive. If this effect is strong enough, $h > 1$ (in absence of solvent) could be modified into $h^* < 1$, i.e., giving preference to the L-form (in the presence of solvent).

Appendices

A. PAIR AND TRIPLET CORRELATIONS BETWEEN EVENTS

We present here two examples where dependence or independence between two events does not imply dependence or independence between three events, and vice versa.

1. Pairwise Independence Does Not Imply Triplewise Independence

Consider a board of total area S being hit by a ball. The probability of hitting a certain region is assumed to be proportional to its area. On this board we draw three regions A, B, and C (Fig. A.1). If the area of the entire board is chosen as unity, and the areas of A, B, and C are 1/10 of S, then we have

$$P(S) = 1, \qquad P(A) = P(B) = P(C) = \frac{1}{10} \tag{A.1}$$

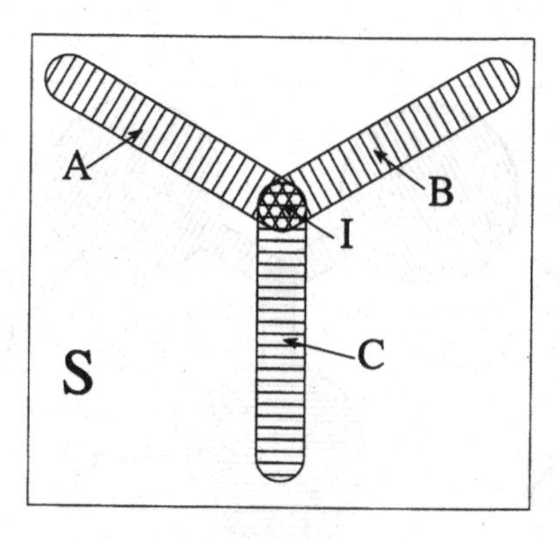

Figure A.1. The three areas A, B, and C and the corresponding intersection I for the first example discussed in the text.

In addition, we know that regions A, B, and C intersect, and that the area of the intersection (region I) is given to be 1/100 of S. Hence, in this system

$$P(A \cdot B) = P(B \cdot C) = P(A \cdot C) = \frac{1}{100} = P(A)P(B) = P(A)P(C) = P(B)P(C) \quad \text{(A.2)}$$

Thus, we have pairwise independence; e.g., the probability of hitting A *and* B is the product of the probabilities $P(A)$ and $P(B)$, hence $g(A, B) = g(A, C) = g(B, C) = 1$. However, in this system

$$P(A \cdot B \cdot C) = \frac{1}{100} \neq P(A)P(B)P(C) = \frac{1}{1000} \quad \text{(A.3)}$$

The probability of hitting A, B, and C is not the product of the probabilities $P(A)$, $P(B)$, and $P(C)$, i.e., $g(A, B, C) \neq 1$.

2. Triplewise Independence Does Not Imply Pairwise Independence

In this example (Fig. A.2) the total area is again S. Also, the area of each region A, B, and C is again 1/10 of S. But now the intersection I of the three regions has an area 1/1000 of S. In this case we have

$$\frac{1}{1000} = P(A \cdot B \cdot C) = P(A)P(B)P(C) \quad \text{(A.4)}$$

The probability of the event $(A \cdot B \cdot C)$ is the product of the three probabilities $P(A)$, $P(B)$, and $P(C)$, i.e., $g(A, B, C) = 1$. On the other hand

$$\frac{1}{1000} = P(A \cdot B) \neq P(A)P(B) = \frac{1}{100} \quad \text{(A.5)}$$

and similarly for both $P(A \cdot C)$ and $P(B \cdot C)$. Hence, in this system there is triplewise independence, $g(A, B, C) = 1$, but not pairwise independence, $g(A, B) = g(A, C) = g(B, C) \neq 1$.

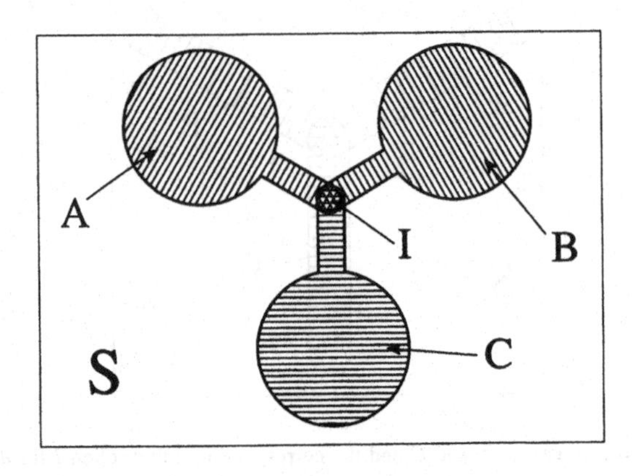

Figure A.2. As in Fig. A.1, but for the second example.

B. LOCALIZATION OF THE ADSORBENT MOLECULES AND ITS EFFECT ON THE BINDING ISOTHERM

Consider a system of M two-site molecules (such as oxalic acid) and N solvent molecules (such as water) in a volume V and at temperature T. The system is closed with respect to the adsorbent molecules and to the solvent molecules, but open with respect to the ligands **L**, maintained at a constant chemical potential μ.

Clearly, if we localize the adsorbent molecules, i.e., we assume that all of the M adsorbent molecules are devoid of translational and rotational degrees of freedom, then we obtain a new system, the thermodynamics of which differ from that of the original system.

For the sake of this appendix we assume that the adsorbent molecules are independent, hence the GPF of our system can be written as

$$\Xi(T, M, N, V, \lambda) = \xi^M/M! \tag{B.1}$$

On the other hand, for the same system but with localized adsorbent molecules we have

$$\Xi_l(T, M, N, V, \lambda) = \xi_l^M \tag{B.2}$$

where

$$\xi(T, \rho, \lambda) = Q(0) + Q(1)\lambda + Q(2)\lambda^2 \tag{B.3}$$

and

$$\xi_l(T, \rho, \lambda) = Q_l(0) + Q_l(1)\lambda + Q_l(2)\lambda^2 \tag{B.4}$$

Here, ξ and ξ_l are the GPFs of a single free and localized adsorbent molecule, respectively, $\rho = N/V$ is the solvent density, and $\lambda = \exp(\beta\mu)$ is the absolute activity of the ligand. Since Ξ differs from Ξ_l, all the thermodynamics of the two systems will, in general, be different. We are now interested in the conditions under which the two systems have the same BI. Clearly, the $M!$ in Eq. (B.1) that is absent in Eq. (B.2) does not affect the BI of the system.

We now write each of the canonical PFs $Q(i)$ as

$$Q(i) = Q_0 Q_{tr}(i) Q_{rot}(i) Q_{ad}(i) Q_{solv}(i) \tag{B.5}$$

where in Q_0 we have collected all factors that do not depend on i. Here, $Q_{tr}(i)$ is the translational PF, depending on i through the mass of the molecule,

$$Q_{tr}(i) = \frac{V}{\Lambda^3} = V\frac{[2\pi m(i)k_BT]^{3/2}}{h^3} \tag{B.6}$$

where $Q_{tr}(0)$ is the translational PF of the empty molecule. The rotational PF, $Q_{rot}(i)$, depends on i through the moment of inertia, which for the linear two-site system is

$$Q_{rot}(i) = \frac{8\pi^2 I(i) k_B T}{h^2 \sigma(i)} \tag{B.7}$$

where $I(i)$ is the moment of inertia and $\sigma(i)$ is a symmetry factor. For our particular example, say oxalic acid, $\sigma(0) = \sigma(2) = 2$ and $\sigma(1) = 1$. The bound ligands contribute the factors

$$\begin{aligned} Q_{ad}(0) &= 1, \quad Q_{ad}(1) = q_0 \exp(-\beta U), \\ Q_{ad}(2) &= q_0^2 \exp[-\beta 2U - \beta U(1,1)] \end{aligned} \tag{B.8}$$

where q_0 is the internal PF of a single ligand molecule. We assume that this does not change upon binding. The quantity U is the binding energy for a single ligand. Finally, the solvation factor is

$$Q_{solv}(i) = \exp[-\beta \Delta\mu^*(i)] \tag{B.9}$$

where $\Delta\mu^*(i)$ is the solvation Gibbs energy of an adsorbent molecule having i ligands.

When the ligand is very small compared with the adsorbent molecule (say oxygen and hemoglobin, or a proton and a dicarboxylic acid; an exception would be proteins bound to DNA, see Section 5.10), we can write

$$Q_{tr}(i) \approx Q_{tr}(0), \quad Q_{rot}(i) \approx Q_{rot}(0)\sigma(0)/\sigma(i), \quad Q_{solv}(i) \approx Q_{solv}(0) \tag{B.10}$$

If we set

$$q = \exp(-\beta U), \qquad S = \exp[-\beta U(1,1)], \qquad \psi = \exp[-\beta \Delta\mu^*(0)], \qquad \lambda = \lambda_0 C$$

then the GPF of a single *free* adsorbent molecule, Eq. (B.3), assumes the form

$$\xi = Q(0)(1 + 2q\lambda_0 C + q^2 S \lambda_0^2 C^2) \tag{B.11}$$

Under the same assumptions, the GPF of a *localized* adsorbent molecule is

$$\xi_l = Q_l(0)(1 + 2q\lambda_0 C + q^2 S \lambda_0^2 C^2) \tag{B.12}$$

Since the factors $Q(0)$ and $Q_l(0)$ do not affect the BI, the two PFs (B.11) and (B.12) would lead to the same BI, even though $Q(0)$ differs from $Q_l(0)$. Note that the factor 2 in Eq. (B.11) originates from $\sigma(0)/\sigma(1)$, but in Eq. (B.12) it originates from the distinguishability of the two sites (since the molecule is localized).

Thus, whenever the ligand is very small compared with the adsorbent molecule, one may neglect its effect on Q_{tr}, Q_{rot}, and Q_{solv}, except for the symmetry factor $\sigma(i)$. This leads to a BI which is the same as that of a localized system, where the sites are distinguishable. When the ligand is not small, such as proteins binding to DNA, then one cannot neglect its effect on the translational, rotational, or the solvational factors in the PF of a single molecule.

C. TRANSITION FROM MICROSTATES TO MACROSTATES

The fundamental relation between the canonical PF and thermodynamics is

$$Q_{mac} = \exp(-\beta A_{mac}) = \sum_i \exp(-\beta E_i) \tag{C.1}$$

where $\beta = (k_B T)^{-1}$, k_B being the Boltzmann constant and T the absolute temperature; A_{mac} is the Helmholtz energy and E_i are the energy levels of the *macroscopic* system. The sum is over all *states* of the macroscopic system.

For systems consisting of M independent subsystems (such as adsorbent molecules), we can write*

$$Q_{mac} = Q^M \tag{C.2}$$

where Q is the canonical PF of a single subsystem. The latter quantity Q is related to the energy levels of a single subsystem by

$$Q = \sum_i \exp(-\beta \varepsilon_i) \tag{C.3}$$

where ε_i are the energy levels of a single subsystem, and the sum over i in Eq. (C.3) is over all the states of the single subsystem. From now on we focus only on a single subsystem which is the adsorbent molecule. We shall discuss a two-site adsorbent molecule having two macrostates. The generalization to any number of sites and any number of macrostates is quite straightforward.

We shall discuss separately the three occupancy states of the system. The empty state introduces the concept of *free-energy levels.* The singly occupied state introduces the concept of *binding free energy,* and the fully occupied state introduces the concept of ligand–ligand *free energy* of *interaction.* In this book, all these three concepts are reduced to: *energy* levels, binding *energies,* and ligand–ligand interaction *energy.*

*We ignore here factors such as $M!$ that do not affect the BI. See also Appendix B.

(a) *The empty molecule*: Suppose we group all the states of the adsorbent molecules into two groups—referred to as L and H—and rewrite the sum in Eq. (C.3) as

$$Q = \sum_i \exp(-\beta\varepsilon_i^{(0)}) = \sum_{i\in L} \exp(-\beta\varepsilon_i^{(0)}) + \sum_{i\in H} \exp(-\beta\varepsilon_i^{(0)}) = Q_L + Q_H \qquad \text{(C.4)}$$

where Q_L and Q_H are defined on the right-hand side of Eq. (C.4); $\varepsilon_i^{(0)}$ are the energy levels of the empty molecule.

The division of all the states into two groups could, in principle, be arbitrary. In practice, such a division is motivated by the existence of two isomers, such as cis and trans, or helix and coil. We define two *free-energy levels* by

$$\exp(-\beta A_L) = Q_L, \qquad \exp(-\beta A_H) = Q_H \qquad \text{(C.5)}$$

and write the free energy of a single molecule, A as

$$Q = \exp(-\beta A) = \exp(-\beta A_L) + \exp(-\beta A_H) \qquad \text{(C.6)}$$

When the molecule is in a solvent, each state of the molecule might possess a different solvation free energy.

Consider again a single adsorbent molecule in a solvent, at some volume V and temperature T. In this case Eq. (C.4) is replaced by

$$Q(0, 0) = C_w \int \cdots \int \sum_i \exp[-\beta(\varepsilon_i^{(0)} + B_i(0, 0))] \exp(-\beta U_N) d\mathbf{X}^N d\mathbf{X}_\mathbf{P} \qquad \text{(C.7)}$$

where C_w is a constant that includes all the internal PFs of the solvent molecules and factors such as $N!$ and $(8\pi^2)^N$, which will cancel out when we take ratios of PFs to form the binding constants or correlation functions. The quantity $B_i(0, 0)$ is the total binding energy* of the empty adsorbent molecule with all solvent molecules being at some specific configuration $\mathbf{X}^N = \mathbf{X}_1 \ldots \mathbf{X}_N$, i.e.,

$$B_i(0, 0) = \sum_{i=1}^{N} U^{(0)}(\mathbf{X}_i, \mathbf{X}_\mathbf{P}) \qquad \text{(C.8)}$$

where $U^{(0)}$ is the pair interaction between the adsorbent molecule ($\mathbf{P}$) and the ith solvent molecule. The quantity U_N is the total solvent–solvent interaction energy at some specific configuration $\mathbf{X}^N = \mathbf{X}_1 \ldots \mathbf{X}_N$. The integration in (C.7) is over all configurations of the solvent molecules and the adsorbent molecule.

*Not to be confused with the binding energy of a ligand to the adsorbent molecule.

The PF of the *pure* solvent, i.e., all N solvent molecules at the same V and T, but in the absence of the adsorbent molecule, is

$$Q_w = C_w \int \cdots \int \exp(-\beta U_N) d\mathbf{X}^N \tag{C.9}$$

Hence, the PF of an empty adsorbent molecule in a solvent can be written as

$$Q(0,0) = Q_w \int \cdots \int \sum_i \exp[-\beta \varepsilon_i^{(0)} - \beta B_i(0,0)] P_0(\mathbf{X}^N) d\mathbf{X}^N d\mathbf{X}_\mathbf{P} \tag{C.10}$$

where $P_0(\mathbf{X}^N)$ is the density distribution of solvent molecules in the absence of $\mathbf{P}$ and given by

$$P_0(\mathbf{X}^N) = \frac{\exp(-\beta U_N)}{\int \cdots \int \exp(-\beta U_N) d\mathbf{X}^N} \tag{C.11}$$

In the two-macrostate approximations, say cis and trans isomers, we assume that $B_i(0, 0)$ can have only two values,

$$B_i(0,0) = \begin{cases} B_L(0,0) \text{ for } i \in L \\ B_H(0,0) \text{ for } i \in H \end{cases} \tag{C.12}$$

in which case $Q(0, 0)$ may be written as

$$\begin{aligned} Q(0,0) &= Q_w Q_L \int \cdots \int \exp[-\beta B_L(0,0)] P_0(\mathbf{X}^N) d\mathbf{X}^N d\mathbf{X}_\mathbf{P} \\ &\quad + Q_w Q_H \int \cdots \int \exp[-\beta B_H(0,0)] P_0(\mathbf{X}^N) d\mathbf{X}^N d\mathbf{X}_\mathbf{P} \\ &= Q_w Q_L \psi_L^{(0)} + Q_w Q_H \psi_H^{(0)} \end{aligned} \tag{C.13}$$

where

$$\psi_L^{(0)} = \exp[-\beta \Delta\mu_L^*(0,0)], \qquad \psi_H^{(0)} = \exp[-\beta \Delta\mu_H^*(0,0)]$$

and $\Delta\mu_\alpha^*(0, 0)$ is the solvation Helmholtz energy of the empty molecule in the macrostate α.

In the majority of this book, except for Chapter 9, we treat A_L and A_H defined in Eq. (C.5) as if they were *energy levels,* although they are in general *free-energy* levels. We do that to stress the emergence of free energies even when we start with *energies*. For instance, the two terms in Eq. (C.13) correspond to the two *free-energy* levels of an adsorbed molecule in a solvent. On the other hand, equations such as (4.5.2) correspond to a system with two *energy* levels.

(b) *The singly occupied molecule*: For the two-site molecule, in the absence of a solvent, the binding process is

$$(0,0) + \mathbf{L} \rightarrow (1,0) \tag{C.14}$$

The corresponding binding *free energy* is defined by

$$\Delta A_B = -k_B T \ln\left[\frac{\Sigma_i \exp(-\beta\varepsilon_i^{(1)})}{\Sigma_i \exp(-\beta\varepsilon_i^{(0)})\Sigma_i \exp(-\beta\varepsilon_i^{\mathrm{L}})}\right] \tag{C.15}$$

For simplicity, we assume that the internal PF of the ligand does not change upon binding, and that all the molecules involved in the process (C.14) are devoid of translational degrees of freedom (localized molecules).

In the two-macrostate approximation we write

$$\Delta A_B = -k_B T \ln\left[\frac{\exp(-\beta A_L - \beta U_L) + \exp(-\beta A_H - \beta U_H)}{\exp(-\beta A_L) + \exp(-\beta A_H)}\right] \tag{C.16}$$

where A_L and A_H are defined as in Eqs. (C.5). The binding free energies to L and H are defined similarly to (C.15), but the sums are over $i \in L$ or $i \in H$, respectively. As in (C.15), both U_L and U_H are, in general, *free energies* of binding. In (C.16) we treat U_L and U_H as if they were *binding energies.*

We see from Eq. (C.16) that even when U_L and U_H are assumed to be *energies,* the overall binding process (C.14) produces a binding *free energy*, i.e.,

$$\Delta A_B = -k_B T \ln[X_L^0 \exp(-\beta U_L) + X_H^0 \exp(-\beta U_H)] \tag{C.17}$$

The PF of a singly occupied molecule is written as

$$Q(1, 0) = Q_L q_L + Q_H q_H \tag{C.18}$$

where

$$q_L = \exp(-\beta U_L), \qquad q_H = \exp(-\beta U_H)$$

If the system is inserted in a solvent, then Eq. (C.18) must be replaced by

$$Q(1, 0) = Q_w(Q_L q_L \psi_L^{(1)} + Q_H q_H \psi_H^{(1)}) \tag{C.19}$$

where $\psi_L^{(1)}$ and $\psi_H^{(1)}$ are related to the solvation free energies $\Delta\mu_L^*(1, 0)$ and $\Delta\mu_H^*(1, 0)$, respectively.

(c) *The doubly occupied molecule*: Assuming, for simplicity, that the two sites are identical and that the internal PF of the ligand does not change upon binding, then the general expression for the ligand–ligand free energy of interaction can be expressed as

$$\Delta A(1, 1) = -k_B T \ln\left[\frac{\Sigma_i \exp(-\beta\varepsilon_i^{(2)})\Sigma_i \exp(-\beta\varepsilon_i^{(0)})}{[\Sigma_i \exp(-\beta\varepsilon_i^{(1)})]^2}\right] \tag{C.20}$$

In the two-macrostate approximation, we write each of the PFs in Eq. (C.20) as

$$\sum_i \exp(-\beta\varepsilon_i^{(2)}) = \exp[-\beta A_L - \beta 2U_L - \beta U_L(1,1)] + \exp[-\beta A_H - \beta 2U_H - \beta U_H(1,1)]$$

$$= Q_L q_L^2 S_L + Q_H q_H^2 S_H \tag{C.21}$$

$$\sum_i \exp(-\beta\varepsilon_i^{(1)}) = Q_L q_L + Q_H q_H \tag{C.22}$$

and

$$\sum_i \exp(-\beta\varepsilon_i^{(0)}) = Q_L + Q_H \tag{C.23}$$

Clearly, even when we assume that A_α, U_α, and $U_\alpha(1, 1)$ are true *energies* ($\alpha = L, H$), the quantity $A(1, 1)$ defined in Eq. (C.20) is a *free energy.*

Applying similar arguments as before, we can write the PF of the doubly occupied molecule in a solvent as

$$Q(1,1) = Q_w(Q_L q_L^2 S_L \psi_L^{(2)} + Q_H q_H^2 S_H \psi_H^{(2)}) \tag{C.24}$$

where

$$\psi_L^{(2)} = \exp[-\beta\Delta\mu_L^*(1,1)] \quad \text{and} \quad \psi_H^{(2)} = \exp[-\beta\Delta\mu_H^*(1,1)]$$

D. FIRST-ORDER CORRECTION TO NONIDEALITY OF THE LIGAND'S RESERVOIR

The general form of the BI is derived from Eq. (2.1.2),

$$\bar{N} = \lambda \frac{\partial \ln \Xi}{\partial \lambda} \tag{D.1}$$

where $\bar{N}$ is the average number of bound ligands on all of the M adsorbent molecules and λ is the absolute activity of the ligand. In most theoretical treatments of the BI two fundamental assumptions were made:

1. The system is very dilute with respect to the adsorbent molecules **P**, hence the GPF of the macroscopic system may be written as

$$\Xi = \xi^M / M! \tag{D.2}$$

where ξ is the GPF of a single adsorbent molecule,

$$\xi = \sum_{l=0}^{m} Q(l)\lambda^l \tag{D.3}$$

2. The reservoir from which the ligands are supplied is maintained at a constant chemical potential $\mu = k_BT \ln \lambda$, and the ligand is very dilute, either in a gaseous phase or in a solvent. In both cases the assumption is made that $\lambda = \lambda_0 C$, where C is the concentration of the ligand in the reservoir and λ_0 is independent of C. With these two assumptions one derives the general form of the equation $\theta = \theta(C)$.

We now examine a first-order deviation from the form of the BI due to nonideality of the reservoir.

First, consider the case of a reservoir consisting of a pure ligand at a chemical potential μ. The general form of the chemical potential in this system is*

$$\mu = k_BT \ln C\Lambda_L^3 q_L^{-1} - k_BT \int_0^C \frac{G}{1+C'G}\, dC' \tag{D.4}$$

where Λ_L^3 and q_L are the momentum and the internal PF of a single ligand molecule, while G is defined by

$$G = \int_0^\infty [g_{LL}(R) - 1] 4\pi R^2 dR \tag{D.5}$$

where $g_{LL}(R)$ is the ligand–ligand pair correlation function in the reservoir.

Since we still assume that the adsorbent molecules are independent, hence Eq. (D.2) is valid, the BI per molecule is

$$n = \lambda \frac{\partial \ln \xi}{\partial \lambda} = \frac{\Sigma_{l=0}^m lQ(l)\lambda^l}{\Sigma_{l=0}^m Q(l)\lambda^l} \tag{D.6}$$

One can define intrinsic binding constants in the same way as before, e.g., Eq. (2.2.25), but now these constants will depend on the ligand concentration C. The relation (D.4) gives an implicit dependence of the absolute activity $\lambda = \exp(\beta\mu)$ on the ligand concentration C. However, since the analytical dependence of G on C is not known, one cannot write the explicit function $\lambda = \lambda(C)$. This may be done to first order in C. Note that when $C \to 0$ the integral on the rhs of Eq. (D.4) is zero, and we have the ideal gas limit of the chemical potential.† If we expand the integral to first order in C, we obtain the first-order deviation with respect to an ideal gas,

*For more details, see Section 3.10 in Ben-Naim (1987) or Chapter 6 in Ben-Naim (1992). Here, we discuss only nonionic ligands.

†Note that at $C = 0$, $\mu = -\infty$. Here, we need only the limiting behavior when $CG \ll 1$. This is the ideal gas limit.

i.e.,

$$\lambda = \exp(\beta\mu) = \Lambda_L^3 q_L^{-1} C \exp(-CG^0)$$

$$= \Lambda_L^3 q_L^{-1} Cf(C) = \lambda_0 Cf(C) \tag{D.7}$$

The BI per adsorbent molecule is thus

$$n = \frac{\Sigma l Q(l)[\lambda_0 Cf(C)]^l}{\Sigma Q(l)[\lambda_0 Cf(C)]^l} \tag{D.8}$$

where $f(C) = \exp(-CG^0)$, and G^0 is related to the second-virial coefficient by*

$$G^0 = \int_0^\infty \{\exp[-\beta U_{LL}(R)] - 1\} 4\pi R^2 dR = -2B_2(T) \tag{D.9}$$

Equation (D.8) gives an explicit form of the BI, i.e., $\bar{n}$ as a function C, for a nonideal reservoir.

The second case is when the reservoir is a solution of the ligand in a solvent, say water. The form of the BI is the same, except for a reinterpretation of the quantity G^0. In Eq. (D.7), G^0 is an integral over the pair correlation function of the ligands in vacuum, i.e., $U_{LL}(R)$ in Eq. (D.9) is the ligand–ligand pair potential. In the case of a small deviation from an ideal-dilute solution, G^0 in Eq. (D.7) is replaced by

$$G^0 = \int_0^\infty [g_{LL}^0(R) - 1] 4\pi R^2 dR \tag{D.10}$$

where now $g_{LL}^0(R)$ is the ligand–ligand pair correlation function in an infinite dilute solution of **L** in a solvent.

When the adsorbent molecules are not independent, we can no longer use the relation (D.2) for the GPF of the system. In this case, we must start from the GPF of the macroscopic system from which we can derive the general form of the BI for any concentration of the adsorbent molecule. The derivation is possible through the McMillan–Mayer theory of solution, but it is long and tedious, even for first-order deviations from an ideal solution. The reason is that, in the general case, the first-order deviations would depend on many second-virial coefficients [the analogue of the quantity $B_2(T)$ in Eq. (D.9)]. For each pair of occupancy states, say i and j, there will be a pair potential $U_{PP}(R, i, j)$, and the corresponding second-virial coefficient

$$B_2(T, i, j) = \int_0^\infty \{\exp[-\beta U_{PP}(R, i, j)] - 1\} 4\pi R^2 dR \tag{D.11}$$

*See Chapters 5 and 6 in Ben-Naim (1992).

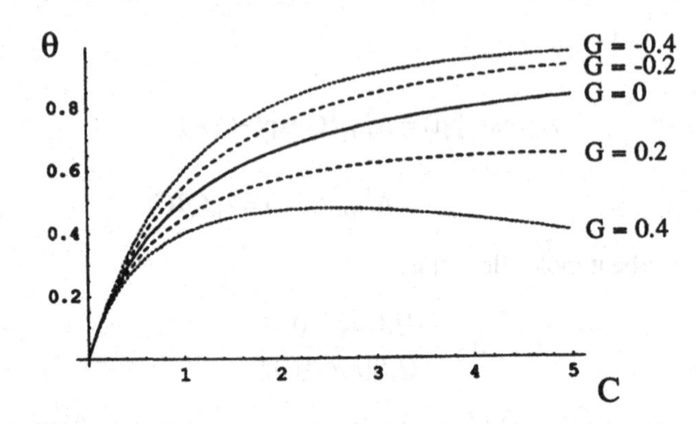

Figure D.1. BIs defined by Eq. (D.8) with different values of G, as indicated next to each curve.

These quantities will affect the absolute activities of all the adsorbent molecules, and hence also the general form of the BI.

In both cases the GPF formalism can, in principle, give the BI of a nonideal system, with respect to either the adsorbent or the ligand molecules. This is not possible if one uses the phenomenological approach as described in Section 2.3.

Figure D.1 shows the BI for a system with $k_1 = 1$, $S = 1$, with different values of G [in Eq. (D.8)]. It is seen from the figure that when $G > 0$ the BI is below the curve for $G = 0$, and when $G < 0$ the BI is above the curve for $G = 0$. Clearly, if we are unaware of the nonideality of the reservoir, we might misinterpret the deviation from the Langmuir curves as due to cooperativities although the system, by construction, is noncooperative.

E. RELATIVE SLOPES OF EQUILIBRATED AND "FROZEN-IN" BIs IN A MULTIMACROSTATE SYSTEM

Consider a binding system with any number of sites and any number of macrostates. The equilibrated and "frozen-in" BIs are

$$\theta^{eq} = \sum_{\alpha} X_{\alpha}^{eq}\theta_{\alpha} \quad \text{and} \quad \theta^{f} = \sum_{\alpha} X_{\alpha}^{f}\theta_{\alpha} \tag{E.1}$$

where the sum is over all macrostates α [a special case where $\alpha = L, H$ are Eqs. (3.5.6) and (3.5.7)]; X_{α}^{eq} is the equilibrium mole fraction of the macrostate α, defined by

$$X_{\alpha}^{eq} = \frac{M_{\alpha}}{M} = \frac{\xi_{\alpha}}{\xi} \tag{E.2}$$

First we show that if we start with an equilibrated system and "freeze-in" the conversion between the species α, then, at that point, the slope of the curve $\theta^{eq}(\lambda)$ is larger than the slope of the curve $\theta^{f}(\lambda)$,

$$\frac{\partial(\theta^{eq}-\theta^{f})}{\partial\lambda}=\sum_{\alpha}\theta_{\alpha}\frac{\partial X_{\alpha}^{eq}}{\partial\lambda}+\sum_{\alpha}(X_{\alpha}^{eq}-X_{\alpha}^{f})\frac{\partial\theta_{\alpha}}{\partial\lambda} \tag{E.3}$$

Clearly, since $X_{\alpha}^{eq}=X_{\alpha}^{f}$ at the point we "froze-in" the equilibrium, the second term on the rhs of Eq. (E.3) is zero. Since $\Sigma X_{\alpha}^{eq}=1$, hence $\Sigma dX_{\alpha}^{eq}=0$, we can rewrite Eq. (E.3), for any θ, as

$$\frac{\partial(\theta^{eq}-\theta^{f})}{\partial\lambda}=\sum_{\alpha}(\theta_{\alpha}-\theta)\frac{\partial X_{\alpha}^{eq}}{\partial\lambda} \tag{E.4}$$

On the other hand, from Eq. (E.2), we have

$$\frac{\partial X_{\alpha}^{eq}}{\partial\lambda}=X_{\alpha}^{eq}\left(\frac{\partial\ln\xi_{\alpha}/\xi}{\partial\lambda}\right)=\frac{X_{\alpha}^{eq}}{\lambda}(\theta_{\alpha}-\theta^{eq}) \tag{E.5}$$

Hence, the difference in the slopes of the two curves is

$$\frac{\partial\theta^{eq}}{\partial\lambda}-\frac{\partial\theta^{f}}{\partial\lambda}=\sum_{\alpha}\frac{(\theta_{\alpha}-\theta^{eq})^{2}X_{\alpha}^{eq}}{\lambda}\geq 0 \tag{E.6}$$

where, on the rhs of Eq. (E.6), we put $\theta=\theta^{eq}$. For a two-state case, discussed in Section 3.5, $\alpha=L, H$, Eq. (E.6) reduces to

$$\frac{\partial\theta^{eq}}{\partial\lambda}-\frac{\partial\theta^{f}}{\partial\lambda}=\frac{X_{L}^{eq}X_{H}^{eq}}{\lambda}(\theta_{L}-\theta_{H})^{2}\geq 0 \tag{E.7}$$

Thus, if we "freeze-in" the equilibrium at any point along the binding process and then continue the process of binding, the slope of the equilibrated curve will always be steeper than that of the "frozen-in" curve.

Note that if we start with $X_{L}^{f}=X_{L}^{0}$ and follow the two curves θ^{eq} and θ^{f}, the difference in the slopes may change sign. In this case,

$$\frac{\partial\theta^{eq}}{\partial\lambda}-\frac{\partial\theta^{f}}{\partial\lambda}=(\theta_{L}-\theta_{H})\frac{\partial X_{L}^{eq}}{\partial\lambda}+(X_{L}^{eq}-X_{L}^{0})\left(\frac{\partial\theta_{L}}{\partial\lambda}-\frac{\partial\theta_{H}}{\partial\lambda}\right) \tag{E.8}$$

and this does not have a permanent sign.

In the special case of single-site molecules $\theta^{eq}-\theta^{f}$ is always positive, i.e.,

θ^{eq} is always above the curve θ^f; indeed

$$\theta^{eq} - \theta^f = \frac{X_L^0 X_H^0 (q_H - q_L)^2 \lambda^2}{(1 + q_H\lambda)(1 + q_L\lambda)(1 + q_H X_H^0 \lambda + q_L X_L^0 \lambda)} \geq 0 \tag{E.9}$$

This cannot be guaranteed for systems with more than one site.

F. SPURIOUS COOPERATIVITY IN SINGLE-SITE SYSTEMS

There are several directions along which one can generalize the concept of spurious cooperativity discussed in Section 3.5. (1) Instead of equal mole fractions $X_L^0 = X_H^0 = 1/2$, one can start with any composition of the two components. (2) Instead of two states L and H, one can have a system with three or more states. (3) Instead of two occupancy states, empty and occupied by a single ligand, one can allow two or more ligands on the same site or on the same adsorbent molecule.

We shall discuss here the first two generalizations.

1. *The case of* $X_L^0 = 2/3$, $X_H^0 = 1/3$: In Sections 3.5 and 4.4 we discussed the case where the initial concentration of the two forms L and H are equal. We found that such a mixture of two components L and H shows a BI indistinguishable from a two-site system with negative cooperativity. We note here that the system discussed in Section 3.5 was obtained by "freezing-in" an equilibrium between two states $L \rightleftharpoons H$. We have pointed out that such a system can have a large negative spurious cooperativity [depending on the difference $(q_L - q_H)^2$] in spite of the fact that $X_L^0 = X_H^0 = 1/2$ (see the footnote on p. 65 of Section 3.5). However, the phenomenon of spurious cooperativity can occur in more general systems of any mixture of two (or more) components, not necessarily derived from an equilibrated system such as $L \rightleftharpoons H$.

We discuss first a simple generalization of the case discussed in Sections 3.5 and 4.4, and then proceed with the more general case. We shall continue to use the notations L and H, but these refer now to any two *different* adsorbent molecules.

Consider a system of $3M$ molecules, $2M$ of which are L with a binding constant k_L and M of which are H with a binding constant k_H. The BI of such a system is (see Section 2.5)

$$\theta = \frac{1}{3}\frac{k_H C}{1 + k_H C} + \frac{2}{3}\frac{k_L C}{1 + k_L C}$$

$$= \frac{1}{3}\frac{(2k_L + k_H)C + 3k_L k_H C^2}{1 + (k_L + k_H)C + k_L k_H C^2} \tag{F.1}$$

This system is not equivalent to a double-site molecule (as discussed in Section 4.4). Instead, we show that the following three systems are equivalent, in the sense of having the *same* BI (Fig. F.1).

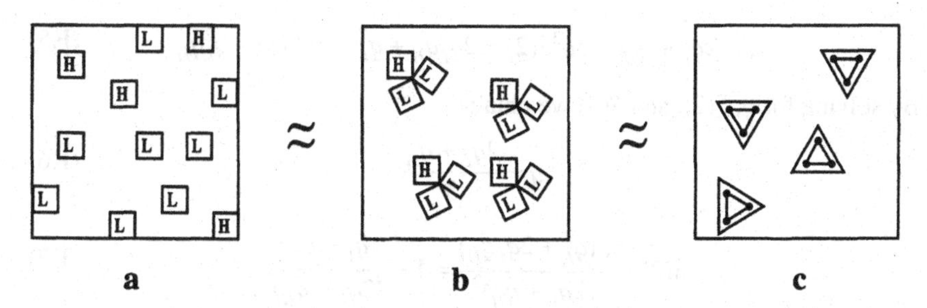

Figure F.1. Schematic illustration of the three equivalent systems **a**, **b**, and **c** corresponding to Eqs. (F.2), (F.3), and (F.4). In **a** and **b** all the sites are independent. In **c** the three sites on each molecule are correlated by pair and triplet correlations.

(**a**) A system of $3M$ *independent* single-site molecules, $2M$ of which are of type L with a binding constant k_L and M of which are of type H with a binding constant k_H. The corresponding GPF is

$$\Xi_{\mathbf{a}} = \frac{(1 + q_L\lambda)^{2M}(1 + q_H\lambda)^M Q_L^{2M} Q_H^M}{(2M)!M!} \tag{F.2}$$

(**b**) A system of M *independent* triple-site molecules. The three sites are *different* and *independent:* two sites with binding constant k_L and one with binding constant k_H. The corresponding GPF is

$$\Xi_{\mathbf{b}} = \frac{[1 + (2q_L + q_H)\lambda + (2q_L q_H + q_L^2)\lambda^2 + q_L^2 q_H \lambda^3]^M Q_b^M}{M!} \tag{F.3}$$

(**c**) A system of M *independent* triple-site molecules. The three sites are *identical* but *dependent,* with the same intrinsic binding constant k $(= q\lambda_0)$, *pair* correlation S(2), and *triplet* correlation S(3). The corresponding GPF is

$$\Xi_{\mathbf{c}} = \frac{[1 + 3q\lambda + 3q^2 S(2)\lambda^2 + q^3 S(3)\lambda^3]^M Q_c^M}{M!} \tag{F.4}$$

The three systems are shown schematically in Fig. F.1. Note again that the three systems are different. However, they have the same BI (factors like Q^M and $M!$ do not affect the BI). The equivalence between **a** and **b** is again trivial. Since in **a** we have 3M independent *sites,* it does not matter how we group them as long as they are still independent, both within the groups and between the groups. The equivalence between **b** and **c** can be obtained by requiring that the two polynomials of degree 3 in λ [Eqs. (F.3) and (F.4)] be identical, i.e., we require the equality of the three coefficients:

$$3q = 2q_L + q_H,\quad 3q^2S(2) = 2q_Lq_H + q_L^2,\quad q^3S(3) = q_L^2q_H \tag{F.5}$$

By solving for q, $S(2)$, and $S(3)$ we obtain

$$q = \frac{2q_L + q_H}{3} \tag{F.6}$$

$$S(2) = \frac{3(q_L^2 + 2q_Lq_H)}{(2q_L + q_H)^2} = 1 - \frac{(q_L - q_H)^2}{(2q_L + q_H)^2} \tag{F.7}$$

$$S(3) = \frac{27q_L^2q_H}{(2q_L + q_H)^3} = 1 - \frac{(q_L - q_H)^2(8q_L + q_H)}{(2q_L + q_H)^2} \tag{F.8}$$

with these transformations, the BIs derived from Eqs. (F.3) and (F.4) are identical.

The most interesting finding is that both $S(2)$ and $S(3)$ in Eq. (F.4) are always smaller than unity, i.e., the system **c** has both pair and triplet *negative* cooperativity. Again from F.7 and F.8 it follows that $S(2) = S(3) = 1$ if and only if $q_L = q_H$. Clearly, these negative cooperativities that are *genuine* in system **c** are only *spurious* in system **a.**

Figure 3.5 shows some binding isotherms for the cases $X_L^0 = 1/2$, $X_L^0 = 2/3$, and $X_L^0 = 3/4$, and for fixed values of $q_L = 1$ and $q_H = 1000$. Note that the slope-curves have only two maxima, corresponding to the maximum slope of θ_L and θ_H. Judging only from the location of the two peaks in Fig. 3.5b, we might conclude that there is only pairwise cooperativity. However, fitting the experimental data to a triple-site BI will reveal also triplet correlation, which is genuine if the system is **c** but spurious if the system is **a.** Similarly, for the case $X_L^0 = 3/4$ and $X_H^0 = 1/4$, we find $S(2) \le 1$, $S(3) \le 1$, and $S(4) \le 1$, i.e., negative pair, triplet, and quadruplet correlations.

2. *The general case*: Consider a system of M independent and single-site molecules, $M_L = X_LM$ of which are of type L and $M_H = X_HM$ of which are of type H, with $X_L + X_H = 1$, and $X_L = r/t$ where r and t are integers ($r < t$).* Since we are interested only in the BI of the system, it is convenient to replace this system with a system of tM molecules (or, equivalently, t independent systems each having M molecules), rM of which are L and $(t - r)M$ are H. The GPF for the three systems **a**, **b**, and **c** are

$$\Xi_{\mathbf{a}} = \frac{[Q_L^r(1 + q_L\lambda)^r]^M}{(rM)!}\,\frac{[Q_H^{(t-r)}(1 + q_H\lambda)^{(t-r)}]^M}{[(t - r)M]!} \tag{F.9}$$

$$\Xi_{\mathbf{b}} = \frac{Q(0)^M}{M!}\left[\sum_{l=0}^{t}\sum_{i+j=l}\binom{r}{i}\binom{t-r}{j}q_L^iq_H^j\lambda^{i+j}\right]^M \tag{F.10}$$

*Note that X_L and X_H are, by definition, rational numbers, hence we can assume that r and t are integers.

$$\Xi_c = \frac{Q(0)^M}{M!}\left[1 + tq\lambda + \sum_{l=2}^{t}\binom{t}{l} q^l S(l)\lambda^l\right]^M \tag{F.11}$$

To prove our assertion that these three systems are equivalent (from the standpoint of having the same BI), we ignore the factors Q and the factorials, and show the identity of the polynomials in λ only. [Physically, this corresponds to a system of localized particles having no internal degrees of freedom, i.e., all Qs are unity, and no factorials in the denominators of expressions (F.9), (F.10), and (F.11).]

The equivalence of **a** and **b** is straightforward. By expanding the two factors in the numerator of Eq. (F.9) and collecting coefficients of the same power of λ, we obtain Eq. (F.10). The equivalence of **b** and **c** is obtained by requiring that the two polynomials of degree t be identical (hence leading to the same BI). This is achieved by imposing the equalities

$$tq = rq_L + (t-r)q_H \tag{F.12}$$

and, for $l \geq 2$,

$$\binom{t}{l} q^l S(l) = \sum_{i+j=l}\binom{r}{i}\binom{t-r}{i} q_L^i q_H^j \tag{F.13}$$

where q is an average of q_L and q_H given by

$$q = \frac{r}{t} q_L + \frac{(t-r)}{t} q_H$$

We obtain the general expression for $S(l)$ by expanding q^l and collecting coefficients with the same powers of $q_L q_H$ (note that $r < t$ and $2 \leq l < t$):

$$S(l) = \frac{\sum_{i=0}^{l}\binom{r}{i}\binom{t-r}{l-i}\binom{t}{l}^{-1} q_L^i q_H^{l-i}}{\sum_{i=0}^{i}\binom{l}{i}\left(\frac{r}{t}\right)^i\left(\frac{t-r}{t}\right)^{l-i} q_L^i q_H^{l-i}} \tag{F.14}$$

When Eqs. (F. 12) and (F.13) are substituted into Eq. (F.11), the systems **b** and **c** become equivalent.

It is next shown that all the cooperativities in system **c** are negative, i.e., for any t, r, and l we have $S(l) \leq 1$.

Extracting q_L^l and setting $h = q_H/q_L$, we rewrite Eq. (F.14) as

$$S(l) = \frac{\sum P_i h^i}{\{r/t + [(t-r)/t]h\}^l} \tag{F.15}$$

It is easy to show that for any t and r, $S(2)$ has the form

$$S(2) = 1 - \frac{(h-1)^2 r(t-r)}{(t-r+hr)^2(t-1)} \tag{F.16}$$

Since $t > r \geq 1$, we have $S(2) \leq 1$ for any t and r [$S(2) = 1$ for $h = 1$].

To prove that $S(l) \leq 1$ for any l, we first note that $S(l) = 1$ for $h = 1$ and any l, t, r. This follows from the fact that P_i is the hypergeometric distribution, hence $\Sigma P_i = 1$. Since $S(l)$ is symmetric with respect to q_L and q_H, it is sufficient to examine the case $q_H > q_L$, i.e., $h > 1$. (If $q_L > q_H$, we may redefine h as q_L/q_H and proceed with the same proof.)

It can easily be shown that $S(l)$ has a maximum as a function of h at $h = 1$. This follows from

$$\left.\frac{\partial S(l)}{\partial h}\right|_{h=1} = 0, \qquad \left.\frac{\partial^2 S(l)}{\partial h^2}\right|_{h=1} = \frac{-(t-r)(l-1)rl}{t^2(t-1)} \tag{F.17}$$

since $t > r \geq 1$ and $l \geq 2$. The second derivative at $h = 1$ is negative. It is also easily shown, after some lengthy algebra [which requires taking the derivative of $S(l)$ in Eq. (F.14) with respect to h, and using well-known expressions for the mean and variance of the hypergeometric distribution), that $S(l, h)$ is a monotonically decreasing function of h for $h > 1$, and a monotonically increasing function of h for $0 < h < 1$. Hence $S(l)$ has only a single maximum at $h = 1$. Figure F.2 shows the typical form of $S(l)$, drawn for the case $t = 2$, $r = 1$, and $l = 2$.

3. *Mixture of three different single-site molecules*: In the previous examples we discussed two-state systems of L and H. These could be either a mixture of two components, or a mixture obtained by "freezing-in" an equilibrium between two states. We extend the discussion to three states, denoted by L, H, and T, with corresponding binding constants k_L, k_H, and k_T.

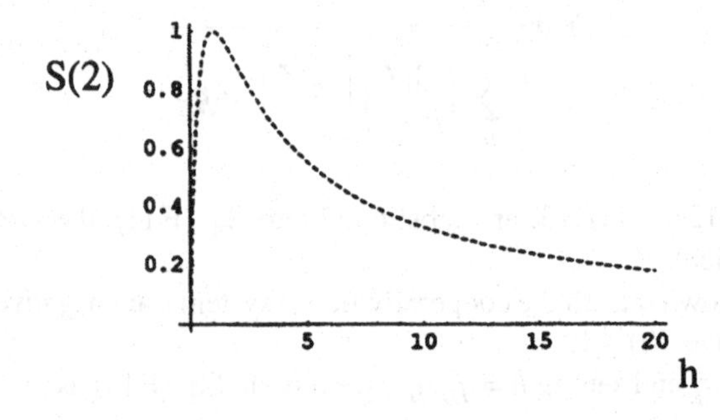

Figure F.2. The form of the spurious pair correlation function $S(2)$ as a function of $h = q_H/q_L$ for the case $t = 2$, $r = 1$.

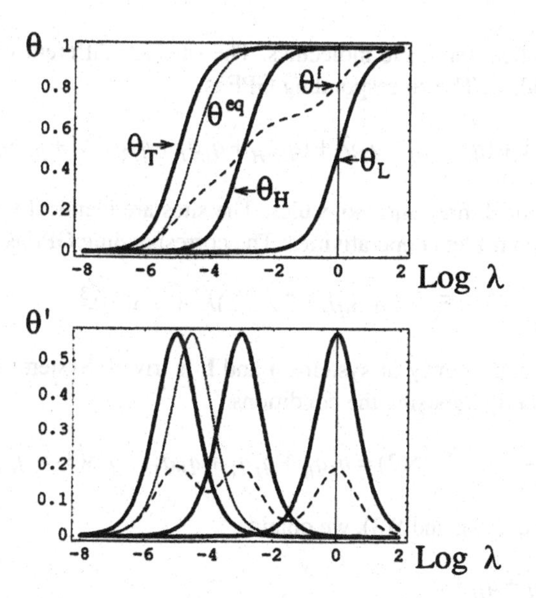

Figure F.3. The BIs θ_L, θ_H, and θ_T, the equilibrated and "frozen-in" BIs, for a system with $q_L = 1$, $q_H = 10^3$, and $q_T = 10^5$.

Figure F.3 shows θ_L, θ_H, and θ_T for such a system with parameters $q_L = 1$, $q_H = 10^3$, and $q_T = 10^5$. The BI of the mixture is defined by

$$\theta^f = \frac{1}{3}(\theta_L + \theta_H + \theta_T) \tag{F.18}$$

We also show the "equilibrated" BI (although the system is not necessarily derived from an equilibrium mixture at three states), defined by

$$\theta^{eq} = X_L^{eq}\theta_L + X_H^{eq}\theta_H + X_T^{eq}\theta_T \tag{F.19}$$

where X_i^{eq} are obtained from the GPF of the system

$$\xi = Q_L + Q_H + Q_T + (Q_L q_L + Q_H q_H + Q_T q_T)\lambda \tag{F.20}$$

Note that while θ_L, θ_H, and θ_T (as well as θ^{eq}, if an equilibrium between the three states exists) are simple Langmuir isotherms, the BI of the mixture θ^f is not.

It is easy to show that the following three systems are equivalent (in the sense of having the same BI).

(a) A mixture of $3M$ single-site molecules, M of which are of type L, M of type H, and M of type T. Ignoring factors like Q^M and $M!$, the GPF is

$$\Xi_a = (1 + q_L\lambda)^M(1 + q_H\lambda)^M(1 + q_T\lambda)^M \tag{F.21}$$

(**b**) M identical three-site molecules. The sites are different with q_L, q_H, and q_T, but independent. The corresponding GPF is

$$\Xi_{\mathbf{b}} = 1 + (q_L + q_H + q_T)\lambda + (q_L q_H + q_L q_T + q_H q_T)\lambda^2 + q_L q_H q_T \lambda^3 \quad \text{(F.22)}$$

(**c**) M identical three-site molecules. The sites are identical with the same q, but with pair and triplet cooperativities. The corresponding GPF is

$$\Xi_{\mathbf{c}} = 1 + 3q\lambda + 3q^2 S(2)\lambda^2 + q^3 S(3)\lambda^3 \quad \text{(F.23)}$$

As before, the equivalency of systems **a** and **b** is trivial. Systems **b** and **c** can be made equivalent by imposing the conditions

$$3q = q_L + q_H + q_H, \quad 3q^2 S(2) = q_L q_H + q_L q_T + q_H q_T, \quad q^3 S(3) = q_L q_H q_T \quad \text{(F.24)}$$

By solving for q, $S(2)$, and $S(3)$, we obtain

$$q = \frac{1}{3}(q_L + q_H + q_T)$$

$$S(2) = \frac{3(q_L q_H + q_L q_T + q_H q_T)}{(q_L + q_H + q_T)^2} = 1 - \frac{1}{2}[(q_H - q_L)^2 + (q_L - q_T)^2 + (q_H - q_T)^2]$$

$$S(3) = \frac{27 q_L q_H q_T}{(q_L + q_H + q_T)^3} = 1 - \frac{1}{2}[(q_H - q_L)^2(7q_T + q_L + q_H)$$

$$+ (q_L - q_T)^2(7q_H + q_L + q_T) + (q_H - q_T)^2(7q_L + q_H + q_T)] \quad \text{(F.25)}$$

Thus, q is the arithmetic average of $q_L + q_H + q_T$, and both $S(2)$ and $S(3)$ are smaller than unity, i.e., the spurious pair and triplet cooperativities are negative. Clearly, when one pair of the qs are identical, this system reduces to the system discussed earlier (in subsection 2). The system will show no spurious cooperativities if and only if $q_L = q_H = q_T$.

G. THE RELATION BETWEEN THE BINDING ISOTHERM AND THE TITRATION CURVE FOR TWO-SITE SYSTEMS

In Section 2.6 we derived a relation between the binding isotherm $\theta = \theta([H])$ and the titration curve $N_B = N_B([H])$, where $[H]$ is the proton concentration. This relation, for the one-site system, is

$$N_B(h) = 1 - \theta(h) - h + K_w/h \quad \text{(G.1)}$$

where, for simplicity of notation, we used $h = [H]$.* Clearly, when $3 \lesssim pH \lesssim 11$ the curves $N_B(h)$ and $1 - \theta(h)$ are nearly identical.† Figure 2.5 shows $N_B(h)$ and $1 - \theta(h)$ for acetic acid, $K_{diss} = 10^{-4.757} = 1.75 \times 10^{-5}$. Measuring K_{diss} either on the titration curve (at $N_B = 1/2$) or on the BI (at $\theta = 1/2$) would give nearly identical results. For dicarboxylic acids, the titration curve is determined by the five equations

$$\left.\begin{aligned} &K_w = [H][OH] = 10^{-14} \\ &K_{1diss} = [HA][H]/[H_2A], \qquad K_{2diss} = [A][H]/[HA] \\ &N_T = [A] + [HA] + [H_2A], \quad 2[A] + [OH] + [HA] = [H] + [N_B] \end{aligned}\right\} \tag{G.2}$$

These should be compared with the four equations (2.6.4) in Section 2.6.

When N_B is solved as a function of $h = [H]$ (with $N_T = 1$), we obtain the titration curve for dicarboxylic acid in terms of the dissociation constants:

$$N_B^{(2)}(h) = \frac{K_w(h^2 + K_{1diss}h + K_{1diss}K_{2diss}) - h^4 + K_{1diss}h^2 - K_{1diss}h^3 + K_{1diss}K_{2diss}(2-h)h}{(h^2 + K_{1diss}h + K_{1diss}K_{2diss})h} \tag{G.3}$$

The corresponding BI in terms of K_{1diss} and K_{2diss} is

$$\theta^{(2)}(h) = \frac{K_{1diss}h + 2h^2}{2(h^2 + K_{1diss}h + K_{1diss}K_{2diss})} \tag{G.4}$$

The two functions are related by

$$N_B^{(2)}(h) = 2 - 2\theta^{(2)}(h) - h + K_w/h \tag{G.5}$$

Again, the difference between $N_B^{(2)}(h)$ and $2 - \theta^{(2)}(h)$ is negligible for $3 \lesssim pH \lesssim 11$.

Figure G.1 shows the titration curve $N_B(h)$ and the BI [plotted as $2 - 2\theta(h)$], with $h = [H]$ for $\alpha - \alpha'$-di-tert-butyl succinic acid in 50% ethanol–water solution. Note that the locations of the two peaks of the derivatives of these curves are nearly identical for the two curves.

*Not to be confused with $h = q_H/q_L$ defined in the rest of the book. We use h for the proton concentration only in this appendix.

†However, these are clearly not identical functions as referred to in some publications [e.g., Wyman and Gill (1990)].

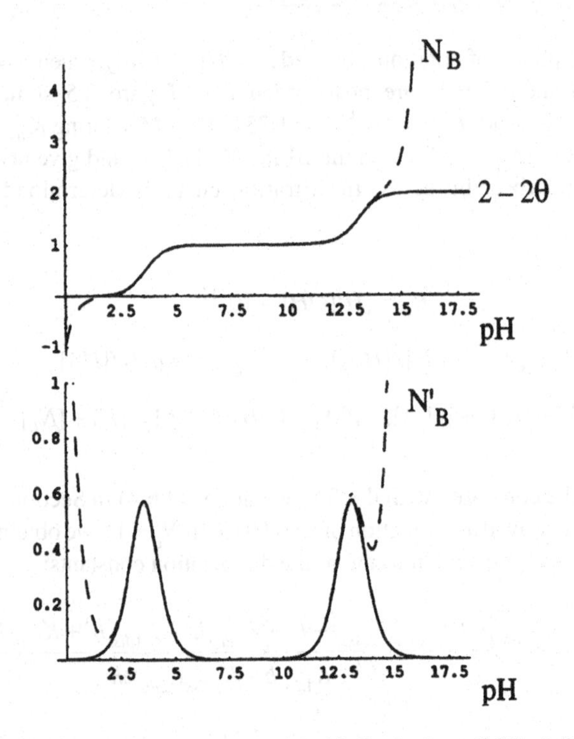

Figure G.1. The titration curve (dashed line), and the binding isotherm (full line) $(2-2\theta)$ for $\alpha-\alpha'$ di-tert-butyl succinic acid (in 50% ethanol solution): $pK_{1diss}=3.58$, $pK_{2diss}=13.12$, and $K_w=7\times10^{-16}$. The lower curves are the derivatives of the upper curves.

H. FITTING SYNTHETIC DATA

We present here an example where "experimental" data are constructed from *exact* binding curves for a three-site system, after which the experimental data are used to calculate the parameters of the system by two different methods.

The exact binding data are obtained with the choice of parameters

$$k_a=10,\quad k_b=1,\quad k_c=10^{-3},\quad g_{ab}=1,\quad g_{ac}=10,\quad g_{bc}=1,\quad g_{abc}=10 \tag{H.1}$$

This system could in principle be an experimental one. Note that in this particular case we chose $g_{abc}=g_{ac}$, i.e., there is *only* long-range correlation and no short-range correlations.

We now choose 30 points in the range $-3\le\log_{10}C\le4$. At each point we create "experimental data" by defining

$$\theta_\alpha^{\exp}=\theta_\alpha^{theor}\pm0.05R \tag{H.2}$$

where θ_α^{theor} is the exact value of the theoretical binding curves with parameters (H.1), and θ_α^{exp} are obtained by adding a random value between ±0.05 (where R is a pseudo-random number between zero and one, selected from a normal distribution).

We now treat these synthetic "experimental data" as if they originated from an experimental source. We first fit three individual binding curves to obtain the seven parameters *without* any restrictions on the correlations. We obtain a very good fit to the data with the parameters

$$k_a = 10.02, \quad k_b = 0.98, \quad k_c = 10^{-3},$$
$$g_{ab} = 0.977, \quad g_{ac} = 10.57 \quad g_{bc} = 0.997, \quad g_{abc} = 9.111 \tag{H.3}$$

Clearly, these are very close to the original parameters in (H.1).

We next fit three curves with the imposed relations, such as those made by Senear *et al.* (1986) (see Section 5.10), namely,

$$g_{ac} = 1 \quad \text{and} \quad g_{abc} = g_{ab} + g_{bc} \tag{H.4}$$

so that only five parameters are to be determined. With these assumptions we obtain a fit to the "experimental data" (with almost the same variance). The resulting parameters are

$$k_a = 10.024, \quad k_b = 1.211, \quad k_c = 0.0005, \quad g_{ab} = 0.815, \quad g_{bc} = 15.0003 \tag{H.5}$$

Clearly, these are quite different from the exact parameters in (H.1). The fitted

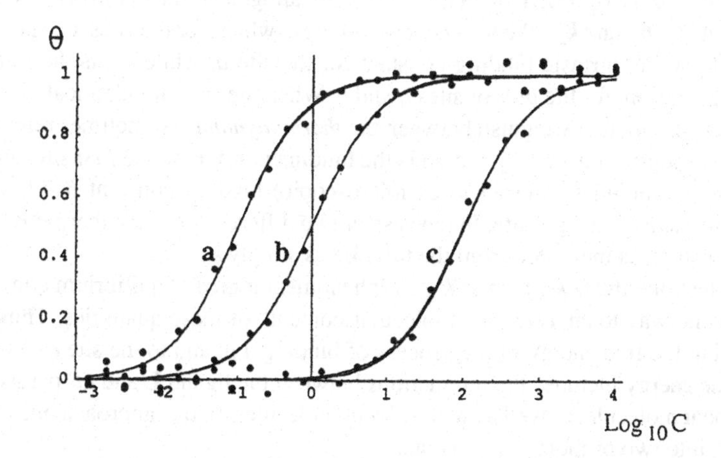

Figure H.1. The BI for the model of Appendix H. The points are obtained from the exact BI by (H.2). The solid lines are the computed BI determined by parameters (H.3) and (H.5).

curves, however, are almost indistinguishable from the exact curves (Fig. H.1). Thus, processing the data with the assumed relations (H.4) would lead us to conclude that there is almost *no correlation* between a and b (i.e., $g_{ab} \approx 1$), and a large *positive* cooperativity between b and c, and by assumption no cooperativity between a and c, in sharp contrast to the *exactly* known properties of the system which, by construction, has no nearest-neighbor correlation but only long-range correlation. This example clearly demonstrates that the binding curves have enough flexibility so that one can obtain a reasonable fit to the experimental data even when unjustified assumptions [such as (H.4)] are introduced in the processing of the data.

I. A COMMENT ON THE NOMENCLATURE

There seems to be some confusion regarding adjectives appended to the various binding constants, such as macroscopic, microscopic, intrinsic, perturbed, and unperturbed. We use here a system of two different sites to clarify these concepts.

For a system of two sites a and b the individual binding isotherms (BI) are

$$\theta_a = \frac{k_a C + k_{ab} C^2}{1 + (k_a + k_b)C + k_{ab} C^2} \tag{I.1}$$

and

$$\theta_b = \frac{k_b C + k_{ab} C^2}{1 + (k_a + k_b)C + k_{ab} C^2} \tag{I.2}$$

The total BI is $\theta_T = \theta_a + \theta_b$. Thus there are altogether three *intrinsic* binding constants k_a, k_b, and k_{ab}. We use the term *intrinsic* whenever the sites are specified. Thus, k_a is the intrinsic binding constant for the site a, while k_{ab} is the intrinsic binding constant for the *pair* of sites a and b. When the sites are identical, then one should take care to distinguish between the *thermodynamic* (sometimes referred to as macroscopic) constant K_1, which is the binding constant to the *first* site, and the *intrinsic* (sometimes referred to as microscopic) binding constant $K_1/2$, which refers to a *specific* single site. In the case of two different sites, the thermodynamic *first* constant, as measured from the total BI, is simply $k_a + k_b$.

When one treats k_a, k_b, and k_{ab} as (phenomenological) equilibrium constants, there is no way to analyze the "molecular content" of these quantities. Thus k_a is related to the free energy of the process of binding a ligand to the site a. As such, this free energy includes the interactions between the ligand **L** and *all* parts of the adsorbent molecule, as well as with solvent molecules. In this approach one cannot split k_a into two or more components.

The situation is entirely different when a molecular or a statistical mechanical approach is adopted. In molecular terms the *three* intrinsic binding constants are defined by

$$k_a = \frac{Q(a, 0)}{Q(0, 0)} \lambda_0, \quad k_b = \frac{Q(0, b)}{Q(0, 0)} \lambda_0, \quad k_{ab} = \frac{Q(a, b)}{Q(0, 0)} \lambda_0^2 \tag{I.3}$$

where $\lambda_0 = \lambda / C$, λ being the absolute activity of the ligand and C its molar concentration in the reservoir. Clearly, each of these constants is related to the free energy of a well-defined binding process.

In some cases one can factor k_a in two or more factors. This can be done, however, only when we use a molecular formulation of the binding constants, and only in some particular cases. We present two such cases, discussed also by Hill (1985).

1. *Purely electrostatic interactions*: Suppose we have a system with two charges at sites **a** and **b** such that the interaction between the (charged) ligand **L** and the adsorbent molecule is purely electrostatic and consists of two contributions only (Fig. I.1),

$$U_a = U_a(\mathbf{L}, \mathbf{a}) + U_a(\mathbf{L}, \mathbf{b}) \tag{I.4}$$

where U_a is the (total) interaction energy of **L** *bound* at **a** with the entire adsorbent molecule, $U_a(\mathbf{L}, \mathbf{a})$ is the electrostatic interaction energy between **L** and the *charge* at site **a**, and similarly $U_a(\mathbf{L}, \mathbf{b})$ is the electrostatic interaction energy between the ligand **L** (bound on **a**) and the charge at **b**. In this case we may write

$$k_a = \frac{Q(0, 0) q_a(\mathbf{L}, \mathbf{a}) q_a(\mathbf{L}, \mathbf{b})}{Q(0, 0)} \lambda_0 = [q_a(\mathbf{L}, \mathbf{a}) \lambda_0] q_a(\mathbf{L}, \mathbf{b}) = k_a'(\mathbf{L}, \mathbf{a}) \cdot \alpha \tag{I.5}$$

where $k_a'(\mathbf{L}, \mathbf{a})$ is the "intrinsic" binding constant to the site **a** when the charge on site **b** is switched off; α includes the interaction of **L** with the other site **b.** The constant $k_a'(\mathbf{L}, \mathbf{a})$ is sometimes referred to as the "intrinsic" (or unperturbed) binding constant, and α is referred to as an interaction (or perturbation) parameter. We stress that such a factorization is possible only when we express k_a in terms of molecular quantities. There is no way of applying such a factorization to k_a when this is determined experimentally. In fact, even in a molecular formulation such a factorization is not always possible. For instance, when we have a fully rotating adsorbent

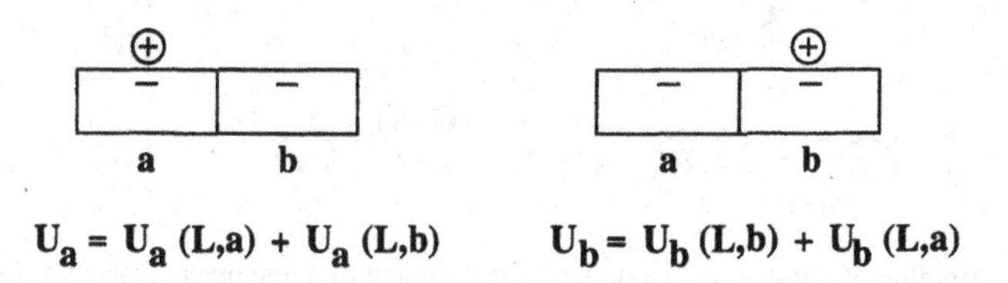

Figure I.1. Binding of a ligand ⊕ to a two-site molecule. Only electrostatic interactions are considered.

molecule, such as succinic acid, then

$$k_a = \frac{\int \exp[-\beta(U_a(\phi) + E(\phi))]d\phi}{\int \exp[-\beta E(\phi)]d\phi} \lambda_0 \tag{I.6}$$

where $U_a(\phi)$ is the ligand–adsorbent interaction energy when the internal rotational angle is ϕ, and $E(\phi)$ is the energy of the unbound adsorbent molecule at ϕ. Clearly, no such factorization as that in Eq. (I.5) is possible in this case. A similar treatment may be given to U_b as defined in Fig. I.1.

2. *Two separable subunits*: Suppose that **a** and **b** are two sites situated at two different subunits A and B (Fig. I.2). One may define the "intrinsic" binding constant to the *isolated* subunit A by

$$k'_A = \frac{Q_A(\mathbf{L})}{Q_A(0)} \lambda_0 \tag{I.7}$$

The intrinsic binding constant to site A on the entire (dimer) adsorbent molecule is now

$$\begin{aligned} k_A &= \frac{Q(A,0)}{Q(0,0)} \lambda_0 = \left[\frac{Q_A(\mathbf{L})}{Q_A(0)} \lambda_0\right] \frac{Q(A,0)}{Q_A(\mathbf{L})} \frac{Q_A(0)}{Q(0,0)} \\ &= k'_A \exp[-\beta(\Delta G(D) - \Delta G(A))] = k'_A \exp[-\beta(\Delta G' - \Delta G^\circ)] \\ &= k'_A \exp(-\beta \Delta\Delta G] = k'_A \alpha \end{aligned} \tag{I.8}$$

where $\Delta\Delta G$ is the difference between the binding free energy to site A on the dimer $\Delta G(D)$ and the binding free energy to the isolated subunit A, $\Delta G(A)$. One may refer to k'_A as an intrinsic binding constant (to the isolated subunit) and to α as a perturbation factor. Again, we stress that such factorization is possible for this particular example.

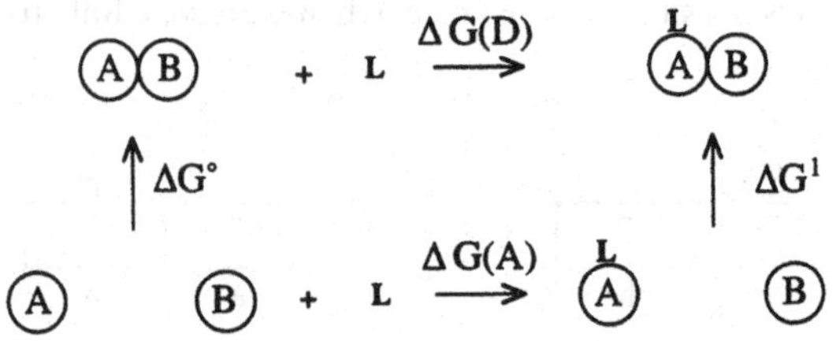

Figure I.2. Binding of a ligand **L** to a dimer (first row) and to a monomer A (second row). ΔG° and $\Delta G'$ are the association free energies for the two monomers, with and without a ligand, respectively.

In general there is no obvious way of separating the adsorbent molecule into two well-defined subunits, nor can one assume that all interactions between the ligand and the adsorbent molecule consist of only two electrostatic interactions. This is clearly the case for succinic acid or for DNA, where there is an infinite number of ways to define an unperturbed intrinsic binding constant. Therefore, the use of such factorization of k_A should be abolished when a phenomenological approach is adopted. On the other hand, when using a molecular approach one may or may not use such factorization. In any case this is certainly not a necessity.

J. AVERAGE BINDING CONSTANTS AND CORRELATION FUNCTIONS

In Chapters 5 and 6 we encountered experimental systems with three and four nonidentical sites. In many cases the individual intrinsic binding constants and the correlations between ligands at specific sites is not known. The only information available is the measurable thermodynamic binding constants. When we do not know the values of the (different) individual binding constants, but we still calculate the correlations as if the system consists of identical sites, we obtain average binding constants and average correlations in the following sense.

For three different sites, denoted by *a, b,* and *c,* the relations between the thermodynamic constants and intrinsic constants are

$$\left.\begin{aligned} K_1 &= k_a + k_b + k_c \ (= 3k_1) \\ K_1K_2 &= k_{ab} + k_{ac} + k_{bc} \ (= 3k_{11}) \\ K_1K_2K_3 &= k_{abc} \ (= k_{111}) \end{aligned}\right\} \tag{J.1}$$

where the equality for the case of strictly identical sites appear in parentheses.

We now use the equalities in the parentheses of Eqs. (J.1), but instead of k_1, k_{11}, and k_{111} we *define* the corresponding average quantities

$$\left.\begin{aligned} 3\langle k_1\rangle &= k_a + k_b + k_c \\ 3\langle k_{11}\rangle &= k_{ab} + k_{ac} + k_{bc} \\ \langle k_{111}\rangle &= k_{abc} \end{aligned}\right\} \tag{J.2}$$

In terms of these average binding constants we define the average correlations, in the same formal manner as we define correlations between specific sites:

$$\left.\begin{aligned} \langle g_{11}\rangle &= \frac{\langle k_{11}\rangle}{\langle k_1\rangle^2} = 3\frac{K_2}{K_1} = 3(g_{ab}x_ax_b + g_{ac}x_ax_c + g_{bc}x_bx_c) \\ \langle g_{111}\rangle &= \frac{\langle k_{111}\rangle}{\langle k_1\rangle^3} = 27\frac{K_1K_2K_3}{K_1^2} = 27g_{abc}x_ax_bx_c \end{aligned}\right\} \tag{J.3}$$

where the "mole fractions" x_i are defined by

$$x_i = \frac{k_i}{\Sigma k_i} = \frac{k_i}{k_a + k_b + k_c} \tag{J.4}$$

When the sites are identical in the weak sense, i.e., $k_1 = k_a = k_b = k_c$, then $x_i = 1/3$ for each i and Eqs. (J.3) reduce to

$$\left.\begin{aligned} \langle k_1 \rangle &= k_1 \\ \langle g_{11} \rangle &= \frac{1}{3}(g_{ab} + g_{ac} + g_{bc}) \\ \langle g_{111} \rangle &= g_{abc} \end{aligned}\right\} \tag{J.5}$$

In this case there is, by definition, only one (first) intrinsic binding constant, hence $\langle k_1 \rangle$ is the same as k_1; but the three pair correlations can be different, hence $\langle g_{11} \rangle$ is the arithmetic average of the three pair correlations.

For three strictly identical sites, Eqs. (J.3) reduce to

$$\langle k_1 \rangle = k_1, \quad \langle g_{11} \rangle = g_{11}, \quad \langle g_{111} \rangle = g_{abc} \tag{J.6}$$

For a four-site system, such as hemoglobin, the sites are different. The analogue of Eqs. (J.1) is

$$\left.\begin{aligned} K_1 &= k_a + k_b + k_c + k_d \, (= 4k_1) \\ K_1K_2 &= k_{ab} + k_{ac} + k_{ad} + k_{bc} + k_{bd} + k_{cd} \, (= 6k_{11}) \\ K_1K_2K_3 &= k_{abc} + k_{abd} + k_{bcd} + k_{acd} \, (= 4k_{111}) \\ K_1K_2K_3K_4 &= k_{abcd} \, (= k_{1111}) \end{aligned}\right\} \tag{J.7}$$

where, again, the equalities for strictly identical sites appear in parentheses. As in Eqs. (J.2) we *define* the average quantities

$$\left.\begin{aligned} 4\langle k_1 \rangle &= k_a + k_b + k_c + k_d \\ 6\langle k_{11} \rangle &= k_{ab} + k_{ac} + k_{ad} + k_{bc} + k_{bd} + k_{cd} \\ 4\langle k_{111} \rangle &= k_{abc} + k_{abd} + k_{bcd} + k_{acd} \\ \langle k_{1111} \rangle &= k_{abcd} \end{aligned}\right\} \tag{J.8}$$

If we introduce new mole fractions

$$x_i = \frac{k_i}{\Sigma k_i} = \frac{k_i}{k_a + k_b + k_c + k_d} \quad \text{(J.9)}$$

then the average correlations can be expressed in the form

$$\left.\begin{aligned}
\langle g_{11}\rangle &= \frac{\langle k_{11}\rangle}{\langle k_1\rangle^2} = \frac{8}{3}\frac{K_2}{K_1} = \frac{8}{3}\,(g_{ab}x_a x_b + g_{ac}x_a x_c + g_{ad}x_a x_d + g_{bc}x_b x_c + g_{cd}x_c x_d)\\
\langle g_{111}\rangle &= \frac{\langle k_{111}\rangle}{\langle k_1\rangle^3} = 16\,\frac{K_1K_2K_3}{K_1^3} =\\
&= 16(g_{abc}x_a x_b x_c + g_{abd}x_a x_b x_d + g_{bcd}x_b x_c x_d + g_{acd}x_a x_c x_d)\\
\langle g_{1111}\rangle &= \frac{\langle k_{1111}\rangle}{\langle k_1\rangle^4} = 64\,\frac{K_1K_2K_3K_4}{K_1^4} = 64 g_{abcd}x_a x_b x_c x_d
\end{aligned}\right\} \quad \text{(J.10)}$$

where the rhs contains the true correlations for the specified sites.

When the sites are identical in the weak sense, then $x_i = 1/4$ for each i and Eqs. (J.10) reduce to

$$\left.\begin{aligned}
\langle g_{11}\rangle &= \tfrac{1}{6}\,(g_{ab} + g_{ac} + g_{ad} + g_{bc} + g_{cd})\\
\langle g_{111}\rangle &= \tfrac{1}{4}\,(g_{abc} + g_{abd} + g_{bcd} + g_{acd})\\
\langle g_{1111}\rangle &= g_{abcd} = g_{1111}
\end{aligned}\right\} \quad \text{(J.11)}$$

So in this case each average correlation is the arithmetic average of all the different correlations of the same order. For strictly identical sites Eqs. (J.11) reduce to

$$\langle g_{11}\rangle = g_{11}, \quad \langle g_{111}\rangle = g_{111}, \quad \langle g_{1111}\rangle = g_{1111} \quad \text{(J.12)}$$

K. UTILITY FUNCTION IN A BINDING SYSTEM

We introduce here the general definition of the utility function for any binding system.

The BI for any binding system can be written as $\theta = \theta(C; \mathbf{a})$, where C is the ligand concentration and $\mathbf{a}$ is a set of parameters that could be molecular (such as the mass or the dipole moment of the ligand) or macroscopic (such as the temperature or concentration of some solutes).

Suppose we use the binding system as a means for transporting the ligand between two stations; we load the system at some fixed ligand concentration C_2 and unload it at a second concentration C_1. For any given values of C_1 and C_2 we define the utility function by the difference

$$U_t = \theta(C_2; \mathbf{a}) - \theta(C_1; \mathbf{a}) \tag{K.1}$$

Clearly, for a fixed set of parameters **a**, we have a single binding curve and the utility function has a fixed value given by Eq. (K.1).

If, on the other hand, we can change one or more of the parameters, we can ask for the value of that parameter for which U_t is maximum, i.e., for which the system will transport the ligand with maximum efficiency.

We consider the following two examples:

1. A one-site system: The BI is

$$\theta = \frac{kC}{1 + kC} \tag{K.2}$$

We now view $\theta(C, k)$, where k is the varying parameter (which may be varied by changing the mass of the ligand, the binding energy, or the temperature). We ask for the maximum of U_t for any fixed values of C_1 and C_2, i.e., we solve

$$\frac{\partial U_t}{\partial k} = 0 \tag{K.3}$$

for which we obtain

$$k_{max} = \frac{1}{\sqrt{C_1 C_2}} \tag{K.4}$$

Figure K.1 shows a family of BIs with $k = 10^i$ ($i = -2$ to $i = 2$). Two vertical lines are drawn at $C_1 = 0.4$ and $C_2 = 0.6$. These lines intersect each of the BIs at two points. The maximum value of U_t is obtained at $k_{max} = (0.4 \times 0.6)^{-1/2} = 2.04124$, or $k_{max} \sim 10^{0.31}$. In the figure the curve for $i = 0.5$ has the largest value of U_t among the curves drawn.

2. A two-site system: For a two-site system the BI is

$$\theta = \frac{kC + k^2SC^2}{1 + 2kC + k^2SC^2} \tag{K.5}$$

Here, we have two parameters k and S. To simplify the examination of the dependence on the cooperativity S, we change variables $x = kC$ and rewrite Eq. (K.5) as

$$\theta = \frac{x + x^2S}{1 + 2x + x^2S} \tag{K.6}$$

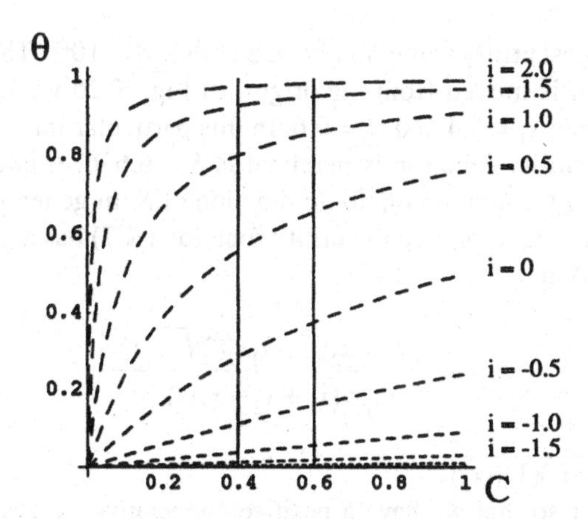

Figure K.1. A family of BIs for a one-site system, with $k = 10^i$ where the values of i are indicated. The two vertical lines at $C_1 = 0.4$ and $C_2 = 0.6$ are the two concentrations between which the ligand is transported.

The utility function is now defined by

$$U_t = \theta(x_2, S) - \theta(x_1, S) \tag{K.7}$$

Figure K.2 shows a family of BI with cooperativities $S = 10^i$ where $-1 \le i \le 2$. It is evident that if we fix the *interval* $\Delta x = 0.2$ near the origin, i.e., between $x_1 \approx 0.001$ and $x_2 = 0.2$, we find that the curve with the largest cooperativity will

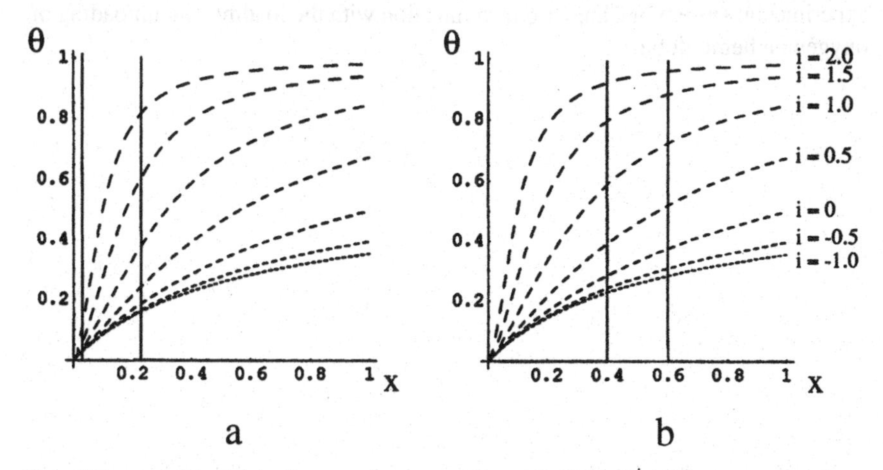

Figure K.2. As in Fig. K.1, but for a two-site system with $x = kC$ and $S = 10^i$. (a) The two concentrations C_1 and C_2 are chosen near the origin. (b) The two concentrations are as in Fig. K.1.

also have the largest utility value (in Fig. K.2a this is $S = 10^2$). This is not true if the same interval is moved from the origin. In Fig. K.2b we have moved the interval to between $x_1 = 0.4$ and $x_2 = 0.6$. In this particular interval the value of S for which the utility function is maximal is $S = 6.636$. It is clear from Fig. K.2b that the utility is not a monotonic function of S. In general, for any given pair of concentrations x_1 and x_2, the utility function (K.7) has a maximum value as a function of S at

$$S = \frac{x_1\sqrt{\ } + x_2\sqrt{\ } + x_1x_2(2\sqrt{\ } - 1)}{x_1x_2(x_1 + x_2 + x_1x_2)} \tag{K.8}$$

where $\sqrt{\ } = \sqrt{(1 + x_1)(1 + x_2)}$.

Clearly, $S \geq 1$ so that we have a positive cooperativity for which the utility function has a maximum value. On the other hand, when we choose $x_1 \approx 0$ then U_t becomes identical with $\theta(x_2, S)$, and this is a monotonically increasing function of $S \geq 1$.

In the above two examples we have varied the parameters k and S and examined the dependence of the utility function on these parameters. The utility function as defined in Eq. (K.1) is the difference between θ for two values of the concentration on a *single* BI. One can conceivably define also utility functions between two or more BIs. For instance, if at the loading terminal the temperature is T_2 and at the unloading terminal it is T_1, then the utility function between C_1 and C_2 is

$$U_t = \theta(C_2, T_2) - \theta(C_1, T_1) \tag{K.9}$$

where $\theta(C, T_2)$ and $\theta(C, T_1)$ are two different BIs. We examine an example of an experimental system in Chapter 6 in connection with the loading and unloading of oxygen on hemoglobin.

Abbreviations Used in the Text

BI	Binding isotherm
BP	Binding polynomial
BPG	D-2,3-bisphosphoglycerate
CPF	Canonical partition function
FG	Functional group
GPF	Grand partition function
IHP	Inositol hexaposphate
lhs	Left-hand side
PF	Partition function
rhs	Right-hand side
HφO	Hydrophobic
HφI	Hydrophilic

References

Ackers, G. K., Johnson, A. D., and Shea, M. A., 1982, *Proc. Natl. Acad. Sci. U.S.A.* **79**:1129.

Adair, G. S., 1925, *J. Biol. Chem.* **63**:529.

Antonini, E., and Brunori, M., 1971, *Hemoglobin and Myoglobin in their Reactions with Ligands*, North-Holland, Amsterdam.

Ben-Naim, A., 1972, in *Water and Aqueous Solutions*, Chapter 11 (R. A. Horne, ed.), Wiley Interscience, New York.

Ben-Naim, A., 1987, *Solvation Thermodynamics*, Plenum Press, New York.

Ben-Naim, A., 1992, *Statistical Thermodynamics for Chemists and Biochemists*, Plenum Press, New York.

Ben-Naim, A., 1997, *J. Chem. Phys.* **107**:10242.

Ben-Naim, A., 1998, *J. Chem. Phys.* **108**:3630 and 6937.

Ben-Naim, A., and Mazo, R., 1993, *J. Phys. Chem.* **97**:10829.

Bjerrum, N., 1923, *Z. Phys. Chem.* **106**:219.

Bohbot, Y., and Ben-Naim, A., 1995, *J. Phys. Chem.* **99**:14544.

Bondi, A., 1968, *Physical Properties of Molecular Crystals, Ligands and Gases*, Wiley, New York.

Bradsley, W. G., and Waight, R. D., 1978, *J. Theor. Biol.* **72**:321.

Briggs, W. E., 1984, *J. Theor. Biol.* **108**:77

Colombo, M. F., Rau, D. C., and Parsegian, A., 1992, *Science* **256**:655.

DiCera, E., 1996, "Thermodynamic theory of site-specific binding processes," in *Biological Macromolecules*, Cambridge University Press.

Eberson, L., 1959, *Acta Chem. Scand.* **13**:211, 224.

Eberson, L., 1992, Chapter 6 in *Carboxylic Acids and Esters* (S. Patai, ed.), Wiley, New York.

Edsall, J. T., and Wyman, J., 1958, *Biophysical Chemistry*, Academic Press, New York.

Eigen, M., 1967, *Kinetics of Reaction Control and Information Transfer in Enzymes and Nucleic Acids*, Proc. 5th Nobel Symp., Interscience, Wiley, New York.

Eliel, E. L., and Wilen, S. H., 1994, *Stereochemistry of Organic Compounds*, Wiley, New York.

Feller, W., 1957, *Introduction to Probability Theory and Its Application*, Vol. I, Wiley, New York.

Gane, R., and Ingold, C. K., 1931, *J. Chem. Soc.*, 2153.

Gerhart, J. C., 1970, *Current Topics in Cellular Regulation*, Vol. 2, Academic Press, New York, p. 275.

Gerhart, J. C., and Pardee, A. B., 1961, *Fed. Proc.* **20**:224.

Gerhart, J. C., and Pardee, A. B., 1962, *J. Biol. Chem.* **237**:891.

Gerhart, J. C., and Pardee, A. B., 1963, *Cold Spring Harbor Symposia on Quantum Biology* **28**:491.

Gerhart, J. C., and Schachman, H. K., 1965, *Biochemistry* **4**:127.

Gerhart, J. C., and Schachman, H. K., 1968, *Biochemistry* **7**:144.

Harned, H. S., and Owen, B. B., 1958, *The Physical Chemistry of Electrolyte Solutions*, Reinhard, New York.
Hill, A. V., 1910, *J. Physiol. (London)* **40**:iv.
Hill, T. L., 1943a, *J. Chem. Phys.* **11**:545.
Hill, T. L., 1943b, *J. Chem. Phys.* **11**:552.
Hill, T. L., 1944a, *J. Chem. Phys.* **12**:56.
Hill, T. L., 1944b, *J. Chem. Phys.* **12**:147.
Hill, T. L., 1960, *An Introduction to Statistical Thermodynamics*, Addison-Wesley, Reading, Mass.
Hill, T. L., 1985, *Cooperativity Theory in Biochemistry, Steady State and Equilibrium Systems*, Springer-Verlag, New York.
Hochschild, A., and Ptashne, M., 1988, *Nature* **336**:353.
Imai, K., 1982, *Allosteric Effects in Haemoglobin*, Cambridge University Press, London.
Imai, K., and Yonetani, T., 1975, *J. Biol. Chem.* **250**:7093.
Johnson, A. D., Meyer, B. J., and Ptashne, M., 1979, *Proc. Natl. Acad. Sci. U.S.A.* **76**:5061.
Jones, J., and Soper, F. G., 1936, *J. Chem. Soc.* 133.
Kantrowitz, E. R., and Lipscomb, W. N., 1988, *Science* **241**:669.
Kantrowitz, E. R., and Lipscomb, W. N., 1990, *Trends in Biochem. Sci.* **15**:53.
Kirkwood, J. G., 1935, *J. Chem. Phys.* **3**:300.
Kirkwood, J. G., and Westheimer, F. H., 1938, *J. Chem. Phys.* **6**:506.
Koblan, K. S., Bain, D. L., Beckett, D., Shea, M. A., and Ackers, G. K., 1992, *Methods in Enzymology*, Academic Press, New York, p. 405.
Kortum, G., Vogel, W., and Andrussow, K., 1961, *Dissociation Constants of Organic Acids in Aqueous Solutions*, Butterworths, London.
Koshland, D. E., 1958, *Proc. Natl. Acad. Sci. Wash.* **44**:98.
Koshland, D. E., 1962, in: *Horizons in Biochemistry*, New York, Academic Press, p. 265.
Koshland, D. E., Nemethy, G., and Filmer, D., 1966, *Biochemistry* **5**:365.
Kondelka, G. B., and Carlson, P., 1992, *Nature* **355**:89.
Kondelka, G. B., Harbury, P., Harrison, S. C., Ptashne, M., 1988, *Proc. Natl. Acad. Sci. U.S.A.* **85**:4633.
Krause, K. L., Volz, K. W., and Lipscomb, W. N., 1985, *Proc. Natl. Acad. Sci. U.S.A.* **82**:1643.
Langmuir, J., 1918, *J. Am. Chem. Soc.* **40**:1361.
Levitzki, A., 1978, *Quantitative Aspects of Allosteric Mechanisms*, Springer-Verlag, New York.
McCoy, L. L., 1967, *J. Am. Chem. Soc.* **89**:1673.
McDaniel, D. H., and Brown, H. C., 1953, *Science* **118**:370.
Minton, A. P., and Saroff, H. A., 1974, *Biophys. Chem.* **2**:296.
Monod, J., Changeux, J. P., and Jacob, F., 1963, *J. Mol. Biol.* **6**:306.
Monod, J., Wyman, J., and Changeux, J. P., 1965, *J. Mol. Biol.* **12**:88.
Newell, O. J., Markby, D. W., and Schachman, H. K., 1989, *J. Biol. Chem.* **264**:2476.
Papoulis, A., 1965, *Probability, Random Variables, and Stochastic Processes*, McGraw-Hill, New York.
Pauling, L., 1935, "The oxygen equilibrium of hemoglobin and its structural interpretation," *Proc. Natl. Acad. Sci. U.S.A.* **21**:186.
Perutz, M. F., 1970, *Nature* **228**:726.
Perutz, M. F., 1990, *Mechanisms of Cooperativity and Allosteric Regulation in Proteins*, Cambridge University Press, New York.
Ptashne, M., 1992, *A Genetic Switch: Gene Control and Phage* λ, Cell Press and Blackwell, Scientific Publ., Cambridge, Mass.
Richards, K. E., and Williams, R. C., 1972, *Biochemistry* **11**:3393.
Robinson, R. A., and Stokes, R. H., 1959, *Electrolyte Solutions*, Butterworths, London.
Saroff, H. A., 1993, *Biopolymers* **33**:1327.
Saroff, H. A., 1987, *Arch .Biochem. Biophys.* **256**:110.
Saroff, H. A., and Minton, A. P., 1972, *Science* **175**:1253.

Schachman, H. K., 1988, *J. Biol. Chem.* **263**:18583.
Senear, D. F., Brenowitz, M., Shea, M. A., and Ackers, G. K., 1986, *Biochemistry* **25**:7344.
Steitz, T. A., 1990, *Quart. Rev. Biophys.* **23**:205.
Warren, S. G., Edwards, B. F. P., Evans, D. R., Wiley, D. C., and Lipscomb, W. N., 1973, *Proc. Natl. Acad. Sci. U.S.A.* **70**:1117.
Westheimer, F. H., and Kirkwood, J. G., 1938, *J. Chem. Phys.* **6**:513.
Westheimer, F. H., and Shookhoff, W. M., 1939, *J. Am. Chem. Soc.* **61**:555.
Whitehead, E. P., 1980, *J. Theor. Biol.* **86**:45.
Whitehead, E. P., 1980, *J. Theor. Biol.* **87**:153.
Wiley, D. C., and Lipscomb, W. N., 1968, *Nature* **218**:1119.
Wiley, D. C., Evans, D. R., Warren, S. G., McMurray, C. H., Edwards, B. F. P., Franks, W. A., and Lipscomb, W. N., 1971, *Cold Spring Harbor Symp. Quant. Biol.* **36**:285.
Wyman, J., 1964, *Adv. Protein Chem.* **19**:223.
Wyman, J., and Gill, S. J., 1990, *Binding Linkage: Functional Chemistry of Biological Macromolecules*, University Science Book, Mill Valley, CA.
Yates, R. A., and Pardee, A. B., 1956, *J. Biol. Chem.* **221**:757.

Index

Absolute activity, 11
Adair equation, 209
Additivity of direct interaction, 145, 175
Allosteric regulation, 264–277
Aspartate transcarbamoylase, 277–280
Average correlation, 164–173, 201–204

Binding constants
  average, 335–337
  for alkylated succinic acid, 131
  for benzen polycarboxylic acids, 174, 204, 206
  conditional, 31, 33
  definition of, 29
  effective, 99, 142
  of equilibrated and frozen-in system, 62–65
  intrinsic, 29
  nonintrinsic, 34
  for normal amines, 47, 48
  for normal carboxylic acids, 44, 46
  and probability, 30
  for protons in amino acids, 121, 123, 193
  for protons in diamines, 120
  for proteins on DNA, 184–188
  of substituted acetic acid, 49
  thermodynamic, 34–37
  thermodynamic interpretation of, 31
  for two protons on dicarboxylic acids, 114, 119
Binding isotherm
  definition of, 25–27
  for equilibrated system, 62
  for frozen-in system, 62
  general form, 25, 26
  for hemoglobin, 212–221
Binding isotherm (*cont.*)
  individual, 26, 30, 32, 177, 188
  Langmuir, 28, 39
  for mixtures of adsorbing molecules, 40
  for mixtures of ligands, 41–43
  for normal amines, 47
  for normal carboxylic acids, 43–45
  and probabilities, 27
  for proteins and DNA, 177
  relation to the partition function, 26
  of system with conformational changes, 56, 104
  in terms of thermodynamic constants, 35
  for three-site system, 146, 147
Binding of protons to a two-site system, 114
Binding polynomial, vi, 37

Chemical potential, 11
Competitive regulation, 263
Concerted model, 112, 211
Conformational states, 12
Cooperativity
  in binding repressor to operator, 184–188
  and correlation, 70, 105
  definition, 68–73
  direct, 73
  in hemoglobin, 207–221
  indirect, 82
  and interaction coefficient, 71
  positive and negative, 72
  solvent effects on, 281
  between two protons, 117
  in two-site systems, 68
Correlations
  in alkylated succinic acid, 131

Correlations (*cont.*)
in α, ω alkane diamines, 120
in α, ω dicarboxylic acids, 119
average, 164–173, 201–204, 336
and conformational changes, 86–91, 105, 149
and cooperativity, 105
and Coulombing interaction, 118, 120
and density of interaction, 201
direct, 73, 145
effective, 99
between four protons, 204–207
for fully rotating model, 127–130
general definition, 23, 69
in hemoglobin, 213–221
indirect, 82, 106, 107, 149
nonadditivity of, 147
in 1-D system, 230
long range, 151–155, 162, 163, 174, 179
nonadditivity, 153–155, 179–184
pair correlation, 7, 24, 280, 309–311
among protons, 175
solvent effect on, 287–293
temperature dependence, 151
between three protons, 173–175
in three-site systems, 148, 162
transmission across boundaries, 155–159
triplet correlation, 8, 234, 309–311
between two protons, 117, 173, 174

Density of interaction, 201, 203
Dipole-dipole interaction, 14
Direct cooperativity, 73

Energy levels, 12

Grand partition function, vi, 17–20
construction of, 18
for four-site system, 194–196
linear model, 197
for localized systems, 311
for mixture of ligands, 41
for non-ideal ligand, 318
and probabilities, 20
for regulation system, 265, 269
square model, 199
for system with conformational changes, 53, 83, 100
tetrahedral model, 200
for three-site systems, 143
for two types of sites, 40

Hemoglobin, 207–222
Adair equation for, 209
A.V. Hill model, 208
binding isotherms, 212–219
cooperativities, 214–219
experimental data, 212
linear model, 198
Pauling model, 210
square model, 199
tetrahedral model, 202
utility function, 218–221
Hill coefficient, 77

Identical sites
in strict sense, 18, 32
in weak sense, 32
Independence between subunits, 102
Induced conformational changes, 57–60
and correlation, 86, 87
extent of, in single-site system, 58–60
in three-site systems, 149, 154
in two-site systems, 82–87
in two subunits, 107–111
Interaction coefficient, 71, 189
Interaction energy
average dipole-dipole, 14
dipole-dipole, 14
direct interaction, 13, 145
pairwise additivity, 13, 145
subunit-subunit, 17
Intrinsic binding constant, 29, 189, 332–335

Koshland, Nemethy, and Filmer model, 113, 211

Langmuir isotherm, 28, 38–40
generalizations of, 40–43
Linear systems
long-range correlation, 248–253
matrix method, 226
partition function of, 213–295
Long-range correlation, 152–155, 162, 163, 248–253

Monod Wyman and Changeux model, 112, 210, 255

Nonadditivity of correlation, 143–154, 175–176, 179–184, 207, 213
Nonideality of the ligand, 317

Occupancy states, 12

Partition function
construction of, 17, 18
for four-site system, 194–197
fully rotating model, 132
general form, 17, 18
relation with thermodynamics, 19
for system with conformational changes, 53
for three-site system, 144
Probabilities
conditional, 22, 27
of disjoint events, 20
of independent events, 22, 309–311
marginal, 54
of molecular events, 20–24
in system with conformational changes, 54

Regulatory curve, 261–276
Regulatory enzymes, 255–258
aspartate transcarbamoylase, 277

Sequential model, 113
Solvation Gibbs energy, 294
Solvent effects, 281–307
Spurious cooperativity, 60–66, 322–328
in alkylated succinic acid, 131–142
in single-site systems, 61
in two-site systems, 77–82, 91–100
Stability condition, 28
States of the system
conformational, 12
occupancy, 12

Thermodynamic binding constants, 34
Titration curve
for carboxylic acid, 44, 46
relation to binding isotherm, 44, 328–330

Utility function, 337–340
for hemoglobin, 218–222
for regulatory enzymes, 262–263